GEOMETRY & TOPOLOGY

GEOMETRY & TOPOLOGY

Editor

Vinicio Zeni

scitus
academics

Geometry & Topology
Edited by **Vinicio Zeni**

This Book is Printed on Acid-Free Paper.

ISBN: 9781681179605
Library of Congress Control Number: 2017949091

SCITUS Academics LLC,
616, Corporate Way, Suite 2, 4766,
Valley Cottage, NY 10989

www.scitusacademics.com

Printed in United States of America.

Preface

Topology is a branch of pure mathematics related to Geometry that describes mathematical spaces, in particular the properties that stem from a space's shape. Topology originates in the early part of the twentieth century, but some isolated results can be traced back several centuries. Many of the shapes topologists deal with are incredibly strange, so much so that practically all everyday objects such as bowls and pets and trees make up a small minority. Topology developed as a field of study out of geometry and set theory, through analysis of concepts such as space, dimension, and transformation. In its most general setting, topology involves objects that are abstract sets of elements. To discuss properties such as continuity of functions between such abstract sets, some additional structure must be imposed on them.

Geometry & Topology provides a glimpse of many research topics in modern algebra, geometry and topology. It presents an array of topics that introduce the reader to key ideas in active areas in geometry and topology. Algebraic geometry and complex manifolds and geometrical aspects of mathematical physics, and relations with manifold topology. This book develops some of the extraordinary richness, beauty, and power of geometry in two and three dimensions, and the strong connection of geometry with topology. The material is presented in a way that both graduate students and researchers should find accessible and enticing.

Table of Contents

CHAPTER 1

Complex Riemannian Geometry—Bianchi Identities and Einstein Tensor

Richard A. Hutchin

Optical Physics Company, Calabasas, CA, USA

ABSTRACT

Riemannian geometry has proved itself to be a useful model of the gravitational phenomena in the universe, but generalizations of it to include other forces have so far not been successful. Here we explore an extension of Riemannian geometry using a complex Hermitian metric tensor. We find that the standard electromagnetic field naturally appears along with two additional fields, which act as mass and charge sources.

A first paper set up the basic geometry and derived the Christoffel symbols plus the E&M field equation. This paper continues development with the generalized Riemann curvature tensor, Bianchi identities and the Einstein tensor, laying the basis for field equations. A final paper will then present the field equations.

Keywords: Unified Field Theory, Gravity, Electromagnetism, Complex Riemannian Geometry, Riemannian Geometry

1. INTRODUCTION

There have been numerous attempts to embed electromagnetism into Riemannian geometry, of which the most well known are Weyl's gauge field [1] , Kaluza's fifth dimension [2] and Eddington's affine geometry [3] . Even exploration of making the metric tensor complex (as we propose here) was investigated almost a hundred years ago by Mie [4] and Reichenbacher [5] . These two did not have general relativity available at that time, and their theories led nowhere.

We explore here the implications of assuming a complex Riemannian geometry as a generalized theory of gravity. We find that the electromagnetic field naturally appears as the imaginary part of the metric tensor and that two other fields also emerge, denoted as S and W. The flow of the development has 5 steps listed below. The first paper [6] set up the complex geometry and completed Step 1. This paper completes Steps 2-4, and a third paper will explore the field equations.

1) Derivation of the Christoffel symbols leading to the E&M field equation;

2) Derivation of the Riemann curvature tensors;

3) Symmetries of the curvature tensors including Bianchi identities;

4) Derivation of the Einstein tensor;

5) Field equations for all four fields.

2. SHORT SUMMARY OF THE FIRST PAPER

We provide here a summary of the first paper [6] , which assumed a complex Riemannian geometry as a generalized theory of gravity. We found that the electromagnetic field naturally appeared as the imaginary part of the metric tensor and that two other fields also emerge, denoted as S and W. That paper presented the concepts of complex Riemannian geometry and derives the Christoffel symbols, the E&M field equation and the two additional fields. While the overall geometry is based on complex numbers, each of the four fields is pure real. The current paper derives the Riemann curvature tensor, Bianchi identities and the generalized Einstein tensor which can be used to provide field equations.

Since general relativity has been a successful theory of gravity, any generalization should reduce to the standard theory in certain approximations. Hence in exploring complex Riemannian geometry, there is strong motivation to include an axiomatic structure which closely parallels the familiar Riemannian geometry. A reference text on General Relativity will be useful to some readers to fill in details of the steps, for the structure of that formulation applies in broad terms. An example of a suitable text would be Introduction to General Relativity by Adler, Bazin and Schiffer (ABS) [7] , but there are many available.

We began with a complex Hermitian metric tensor $c_{\alpha\beta}$ which defined an inner product between covariant complex vectors. The barred index is meant to indicate a complex-conjugate index, so that the inner product of a vector with itself will be real as defined in Equation (2.1), where *indicates complex conjugation.

$$|A|^2 = A^\alpha c_{\alpha\bar{\beta}} \left(A^\beta\right)^* \tag{2.1}$$

This form of $c_{\alpha\bar{\beta}}$ can be written as a sum of its real and imaginary parts as shown in Equation (2.2a) with its Hermitian inverse $c^{\bar{\beta}\alpha}$ also decomposed into real and imaginary parts as shown in Equation (2.2b). We note that since the two indices are typed for the complex metric tensors $c_{\alpha\bar{\beta}}$ and $c^{\bar{\beta}\alpha}$, the indices can be in any order without confusion such as $c^{\bar{\beta}\alpha}$ and $c_{\alpha\bar{\beta}}$ —a convention that will simplify some of the later equations. We will also use the same convention for $f_{\alpha\bar{\beta}} = f_{\bar{\beta}\alpha}$ and its raised form $F^{\bar{\beta}\alpha} = F^{\alpha\bar{\beta}}$. Since the real part of $c^{\bar{\beta}\alpha}$ and $c_{\alpha\bar{\beta}}$ are symmetric, typing is not required for those two ($g_{\alpha\beta}$ and $G^{\beta\alpha}$).

$$c_{\alpha\bar{\beta}} = \mathbf{g}_{\alpha\beta} + i\mathbf{f}_{\alpha\bar{\beta}} \tag{2.2a}$$

$$c^{\bar{\beta}\alpha} = G^{\beta\alpha} - iF^{\bar{\beta}\alpha} \tag{2.2b}$$

Here g is the usual symmetric metric tensor and f is an antisymmetric tensor. We will use bold font for those two basic fields, ($\mathbf{g}_{\alpha\beta}$ and $\mathbf{f}_{\alpha\beta}$) as well as the two others (S and W), which naturally appear later. Thus it will be easy for the reader to identify the key fields. One may correctly surmise from this notation that f will be identified with electromagnetism. Since $G^{\beta\alpha}$ is not the inverse of $\mathbf{g}_{\alpha\beta}$ due to the impact of the imaginary part $f_{\alpha\beta}$, we will use the standard notation $g^{\beta\alpha}$ for the inverse of $\mathbf{g}_{\alpha\beta}$. With this definition, we list some algebraic relationships in Equations (2.2c)-(2.2g). By expanding the equation $c^{\bar{\alpha}\beta}c_{\beta\bar{\gamma}} = c_{\gamma\bar{\beta}}c^{\bar{\beta}\alpha} = \delta^{\alpha}{}_{\gamma}$ into real and imaginary parts, we can derive (2.2c)-(2.2e). Multiply Equation (2.2c) by $g^{\gamma\beta}$ to get Equation (2.2f) and likewise multiply Equation (2.2d) by $g^{\gamma\beta}$ to get Equation (2.2g).

Equation (2.2f) shows that $G^{\bar{\alpha}\beta}$ equals $g_{\alpha\beta}$ except for terms second order and higher in $f_{\alpha\beta}$. When we limit our consideration to second order terms, $G^{\bar{\alpha}\beta}$ can be replaced by $g_{\alpha\beta}$.

Using this result, Equation (2.2g) shows that $F^{\bar{\alpha}\beta}$ equals $g^{\alpha\mu}f_{\mu\bar{\nu}}g^{\nu\beta}$ except for terms third order and higher in $f_{\alpha\beta}$. Thus when we limit our consideration to second order terms, $F^{\bar{\alpha}\beta}$ can be replaced by the familiar raised form of the E&M tensor, $f_{\alpha\beta}$.

$$G^{\bar{\alpha}\mu}g_{\mu\gamma} + F^{\bar{\alpha}\mu}f_{\mu\gamma} = g_{\gamma\mu}G^{\bar{\mu}\alpha} + f_{\gamma\mu}F^{\bar{\mu}\alpha} = \delta^{\alpha}{}_{\gamma} \tag{2.2c}$$

$$F^{\bar{\alpha}\mu}g_{\mu\gamma} = G^{\bar{\alpha}\mu}f_{\mu\gamma} \tag{2.2d}$$

$$g_{\alpha\mu}F^{\bar{\mu}\gamma} = f_{\alpha\mu}G^{\bar{\mu}\gamma} \tag{2.2e}$$

$$G^{\bar{\alpha}\beta} = g^{\bar{\alpha}\beta} - F^{\bar{\alpha}\mu}f_{\mu\bar{\tau}}g^{\tau\beta} = g^{\alpha\beta} - g^{\alpha\mu}f_{\mu\bar{\nu}}G^{\bar{\nu}\rho}f_{\rho\bar{\tau}}g^{\tau\beta} \tag{2.2f}$$

$$F^{\bar{\alpha}\beta} = g^{\alpha\mu}f_{\mu\bar{\nu}}G^{\bar{\nu}\beta} = G^{\bar{\alpha}\mu}f_{\mu\bar{\nu}}g^{\nu\beta} \tag{2.2g}$$

Raising and Lowering Indices: The metric tensor is customarily used for raising and lowering indices of vectors and tensors, and this property also applies to the Hermitian metric with one caveat—the conjugate quality of the index switches. Thus $c_{\alpha\bar{\beta}}$ applied to A^{α} gives $A_{\bar{\beta}}$ and vice versa as summarized in Equations (2.3)-(4). One implication is that barred indices must be contracted with barred indices, and unbarred with unbarred. We identify $c^{\bar{\beta}\alpha}$ as the matrix inverse of $c_{\alpha\bar{\beta}}$, one being used for raising indices and the other for lowering. This definition makes raising and lowering indices inverse operations as usual.

$$A_{\bar{\beta}} = A^{\alpha} c_{\alpha\bar{\beta}} \tag{2.3a}$$

$$A^{\alpha} = A_{\bar{\beta}} c^{\bar{\beta}\alpha} \tag{2.3b}$$

We also assume that one can convert between barred and unbarred indices by complex conjugation, i.e. you can create a new tensor with indices of flipped type by taking the complex conjugation of a previous tensor.

$$A_{\bar{\beta}} = A_{\beta}^{*} \tag{2.4a}$$

$$A^{\alpha}{}_{\bar{\beta}} = A^{\bar{\alpha}}{}_{\beta}{}^{*} \tag{2.4b}$$

3. COVARIANT DERIVATIVE

We begin with scalar differentiation. In classical relativity, the partial derivative and the covariant derivative of a scalar field ϕ are the same numerically and are denoted respectively as $\phi_{|\gamma}$ and $\phi_{\|\gamma}$, according to a familiar notation. We keep the same assumption here. Both types of scalar derivatives are numerically equal, as shown in Equation (3.1), but will differ on subsequent covariant derivatives according to their type.

$$\phi_{|\gamma} = \phi_{\|\bar{\gamma}} = \phi_{\|\gamma} \tag{3.1}$$

Vector Differentiation: Complex geometry also introduces covariant differentiation with Christoffel symbols in the standard fashion, where there are now two different types of indices for the vector and two for the derivative resulting in four different types of Christoffel symbols shown in Equations (3.2a)-(3.2d).

$$A^{\alpha}{}_{\|\gamma} = A^{\alpha}{}_{|\gamma} - \Gamma^{\alpha}{}_{\nu\gamma} A^{\nu} \tag{3.2a}$$

$$A^{\alpha}{}_{\|\bar{\gamma}} = A^{\alpha}{}_{|\gamma} - \Gamma^{\alpha}{}_{\nu\bar{\gamma}} A^{\nu} \tag{3.2b}$$

$$A^{\bar{\alpha}}{}_{\|\gamma} = A^{\bar{\alpha}}{}_{|\gamma} - \Gamma^{\bar{\alpha}}{}_{\bar{\nu}\gamma} A^{\bar{\nu}} \tag{3.2c}$$

$$A^{\bar{\alpha}}{}_{\|\bar{\gamma}} = A^{\bar{\alpha}}{}_{|\gamma} - \Gamma^{\bar{\alpha}}{}_{\bar{\nu}\bar{\gamma}} A^{\bar{\nu}} \tag{3.2d}$$

We note that the Christoffel symbols for the lowered indices can be derived from Equations (3.2a)-(3.2d) using $\phi = A^{\nu} A_{\nu}$ and Equation (3.1) for the covariant derivative of a scalar function—just as is frequently done in classical Riemannian geometry. The result from the first paper [6] is summarized below.

$$B_{\alpha\|\gamma} = B_{\alpha|\gamma} + \Gamma^{\nu}{}_{\alpha\gamma} B_{\nu} \tag{3.3a}$$

$$B_{\alpha\|\bar{\gamma}} = B_{\alpha|\bar{\gamma}} + \Gamma^{\nu}{}_{\alpha\bar{\gamma}} B_{\nu} \tag{3.3b}$$

$$B_{\bar{\alpha}\|\gamma} = B_{\bar{\alpha}|\gamma} + \Gamma^{\bar{\nu}}{}_{\bar{\alpha}\gamma} B_{\bar{\nu}} \tag{3.3c}$$

$$B_{\bar{\alpha}\|\bar{\gamma}} = B_{\bar{\alpha}|\bar{\gamma}} + \Gamma^{\bar{\nu}}{}_{\bar{\alpha}\bar{\gamma}} B_{\bar{\nu}} \tag{3.3d}$$

Note that with all derivatives of a vector, the vector index is always contracted with one of the first two indices of the Christoffel symbol, which two indices are always of the same type (barred or unbarred). The last index is always the derivative index.

We also require that covariant differentiation maintain the conjugation symmetry. This last statement means that:

$$\left(A^{\alpha}{}_{\|\gamma}\right)^{*} = A^{\bar{\alpha}}{}_{\|\bar{\gamma}} \tag{3.4a}$$

and

$$\left(A^{\alpha}{}_{\|\bar{\gamma}}\right)^{*} = A^{\bar{\alpha}}{}_{\|\gamma}. \tag{3.4b}$$

Equations (3.4a) and (3.4b) imply conjugation symmetry on the Christoffel symbols, namely:

$$\left(\Gamma^{\alpha}{}_{\rho\gamma}\right)^{*} = \Gamma^{\bar{\alpha}}{}_{\bar{\rho}\bar{\gamma}} \tag{3.5a}$$

$$\left(\Gamma^{\bar{\alpha}}{}_{\bar{\rho}\gamma}\right)^{*} = \Gamma^{\alpha}{}_{\rho\bar{\gamma}}. \tag{3.5b}$$

We will frequently need to use the explicit difference S between the barred and unbarred Christoffel symbols and thus give it its own symbol as defined in Equation (3.6a) and (3.6b). Since the difference between two Christoffel symbols transforms as a tensor under coordinate transformations, we know that S must be a tensor, and we proved in the first paper [6] that S is pure real. We also showed that S switches sign when the type of one of its indices switches, as it does in Equation (3.6).

$$2S^{\alpha}{}_{\beta\gamma} = 2\Gamma^{\alpha}{}_{\beta\gamma} - 2\Gamma^{\alpha}{}_{\beta\bar{\gamma}} = -2S^{\alpha}{}_{\beta\bar{\gamma}} \tag{3.6}$$

We note one immediate corollary to this definition of S in (3.7a) or equivalently in Equation (3.7b).

$$A^{\alpha}{}_{\|\gamma} - A^{\alpha}{}_{\|\bar{\gamma}} = -\left(\Gamma^{\alpha}{}_{\beta\gamma} - \Gamma^{\alpha}{}_{\beta\bar{\gamma}}\right)A^{\beta} = -2S^{\alpha}{}_{\beta\gamma}A^{\beta} \tag{3.7a}$$

$$A^{\bar{\alpha}}{}_{\|\gamma} - A^{\bar{\alpha}}{}_{\|\bar{\gamma}} = -\left(\Gamma^{\bar{\alpha}}{}_{\bar{\beta}\gamma} - \Gamma^{\bar{\alpha}}{}_{\bar{\beta}\bar{\gamma}}\right)A^{\bar{\beta}} = -2S^{\bar{\alpha}}{}_{\bar{\beta}\gamma}A^{\bar{\beta}} \tag{3.7b}$$

Christoffel Symmetry Condition: Since we want this new geometry to reduce to classical Riemannian geometry when the metric tensor is pure real, we will want to impose symmetry conditions on the Christoffel symbols which map to the classical symmetry in the last two indices. We chose the condition shown in Equations (3.8a)-(3.8c), where the Christoffel is unchanged when the last two indices switch. As will be shown below, this condition leads to generalized Bianchi conditions and resulting Einstein tensor. An alternate symmetry, which used complex conjugation in Equation (3.8c), was also explored, but it was found to produce only a constant E&M field—useless for this geometry.

$$\Gamma^{\alpha}{}_{\beta\gamma} = \Gamma^{\alpha}{}_{\gamma\beta} \tag{3.8a}$$

$$\Gamma^{\bar{\alpha}}{}_{\bar{\beta}\bar{\gamma}} = \Gamma^{\bar{\alpha}}{}_{\bar{\gamma}\bar{\beta}} \tag{3.8b}$$

$$\Gamma^{\alpha}{}_{\beta\bar{\gamma}} = \Gamma^{\alpha}{}_{\gamma\bar{\beta}} \tag{3.8c}$$

Zero Derivatives for the Metric Tensor: The last requirement we want to make on the Christoffel symbols is that the covariant derivative of the metric tensor vanishes as usual. Since there are two types of covariant derivative, we get two equations—which are just complex conjugates of each other (with a flip of indices).

$$c_{\rho\bar{\nu}\|\gamma} = \Gamma^{\alpha}{}_{\rho\gamma} c_{\alpha\bar{\nu}} + c_{\rho\bar{\alpha}} \Gamma^{\bar{\alpha}}{}_{\bar{\nu}\gamma} + c_{\rho\bar{\nu}|\gamma} = 0 \tag{3.9a}$$

$$c_{\rho\bar{\nu}\|\bar{\gamma}} = \Gamma^{\alpha}{}_{\rho\bar{\gamma}} c_{\alpha\bar{\nu}} + c_{\rho\bar{\alpha}} \Gamma^{\bar{\alpha}}{}_{\bar{\nu}\bar{\gamma}} + c_{\rho\bar{\nu}|\bar{\gamma}} = 0 \tag{3.9b}$$

By taking the average of these two equations, we note that there is a third derivative form in Equation (3.9c) that also gives a zero derivative for the metric tensor and thus may have a place in the geometry. Since this derivative is the average of barred and unbarred derivatives, we use a different slashed symbol. This form of derivative will be essential to derive the Riemann curvature tensor. All the forms of the barred, unbarred and slashed Christoffel symbols are summarized in Equation (3.13).

We note as a caution that the derivation of the Christoffel symbols in the first paper [6] required the selection of a fully real Christoffel symbol to make the complex Christoffel symbols correctly invariant under coordinate transformations. We chose the classical Christoffel symbols for convenience and because they gave an elegant non-recursive formula for the Christoffel symbols. However, as a note for the future, we could also have chosen the fully slashed Christoffel symbols, which are also pure real and are more naturally embedded into this complex geometry. This would change some of the higher order terms in the field equations. Assuming this basic unification process is correct, then only experimental observations will be able to determine which path to take.

$$\Gamma^{a}{}_{b\not\gamma} = \frac{1}{2}\left(\Gamma^{a}{}_{b\gamma} + \Gamma^{a}{}_{b\bar{\gamma}}\right) \tag{3.10a}$$

$$c_{\rho\bar{\nu}\|\not\gamma} = \Gamma^{\alpha}{}_{\rho\not\gamma}\, c_{\alpha\bar{\nu}} + c_{\rho\bar{\alpha}}\Gamma^{\bar{\alpha}}{}_{\bar{\nu}\not\gamma} + c_{\rho\bar{\nu}|\gamma} = 0 \tag{3.10b}$$

The E&M Field Equation:

Once we have the transpose symmetry for the Christoffel symbols (Equations (3.8)) and the zero derivative Equations (3.9), we can derive the desired E&M field equation by antisymmetrizing Equation (3.9a) i.e. summing over all signed permutations of ρ, ν and γ. (This is denoted $\{\ldots\ldots\}_{(\rho,\nu,\gamma)}$.) Since the Christoffel symbols are all symmetric in the lower indices, all the Christoffel terms vanish, leaving only one term in (3.11).

$$\left\{\Gamma^{\alpha}{}_{\rho\gamma} c_{\alpha\bar{\nu}} + c_{\rho\bar{\alpha}}\Gamma^{\bar{\alpha}}{}_{\bar{\nu}\gamma} + c_{\rho\bar{\nu}|\gamma}\right\}_{(\rho,\nu,\gamma)} = \left\{c_{\rho\bar{\nu}|\gamma}\right\}_{(\rho,\nu,\gamma)} = 0 \tag{3.10}$$

Expanding that metric tensor in (3.11), we get further simplification in Equation (3.12) since the symmetric $g_{\rho\underline{\nu}}$ term drops out in the antisymmetrization. What is left is i times the standard E&M field equation.

$$\left\{c_{\rho\bar{\nu}|\gamma}\right\}_{(\rho,\gamma,\nu)} = \left\{\boldsymbol{g}_{\rho\bar{\nu}|\gamma} + i\boldsymbol{f}_{\rho\bar{\nu}|\gamma}\right\}_{(\rho,\gamma,\nu)} = i\left\{\boldsymbol{f}_{\rho\bar{\nu}|\gamma}\right\}_{(\rho,\gamma,\nu)} = 0 \tag{3.11}$$

We note that this field Equation (3.11) is well known to imply that the E&M field tensor $f_{\rho\underline{\nu}}$ can be expressed as the antisymmetric derivative of a vector potential A_{ρ} as shown in Equation (3.12) [7] . Thus we could have this E&M field Equation (3.11) satisfied simply by using Equation (3.12) instead of $f_{\rho\underline{\nu}}$. This is often done in modern E&M calculations.

$$\boldsymbol{f}_{\rho\bar{\nu}} = \boldsymbol{A}_{\rho|\nu} - \boldsymbol{A}_{\nu|\rho} \tag{3.12}$$

As a note on the indexing convention used here: A Greek index carries a type, either barred or unbarred, while a roman index indicates a more general index of either type. This convention makes many of the equations much simpler. For instance, the general solution for a lowered Christoffel symbol is shown in Equation (3.13a), with the only condition being that indices a and b must have the opposite type. The bold green Christoffel symbols are the standard ones from classical general relativity, and the bold green covariant derivatives use the classical covariant derivatives. With the classical derivatives and the barred and unbarred derivatives, this complex Riemannian geometry has a total of three different types of derivatives—each of which has its purpose. It is fascinating that the classical Christoffel symbols evolve intrinsically in this complex geometry—as explained in the first paper [6] .

The same expression for the raised Christoffel symbols is given in Equation (3.13b). Note that the barred or unbarred type of index a is used to determine the sign of the f terms. In Equation (3.13a), if index a is barred, then that index position is barred in the two f terms to give $i\left(f_{b\bar{a}\|c}+f_{c\bar{a}\|b}\right)/3$. If a is unbarred, then these terms become $i\left(f_{\bar{b}a\|c}+f_{\bar{c}a\|b}\right)/3$, which is minus the previous term. In Equation (3.13b), the type of index a determines the type of index d, which then determines the sign of the f terms as in the previous sentence. Equations (3.13c), (3.13d) and (3.13e) summarize the slashed Christoffel symbols.

$$\Gamma_{abc}=c_{ad}\boldsymbol{\Gamma}^{d}{}_{bc}-S_{abc}+i\left(f_{ba\|c}+f_{ca\|b}\right)/3+iW_{abc} \tag{3.13a}$$

$$\Gamma^{a}{}_{bc}=\boldsymbol{\Gamma}^{a}{}_{bc}+c^{ad}S_{dbc}+ic^{ad}\left(f_{bd\|c}+f_{cd\|b}\right)/3+ic^{ad}W_{dbc} \tag{3.13b}$$

$$\Gamma^{\not{a}}{}_{\not{b}c}=\boldsymbol{\Gamma}^{a}{}_{bc}-iF^{ad}S_{dbc}+F^{ad}\left(f_{bd\|c}+f_{cd\|b}\right)/3+F^{ad}W_{dbc} \tag{3.13c}$$

$$\Gamma^{a}{}_{b\not{c}}=\boldsymbol{\Gamma}^{a}{}_{bc}+ic^{ad}\left(f_{bd\|c}+f_{cd\|b}\right)/3+ic^{ad}W_{dbc} \tag{3.13d}$$

$$\Gamma^{\not{a}}{}_{\not{b}\not{c}}=\boldsymbol{\Gamma}^{a}{}_{bc}+F^{ad}\left(f_{bd\|c}+f_{cd\|b}\right)/3+F^{ad}W_{abc} \tag{3.13e}$$

For easy reference we summarize the symmetries of the S and W tensors here, derived in the previous paper.

The S tensor changes sign if the type of the first two indices flip (which must be opposite types) or the third index flips.

$$S_{\bar{\alpha}\beta\bar{\gamma}}=-S_{\bar{\alpha}\beta\gamma}=-S_{\alpha\bar{\beta}\gamma}=S_{\alpha\overline{\beta\gamma}} \tag{3.14}$$

If types are fixed, then S is invariant under all 6 permutations of the index values.

$$S_{\bar{\alpha}\beta\bar{\gamma}}=S_{\bar{\alpha}\gamma\bar{\beta}}=S_{\bar{\beta}\alpha\bar{\gamma}}=S_{\bar{\beta}\gamma\bar{\alpha}}=S_{\bar{\gamma}\alpha\bar{\beta}}=S_{\bar{\gamma}\beta\bar{\alpha}} \tag{3.15}$$

The W tensor changes sign if the type of the first two indices switch types, but index type makes no difference to the sign in the last index for W.

$$W_{\bar{\alpha}\beta\bar{\gamma}}=W_{\bar{\alpha}\beta\gamma}=-W_{\alpha\overline{\beta\gamma}}=-W_{\alpha\bar{\beta}\gamma} \tag{3.16}$$

Just like S, if types are fixed, then W is invariant under all 6 permutations of the index values

$$W_{\bar{\alpha}\beta\bar{\gamma}}=W_{\bar{\alpha}\gamma\bar{\beta}}=W_{\bar{\beta}\alpha\bar{\gamma}}=W_{\bar{\beta}\gamma\bar{\alpha}}=W_{\bar{\gamma}\alpha\bar{\beta}}=W_{\bar{\gamma}\beta\bar{\alpha}} \tag{3.17}$$

With these symmetries, we note that the sum of two Christoffel symbols which switch type in the first two indices is twice the classical Christoffel symbol.

4. SCALAR CURVATURE TENSOR

Now we can deduce some general properties of this type of geometry. We start with the scalar curvature tensor where a scalar function is twice differentiated. When both derivatives are of the same type, then derivatives commute as usual as shown in Equations (4.1a) and (4.1b). However, when they are of different types, then they don't commute in general as shown in Equation (4.1c). Note that since all derivatives of a scalar field are identical, we can write Equation (4.1c) using the slashed derivative, which is obviously a tensor equation since the difference between two Christoffel symbols is a tensor under coordinate transformations.

$$\Phi_{\|\beta\|\gamma}-\Phi_{\|\gamma\|\beta}=\left(\Gamma^{\nu}{}_{\beta\gamma}-\Gamma^{\nu}{}_{\gamma\beta}\right)\Phi_{\|\nu}=0 \tag{4.1a}$$

$$\Phi_{\|\bar{\beta}\|\bar{\gamma}} - \Phi_{\|\bar{\gamma}\|\bar{\beta}} = \left(\Gamma^{\bar{\nu}}{}_{\bar{\beta}\gamma} - \Gamma^{\bar{\nu}}{}_{\gamma\bar{\beta}}\right)\Phi_{\|\bar{\nu}} = 0 \tag{4.1b}$$

$$\Phi_{\|\beta\|\bar{\gamma}} - \Phi_{\|\bar{\gamma}\|\beta} = \left(\Gamma^{\nu}{}_{\beta\bar{\gamma}}\Phi_{\|\nu} - \Gamma^{\bar{\nu}}{}_{\bar{\gamma}\beta}\Phi_{\|\bar{\nu}}\right) = \left(\Gamma^{\nu}{}_{\beta\bar{\gamma}} - \Gamma^{\bar{\nu}}{}_{\bar{\gamma}\beta}\right)\Phi_{\|\nu} = K^{\nu}{}_{\beta\bar{\gamma}}\Phi_{\|\nu} \tag{4.1c}$$

The general result is given in Equations (4.2a) and (4.2b), where for convenience in writing these equations we also define the scalar curvature tensor $K^{m}{}_{ab} = \Gamma^{m}{}_{ab} - \Gamma^{m}{}_{ba}$, which vanishes unless the indices a and b are different types. With the symmetry in the lower two indices of a Christoffel symbol and the complex conjugation symmetry when the types are switched, $K^{m}{}_{ab}$ is either identically 0 or pure imaginary, and thus can be treated always as pure imaginary—obviously vanishing in the classical limit where all fields are real.

$$\Phi_{\|a\|b} - \Phi_{\|b\|a} = K^{m}{}_{ab}\Phi_{\|m} \tag{4.2a}$$

where

$$K^{m}{}_{ab} = \Gamma^{m}{}_{ab} - \Gamma^{m}{}_{ba} \text{ (pure imaginary)} \tag{4.2b}$$

Since the slashed derivative has been established as a third valid type of covariant derivative with its own Christoffel symbols, we point out that indices a and b in Equation (4.2b) can be slashed as well as barred or unbarred. The slashed scalar curvature tensor is simply and intuitively related to the unslashed scalar curvature tensor as shown in Equation (4.3)

$$K^{m}{}_{ab} = \Gamma^{m}{}_{ab} - \Gamma^{m}{}_{ba} = \frac{1}{2}\left(\Gamma^{m}{}_{ab} + \Gamma^{m}{}_{a\bar{b}} - \Gamma^{m}{}_{ba} - \Gamma^{\bar{m}}{}_{\bar{b}a}\right) = \frac{1}{2}\left(K^{m}{}_{ab} + K^{m}{}_{a\bar{b}}\right) \tag{4.3}$$

5. RIEMANN CURVATURE TENSOR

We next derive the Riemann curvature tensor and explore its symmetries. It is defined as usual by taking the antisymmetrized double derivative of a vector—but in this derivative the vector may be barred or unbarred and each derivative may be barred or unbarred thereby creating 8 different curvature tensors. While this may seem like a large number, the symmetries will reduce them to 3 types.

Derivatives of the Same Type: To make this process explicit, we start with the simplest curvature tensor derivation, where all indices are of the same type. Begin with the definition of covariant derivative of an arbitrary twice differentiable vector field A^{α} in Equation (5.1a), then differentiate again in (5.1b), reverse the order of differentiation in (5.1c) and subtract the two in (5.1d). The terms involving A^{β} subtract away, and the coefficient of the $A^{\alpha}{}_{|\nu}$ factor is designated the Riemann curvature tensor as usual. Its final equation summarized in (5.1f) looks identical to the standard expression for the Riemann curvature tensor in textbooks, although the Christoffel symbols here are now complex.

$$A^{\alpha}{}_{\|\gamma} = A^{\alpha}{}_{|\gamma} - \Gamma^{\alpha}{}_{\beta\gamma} A^{\beta} \tag{5.1a}$$

$$A^{\alpha}{}_{\|\gamma\|\delta} = A^{\alpha}{}_{|\gamma|\delta} - \Gamma^{\alpha}{}_{\beta\gamma|\delta} A^{\beta} - \Gamma^{\alpha}{}_{\beta\gamma} A^{\beta}{}_{|\delta} - \Gamma^{\alpha}{}_{\beta\delta} A^{\beta}{}_{|\gamma} + \Gamma^{\nu}{}_{\gamma\delta} A^{\alpha}{}_{|\nu} + \Gamma^{\alpha}{}_{\nu\delta} \Gamma^{\nu}{}_{\beta\gamma} A^{\beta} - \Gamma^{\nu}{}_{\gamma\delta} \Gamma^{\alpha}{}_{\beta\nu} A^{\beta} \tag{5.1b}$$

$$A^{\alpha}{}_{\|\delta\|\gamma} = A^{\alpha}{}_{|\delta|\gamma} - \Gamma^{\alpha}{}_{\beta\delta|\gamma} A^{\beta} - \Gamma^{\alpha}{}_{\beta\delta} A^{\beta}{}_{|\gamma} - \Gamma^{\alpha}{}_{\beta\gamma} A^{\beta}{}_{|\delta} + \Gamma^{\nu}{}_{\delta\gamma} A^{\alpha}{}_{|\nu} + \Gamma^{\alpha}{}_{\nu\gamma} \Gamma^{\nu}{}_{\beta\delta} A^{\beta} - \Gamma^{\nu}{}_{\delta\gamma} \Gamma^{\alpha}{}_{\beta\nu} A^{\beta} \tag{5.1c}$$

$$A^{\alpha}{}_{\|\gamma\|\delta} - A^{\alpha}{}_{\|\delta\|\gamma} = \left(-\Gamma^{\alpha}{}_{\beta\gamma|\delta} + \Gamma^{\alpha}{}_{\beta\delta|\gamma} - \Gamma^{\alpha}{}_{\nu\gamma} \Gamma^{\nu}{}_{\beta\delta} + \Gamma^{\alpha}{}_{\nu\delta} \Gamma^{\nu}{}_{\beta\gamma} \right) A^{\beta} \tag{5.1d}$$

$$A^{\alpha}{}_{\|\gamma\|\delta} - A^{\alpha}{}_{\|\delta\|\gamma} = R^{\alpha}{}_{\beta\gamma\delta} A^{\beta} \tag{5.1e}$$

$$R^{\alpha}{}_{\beta\gamma\delta} = -\Gamma^{\alpha}{}_{\beta\gamma|\delta} + \Gamma^{\alpha}{}_{\beta\delta|\gamma} - \Gamma^{\alpha}{}_{\nu\gamma} \Gamma^{\nu}{}_{\beta\delta} + \Gamma^{\alpha}{}_{\nu\delta} \Gamma^{\nu}{}_{\beta\gamma} \tag{5.1f}$$

We note that the type of index α had no impact on the structure of the curvature tensor derivation above. Thus the same equation for the Riemann curvature tensor will work if α is barred or unbarred (and always the same type as β). Also γ and δ above can be any two indices of the same type and the algebra will work the same. Thus Equation (5.1f) applies to all index combinations where γ and δ have the same type.

Derivatives of Different Types: The derivation with mixed types of derivatives begins the same way in Equations (5.2a)-(5.2c), but the difference in Equation (5.2d) has two extra terms, marked in yellow because there are two types of derivatives—one barred and one unbarred. We must keep track of those two terms when doing mixed derivatives.

$$A^{\alpha}{}_{\|\gamma} = A^{\alpha}{}_{|\gamma} - \Gamma^{\alpha}{}_{\beta\gamma} A^{\beta} \tag{5.2a}$$

$$A^{\alpha}{}_{\|\gamma\|\bar{\delta}} = A^{\alpha}{}_{|\gamma|\bar{\delta}} - \Gamma^{\alpha}{}_{\beta\gamma|\bar{\delta}} A^{\beta} - \Gamma^{\alpha}{}_{\beta\gamma} A^{\beta}{}_{|\bar{\delta}} - \Gamma^{\alpha}{}_{\beta\bar{\delta}} A^{\beta}{}_{|\gamma} - \Gamma^{\nu}{}_{\gamma\bar{\delta}} A^{\alpha}{}_{\|\nu} + \Gamma^{\alpha}{}_{\nu\bar{\delta}} \Gamma^{\nu}{}_{\beta\gamma} A^{\beta} \tag{5.2b}$$

$$A^{\alpha}{}_{\|\bar{\delta}\|\gamma} = A^{\alpha}{}_{|\bar{\delta}|\gamma} - \Gamma^{\alpha}{}_{\beta\bar{\delta}|\gamma} A^{\beta} - \Gamma^{\alpha}{}_{\beta\bar{\delta}} A^{\beta}{}_{|\gamma} - \Gamma^{\alpha}{}_{\beta\gamma} A^{\beta}{}_{|\bar{\delta}} - \Gamma^{\bar{\nu}}{}_{\bar{\delta}\gamma} A^{\alpha}{}_{\|\bar{\nu}} + \Gamma^{\alpha}{}_{\nu\gamma} \Gamma^{\nu}{}_{\beta\bar{\delta}} A^{\beta} \tag{5.2c}$$

$$\left(\Gamma^{\bar{\nu}}{}_{\bar{\delta}\gamma} A^{\alpha}{}_{\|\bar{\nu}} - \Gamma^{\nu}{}_{\gamma\bar{\delta}} A^{\alpha}{}_{\|\nu}\right) \tag{5.2d}$$

To find the curvature tensor we need to separate the right side of Equation (5.2d) into a tensor component multiplying A^{β} and a tensor term multiplying the covariant derivative of A^{α}. The part multiplying A^{β} will then be the generalized Riemann curvature tensor.

To make this work, we need to use the slashed derivative as we did in Equation (4.1c), which is half way between the barred and unbarred derivatives. The algebra is shown in Equations (5.3a)-(5.3c). In Equation (5.3a) we expand the covariant derivatives. In Equation (5.3b), we add and subtract slashed Christoffel terms, which we regroup in Equation (5.3c). Finally we substitute that result into Equation (5.2d) to get Equation (5.3d), which has clearly separated out a covariant component (the same difference of Christoffel symbols we found in Equation (4.1c)) that multiplies the covariant derivative, leaving the rest to be the Riemann curvature tensor term.

For compactness we defined $\Delta\Gamma^{a}{}_{bc} = \Gamma^{a}{}_{bc} - \Gamma^{a}{}_{b\not{c}}$, which is itself a tensor under coordinate transformations since it is the difference between two Christoffel symbols. As shown in Equation (5.3e) (referencing Equation (3.6a)), it is exactly equal to the S tensor $S^{a}{}_{bc}$, which also equals $-S^{a}{}_{b\bar{c}}$.

$$\Gamma^{\bar{\nu}}{}_{\bar{\delta}\gamma} A^{\alpha}{}_{\|\bar{\nu}} - \Gamma^{\nu}{}_{\gamma\bar{\delta}} A^{\alpha}{}_{\|\nu} = \Gamma^{\bar{\nu}}{}_{\bar{\delta}\gamma}\left(A^{\alpha}{}_{|\nu} - \Gamma^{\alpha}{}_{\beta\bar{\nu}} A^{\beta}\right) - \Gamma^{\nu}{}_{\gamma\bar{\delta}}\left(A^{\alpha}{}_{|\nu} - \Gamma^{\alpha}{}_{\beta\nu} A^{\beta}\right) \tag{5.3a}$$

$$\begin{aligned} \Gamma^{\bar{\nu}}{}_{\bar{\delta}\gamma} A^{\alpha}{}_{\|\bar{\nu}} - \Gamma^{\nu}{}_{\gamma\bar{\delta}} A^{\alpha}{}_{\|\nu} &= \Gamma^{\bar{\nu}}{}_{\bar{\delta}\gamma}\left(A^{\alpha}{}_{|\nu} - \Gamma^{\alpha}{}_{\beta\not{\nu}} A^{\beta}\right) - \Gamma^{\nu}{}_{\gamma\bar{\delta}}\left(A^{\alpha}{}_{|\nu} - \Gamma^{\alpha}{}_{\beta\nu} A^{\beta}\right) \\ &\quad + \Gamma^{\bar{\nu}}{}_{\bar{\delta}\gamma}\left(\Gamma^{\alpha}{}_{\beta\not{\nu}} - \Gamma^{\alpha}{}_{\beta\bar{\nu}}\right) A^{\beta} - \Gamma^{\nu}{}_{\gamma\bar{\delta}}\left(\Gamma^{\alpha}{}_{\beta\not{\nu}} - \Gamma^{\alpha}{}_{\beta\nu}\right) A^{\beta} \end{aligned} \tag{5.3b}$$

$$\Gamma^{\bar{\nu}}{}_{\bar{\delta}\gamma} A^{\alpha}{}_{\|\bar{\nu}} - \Gamma^{\nu}{}_{\gamma\bar{\delta}} A^{\alpha}{}_{\|\nu} = \left(\Gamma^{\bar{\nu}}{}_{\bar{\delta}\gamma} - \Gamma^{\nu}{}_{\gamma\bar{\delta}}\right) A^{\alpha}{}_{\|\not{\nu}} + \Gamma^{\bar{\nu}}{}_{\bar{\delta}\gamma}\left(\Gamma^{\alpha}{}_{\beta\not{\nu}} - \Gamma^{\alpha}{}_{\beta\bar{\nu}}\right) A^{\beta} - \Gamma^{\nu}{}_{\gamma\bar{\delta}}\left(\Gamma^{\alpha}{}_{\beta\not{\nu}} - \Gamma^{\alpha}{}_{\beta\nu}\right) A^{\beta} \tag{5.3c}$$

$$A^{\alpha}{}_{\|\gamma\|\bar{\delta}} - A^{\alpha}{}_{\|\bar{\delta}\|\gamma} = \left(-\Gamma^{\alpha}{}_{\beta\gamma|\bar{\delta}} + \Gamma^{\alpha}{}_{\beta\bar{\delta}|\gamma} - \Gamma^{\alpha}{}_{\nu\gamma}\Gamma^{\nu}{}_{\beta\bar{\delta}} + \Gamma^{\alpha}{}_{\nu\bar{\delta}}\Gamma^{\nu}{}_{\beta\gamma} + \Gamma^{\nu}{}_{\gamma\bar{\delta}}\Delta\Gamma^{\alpha}{}_{\beta\nu} - \Gamma^{\bar{\nu}}{}_{\bar{\delta}\gamma}\Delta\Gamma^{\alpha}{}_{\beta\bar{\nu}}\right)A^{\beta} + K^{\not\nu}{}_{\bar{\delta}\gamma}A^{\alpha}{}_{\|\not\nu} \tag{5.3d}$$

where

$$\Delta\Gamma^{a}{}_{bc} = \Gamma^{a}{}_{bc} - \Gamma^{a}{}_{b\not c} = \Gamma^{a}{}_{bc} - \frac{1}{2}\left(\Gamma^{a}{}_{bc} + \Gamma^{a}{}_{b\bar{c}}\right) = \frac{1}{2}\left(\Gamma^{a}{}_{bc} - \Gamma^{a}{}_{b\bar{c}}\right) = S^{a}{}_{bc} \tag{5.3e}$$

Since the left side and the last term in Equation (5.e) are obviously tensors, then so is the first term which multiples A^{β}. Then since A^{β} is an arbitrary differentiable vector field, its coefficient (the Riemann curvature tensor) must also be a tensor. Using the simplification in Equation (5.3e), we summarize the final result that applies to all eight possible barred and unbarred index combinations in Equations (5.4a) and (5.4b), which use the roman indices that can be barred or unbarred—all combinations giving the correct result. The slashed indices are simply averages of barred and unbarred indices and do not add any more degrees of freedom. Note that indices a and b must have the same type, and that the two internal m and n indices must have types consistent with the types of c and d respectively.

$$A^{a}{}_{\|c\|d} - A^{a}{}_{\|d\|c} = R^{a}{}_{bcd}A^{b} + K^{\not m}{}_{cd}A^{a}{}_{\|\not m} \tag{5.4a}$$

$$R^{a}{}_{bcd} = -\Gamma^{a}{}_{bc|d} + \Gamma^{a}{}_{bd|c} - \Gamma^{a}{}_{ec}\Gamma^{e}{}_{bd} + \Gamma^{a}{}_{ed}\Gamma^{e}{}_{bc} + \left(\Gamma^{m}{}_{cd}S^{a}{}_{bm} - \Gamma^{n}{}_{dc}S^{a}{}_{bn}\right) \tag{5.4b}$$

We also note a second alternative form of the curvature tensor in Equation (5.4c) which uses covariant derivatives rather than partial derivatives. It is easily verified by expanding the covariant derivative, and is sometimes more useful in derivations because it immediately gives covariant derivatives. It also allows the Riemann curvature to be written with all indices lowered without much change in the form as shown in Equation (5.4d) and then expanded out with simple derivatives in Equation (5.4e). This same technique works with the classical Riemann curvature tensor as well.

$$R^{a}{}_{bcd} = -\Gamma^{a}{}_{bc\|d} + \Gamma^{a}{}_{bd\|c} + \Gamma^{a}{}_{ec}\Gamma^{e}{}_{bd} - \Gamma^{a}{}_{ed}\Gamma^{e}{}_{bc} + \left(\Gamma^{m}{}_{cd}S^{a}{}_{bm} - \Gamma^{n}{}_{dc}S^{a}{}_{bn}\right) \tag{5.4c}$$

$$R_{abcd} = -\Gamma_{abc\|d} + \Gamma_{abd\|c} + \Gamma_{aec}\Gamma^{e}{}_{bd} - \Gamma_{aed}\Gamma^{e}{}_{bc} + \left(\Gamma^{m}{}_{cd}S_{abm} - \Gamma^{n}{}_{dc}S_{abn}\right) \tag{5.4d}$$

$$R_{abcd} = -\Gamma_{abc|d} + \Gamma_{abd|c} - \Gamma_{ebc}\Gamma^{e}{}_{ad} + \Gamma_{ebd}\Gamma^{e}{}_{ac} + \left(\Gamma^{m}{}_{cd}S_{abm} - \Gamma^{n}{}_{dc}S_{abn}\right) \tag{5.4e}$$

One immediate question that arises is how many independent curvature tensors do we have? Since indices a, c and d can be barred or unbarred, we begin with 8. (The type of index b depends on the type of index a and thus is not an independent degree of freedom.) Complex conjugation switches barred and unbarred types and thus divides the number of degrees of freedom in half to 4. Finally, if the last indices are of different type, switching their order simply changes the overall sign of the tensor. This reduces the number of different curvature tensors to 3.

Curvature Tensors for Lowered Indices: As customary, lowered indices also have their own mixed derivative relationships, which use the same curvature tensors as defined in (5.4). To get the lowered curvature tensors, we simply twice differentiate the scalar function $\Phi = B^{a}A_{a}$, using Equation (4.1c), where A and B are arbitrary twice differentiable vector fields as shown in Equation (5.5a).

$$\left(B^{a}A_{a}\right)_{\|c\|d} - \left(B^{a}A_{a}\right)_{\|d\|c} = \left(\Gamma^{m}{}_{cd} - \Gamma^{m}{}_{dc}\right)\left(B^{a}A_{a}\right)_{\|m} \tag{5.5a}$$

After some algebra, we get the familiar result in Equation (5.5d) with the expected additional term using the scalar curvature tensor.

$$\begin{aligned}&\left(B^{a}{}_{\|c\|d}A_{a} + B^{a}{}_{\|c}A_{a\|d} + B^{a}{}_{\|d}A_{a\|c} + B^{a}A_{a\|c\|d}\right) - \left(B^{a}{}_{\|d\|c}A_{a} + B^{a}{}_{\|d}A_{a\|c} + B^{a}{}_{\|c}A_{a\|d} + B^{a}A_{a\|d\|c}\right)\\ &= \left(B^{a}{}_{\|c\|d} - B^{a}{}_{\|d\|c}\right)A_{a} + B^{a}\left(A_{a\|c\|d} - A_{a\|d\|c}\right) = K^{m}{}_{cd}\left(B^{a}A_{a}\right)_{\|m}\end{aligned} \tag{5.5b}$$

Using the definition of the Riemann curvature above in Equation (5.3a) and reordering terms, we get Equation (5.5c).

$$\begin{aligned}B^{a}\left(A_{a\|c\|d} - A_{a\|d\|c}\right) &= -R^{b}{}_{acd}B^{a}A_{b} - K^{m}{}_{cd}B^{a}{}_{\|m} + K^{m}{}_{cd}\left(B^{a}A_{a}\right)_{\|m}\\ &= -R^{b}{}_{acd}A_{b}B^{a} + K^{m}{}_{cd}A_{a\|m}B^{a}\end{aligned} \tag{5.5c}$$

Since B^{a} is an arbitrary twice differentiable field, we can remove it from the equality to get Equation (5.4d). This equation is identical to the classical equation except for the addition of the scalar curvature tensor term K.

$$A_{a\|c\|d} - A_{a\|d\|c} = -R^{b}{}_{acd}A_{b} + K^{m}{}_{cd}A_{a\|m} \tag{5.5d}$$

Thus we see that the curvature tensor for lowered indices is the same curvature tensor as for raised indices with a switch in sign—the familiar classical result.

6. SYMMETRIES OF THE RIEMANN CURVATURE TENSOR

Now we get to the critical discussion of the symmetries on the Riemann curvature tensor which will allow us to construct the Einstein tensor and field equations. Symmetries come in two versions. One version has the types moving with the indices, and the other version has types remaining in their fixed position with the indices moving. In classical relativity, all indices are the same type, so the difference never appears.

Last two indices: The most obvious symmetry for the Riemann curvature tensor is the antisymmetry in the last two indices, which trivially comes from the definition of the curvature tensor. This symmetry requires that types move with indices. Thus this symmetry is expressed simply as Equation (6.1).

$$R_{abcd} = -R_{abdc} \text{ (typed indices)} \tag{6.1}$$

First two indices: We can derive another symmetry relationship for the complex Riemann tensor using techniques similar to ABS. Begin with the scalar function ϕ defined as the dot product between two arbitrary twice differentiable vector fields $A^{\nu} \leftarrow$ and $B^{\underline{\tau}}$ in Equation (6.2a), and then take the difference of double derivatives as shown in Equations (6.2b). For a scalar function this difference is zero, but it can also be calculated using the derivatives of the vectors as shown in Equation (6.2c). Using Equations (5.3a) and e to write the vector differences in terms of the curvature tensor we get Equation (6.2d).

$$\phi = A^{\nu} c_{\nu\bar{\bar{\tau}}} B^{\bar{\bar{\tau}}} \quad (\qquad 6.2a)$$

$$\phi_{\|\alpha\|\beta} - \phi_{\|\beta\|\alpha} = 0 \tag{6.2b}$$

$$= \left(A^{\nu}{}_{\|\alpha\|\beta} - A^{\nu}{}_{\|\beta\|\alpha} \right) c_{\nu\bar{\bar{\tau}}} B^{\bar{\bar{\tau}}} + A^{\nu} c_{\nu\bar{\bar{\tau}}} \left(B^{\bar{\bar{\tau}}}{}_{\|\alpha\|\beta} - B^{\bar{\bar{\tau}}}{}_{\|\beta\|\alpha} \right) \tag{6.2c}$$

$$= R^{\nu}{}_{\rho\alpha\beta} A^{\rho} c_{\nu\bar{\tau}} B^{\bar{\tau}} + A^{\nu} c_{\nu\bar{\tau}} R^{\bar{\tau}}{}_{\bar{\rho}\alpha\beta} B^{\bar{\rho}} \tag{6.2d}$$

Shuffling some indices and combining terms, we get the Equation (6.3a), which becomes the familiar antisymmetric relationship in Equation (6.3b) when we realize that A^{ν} and $B^{\bar{\tau}}$ are arbitrary vector fields.

$$\left(R_{\bar{\tau}\nu\alpha\beta} + R_{\nu\bar{\tau}\alpha\beta}\right) A^{\nu} B^{\bar{\tau}} = 0 \tag{6.3a}$$

$$R_{\bar{\tau}\nu\alpha\beta} = -R_{\nu\bar{\tau}\alpha\beta} \tag{6.3b}$$

The exact same logic holds when α and β are both barred indices resulting in Equation (6.3c).

$$R_{\bar{\tau}\nu\bar{\alpha}\bar{\beta}} = -R_{\nu\bar{\tau}\bar{\alpha}\bar{\beta}} \tag{6.3c}$$

We also see that the same symmetry holds with mixed type derivatives. We define the scalar function ϕ again in Equation (6.2a), and then apply the mixed derivative formula for ϕ, using Equation (4.1c), to give Equation (6.2b).

$$\phi = A^{\nu} c_{\nu\bar{\tau}} B^{\bar{\tau}} \tag{6.2a}$$

$$\phi_{\|\alpha\|\bar{\beta}} - \phi_{\|\bar{\beta}\|\alpha} = K^{\rho}{}_{\alpha\bar{\beta}} \Phi_{\|\rho} = K^{\rho}{}_{\alpha\bar{\beta}} \left(A^{\nu} c_{\nu\bar{\tau}} B^{\bar{\tau}}\right)_{\|\rho} = K^{\rho}{}_{\alpha\bar{\beta}} A^{\nu}{}_{\|\rho} c_{\nu\bar{\tau}} B^{\bar{\tau}} + A^{\nu} c_{\nu\bar{\tau}} K^{\rho}{}_{\alpha\bar{\beta}} B^{\bar{\tau}}{}_{\|\rho} \tag{6.2b}$$

Next we expand ϕ and differentiate the function piece by piece using the curvature tensor formula in Equation (5.4d) to give Equation (6.2c). Note that all first derivative terms simply subtract away trivially, and that all covariant derivatives of the metric tensor $c_{\nu\bar{\tau}}$ vanish automatically.

$$\begin{aligned} &= \left(A^{\nu}{}_{\|\alpha\|\bar{\beta}} - A^{\nu}{}_{\|\bar{\beta}\|\alpha}\right) c_{\nu\bar{\tau}} B^{\bar{\tau}} + A^{\nu} c_{\nu\bar{\tau}} \left(B^{\bar{\tau}}{}_{\|\alpha\|\bar{\beta}} - B^{\bar{\tau}}{}_{\|\bar{\beta}\|\alpha}\right) \\ &= R^{\nu}{}_{\rho\alpha\bar{\beta}} A^{\rho} c_{\nu\bar{\tau}} B^{\bar{\tau}} + A^{\nu} c_{\nu\bar{\tau}} R^{\bar{\tau}}{}_{\bar{\rho}\alpha\bar{\beta}} B^{\bar{\rho}} + K^{\rho}{}_{\alpha\bar{\beta}} A^{\nu}{}_{\|\rho} c_{\nu\bar{\tau}} B^{\bar{\tau}} + K^{\rho}{}_{\alpha\bar{\beta}} A^{\nu} c_{\nu\bar{\tau}} B^{\bar{\tau}}{}_{\|\rho} \end{aligned} \tag{6.2c}$$

Realizing that the K terms subtract away and then switching index names to combine terms, we get the Equation (6.5a), which becomes the familiar antisymmetric symmetry relationship in Equation (6.5b) when we realize that A^{ν} and $B^{\bar{\tau}}$ are arbitrary vector fields.

$$\left(R_{\bar{\tau}\nu\alpha\beta} + R_{\nu\bar{\tau}\alpha\beta}\right) A^{\nu} B^{\bar{\tau}} = 0 \tag{6.5a}$$

$$R_{\bar{\tau}\nu\alpha\beta} = -R_{\nu\bar{\tau}\alpha\bar{\beta}} \tag{6.5b}$$

The final summarized result is that the antisymmetry in the first two indices holds for all possible combinations of barred and unbarred indices as stated in Equation (6.6a).

$$R_{abcd} = -R_{bacd} \text{ (typed indices)} \tag{6.6a}$$

Three-fold symmetry: Another immediately obvious symmetry is a familiar three-fold symmetry but (caution!) here the types do not move with the indices. Such symmetries have not been found to be very useful compared to symmetries where the types move with the indices, but we list it here for completeness. This symmetry follows immediately from inspection of the expressions for the curvature tensor in (5.3c). Since every Christoffel symbol is symmetric in the last two indices, every term immediately zeroes out under the signed sum of permutations. To differentiate clearly between the cases where the types are fixed to positions and where the types are fixed to the indices, we annotate each equation and also use the equation format $\{\ldots\}_{(a,b,c)}$ for typed positions and $\{\ldots\}_{(a,b,c)}$ for typed indices.

$$\left\{R^{a}{}_{bcd}\right\}_{(b,c,d)} = 0 \text{ (typed positions)} \tag{6.6b}$$

$$\left\{R_{abcd}\right\}_{(b,c,d)} = 0 \text{ (typed positions)} \tag{6.6c}$$

$$\left\{R_{bacd}\right\}_{(b,c,d)} = 0 \text{ (typed positions)} \tag{6.6d}$$

Bianchi Identities: From here on, all the symmetries will use typed indices. The Bianchi identities can now be gotten using the following straightforward technique. We simply begin in Equation (6.8a) and (6.8b) with two trivial identities, where the first term of each equation produces the same 6 terms as the second but with opposite signs since two of the indices are switched.

$$\left\{A_{a\|b\|c\|d}\right\}_{[b,c,d]}+\left\{A_{a\|b\|d\|c}\right\}_{[b,c,d]}=0$$ (typed indices) (6.8a)

$$\left\{A_{a\|b\|d\|c}\right\}_{[b,c,d]}+\left\{A_{a\|d\|b\|c}\right\}_{[b,c,d]}=0$$ (typed indices) (6.8b)

Subtracting Equation (6.8b) from Equation (6.8a), we get Equation (6.8c). This becomes Equation (6.8d) when we use the Riemann curvature Equation (5.4d).

$$\left\{A_{a\|b\|c\|d}-A_{a\|b\|d\|c}\right\}_{[b,c,d]}+\left\{A_{a\|b\|d\|c}-A_{a\|d\|b\|c}\right\}_{[b,c,d]}=0$$

(typed indices) (6.8c)

$$\left\{-R^{n}{}_{acd}A_{n\|b}-R^{n}{}_{bcd}A_{a\|n}+K^{n}{}_{cd}A_{a\|b\|n}\right\}_{[b,c,d]}+\left\{\left(-R^{n}{}_{abd}A_{n}+K^{n}{}_{bd}A_{a\|n}\right)_{\|c}\right\}_{[b,c,d]}=0$$

(typed indices) (6.8d)

Now we expand the $\|c$ derivative in the second term in Equation (6.8e), and switch b and c in the term $K^{n}{}_{cd}A_{a\|b\|n}$ along with a sign switch (OK since we have the antisymmetrized $[b,c,d]$ sum applied).

$$\left\{-R^{n}{}_{acd}A_{n\|b}-R^{n}{}_{bcd}A_{a\|n}-K^{n}{}_{bd}A_{a\|c\|n}\right\}_{[b,c,d]}$$
$$+\left\{-R^{n}{}_{abd\|c}A_{n}-R^{n}{}_{abd}A_{n\|c}+K^{n}{}_{bd\|c}A_{a\|n}+K^{n}{}_{bd}A_{a\|n\|c}\right\}_{[b,c,d]}=0$$ (6.8e)

We note that the first term in the first line subtracts out the second term in the second line using the antisymmetric sum $[b,c,d]$. We also note that the two double derivative terms combine to give another term with a curvature tensor times A_n. Doing these simplifications and collecting terms, we have Equation (6.8f).

$$\left\{R^{n}{}_{bcd}\right\}_{[b,c,d]} A_{a\|n} + \left\{R^{n}{}_{abd\|c} + K^{m}{}_{bd}R^{n}{}_{acm}\right\}_{[b,c,d]} A_{n} - \left\{K^{n}{}_{bd\|c} + K^{m}{}_{bd}K^{n}{}_{cm}\right\}_{[b,c,d]} A_{a\|n} = 0 \tag{6.8f}$$

Note that if indices b, c and d were all the same type, then all the K terms would vanish and we would have only two terms left as shown in Equation (6.8g).

$$\text{If } (b,c,d) \text{ have the same type} \rightarrow \left\{R^{n}{}_{bcd}\right\}_{[b,c,d]} A_{a\|n} + \left\{R^{n}{}_{abd\|c}\right\}_{[b,c,d]} A_{n} = 0 \tag{6.8g}$$

Since an arbitrary field and its covariant derivative are two independent degrees of freedom, both components must separately vanish to give the familiar Bianchi identities as the coefficient of $A_n$and the familiar three-index symmetry of the Riemann curvature tensor as the coefficient of $A_{a\|n}$. Thus the only issue here is how to extend these results to include indices of different types, and the main issue is how to deal with the $A_{a\|n}$ and the $A_{a\|n}$ terms. Are they independent degrees of freedom or not? We believe that they have the effect of being independent degrees of freedom, which means that all the terms in Equation (6.8f) which multiply either $A_{a\|n}$, A_n or $A_{a\|n}$ are each individually zero.

The logic supporting this principle goes as follows: The first insight is to point out that the partial derivative $A_{a|n}$ is equally present in both the $A_{a\|n}$ and the $A_{a\|n}$ terms and can take on any value independent of A_a. This means that the sum of those two terms must vanish and also that the middle term multiplying the A_n field must independently vanish since A_n is an independent degree of freedom. This is enough to prove the generalized Bianchi identities in Equation (6.9a) and the vanishing of the sum in Equation (6.9b).

$$\text{Bianchi Identities: } \left\{R^{n}{}_{abd\|c} + K^{m}{}_{bd}R^{n}{}_{acm}\right\}_{[b,c,d]} = 0 \tag{6.9a}$$

Triplet Symmetry:

$$\left\{R^{n}{}_{bcd}\right\}_{[b,c,d]} - \left\{K^{n}{}_{bd\|c} + K^{m}{}_{bd}K^{n}{}_{cm}\right\}_{[b,c,d]} = 0 \tag{6.9b}$$

Dropping the now proven zero middle term in Equation (6.8f), we can rewrite Equation (6.8f) as (6.9c). Then using Equation (6.9b), the terms multiplying the derivatives $A_{a|n}$ also drop out as zero to give Equation (6.9d).

$$\left\{R^{n}{}_{bcd}\right\}_{[b,c,d]} A_{a\|n} - \left\{K^{n}{}_{bd\|c} + K^{m}{}_{bd} K^{n}{}_{cm}\right\}_{[b,c,d]} A_{a\|n} = 0 \tag{6.9c}$$

$$\left(\left\{R^{n}{}_{bcd}\right\}_{[b,c,d]} \Gamma^{e}{}_{an} - \left\{K^{n}{}_{bd\|c} + K^{m}{}_{bd} K^{n}{}_{cm}\right\}_{[b,c,d]} \Gamma^{e}{}_{an}\right) A_{e} = 0 \tag{6.9d}$$

We then add and subtract terms in Equation (6.9d) to get Equation (6.9e).

$$\left(\left\{R^{n}{}_{bcd}\right\}_{[b,c,d]} - \left\{K^{n}{}_{bd\|c} + K^{m}{}_{bd} K^{n}{}_{cm}\right\}_{[b,c,d]}\right) \Gamma^{e}{}_{an} A_{e} + \left(\left\{R^{n}{}_{bcd}\right\}_{[b,c,d]}\right)\left(\Gamma^{e}{}_{an} - \Gamma^{e}{}_{an}\right) A_{e} = 0 \tag{6.9e}$$

Since we have already proven that the first term is identically zero in Equation (6.9b), so is the second term. Also recognizing the difference of the Christoffel symbols from Equation (5.3e) as $S^{e}{}_{an}$, we simplify Equation (6.9e) to Equation (6.9f).

$$\left(\left\{R^{n}{}_{bcd}\right\}_{[b,c,d]}\right) S^{e}{}_{an} A_{e} = 0 \tag{6.9f}$$

Given that A_e is an arbitrary field and a is an arbitrary index, as long as the S field is even microscopically non-zero everywhere, then we conclude the very simple result in Equation (6.9g).

$$\left\{R^{n}{}_{bcd}\right\}_{[b,c,d]} = 0 \tag{6.9g}$$

The final results of this analysis are given in Equations (6.10).

Bianchi Identities: $\left\{R^{n}{}_{abd\|c} + K^{m}{}_{bd} R^{n}{}_{acm}\right\}_{[b,c,d]} = 0$ (typed indices) (6.10a)

Triplet R Symmetry: $\left\{R^{n}{}_{bcd}\right\}_{[b,c,d]} = 0$ (typed indices) (6.10b)

Triplet K Symmetry: $\left\{K^{n}{}_{bd\|c} + K^{m}{}_{bd}K^{n}{}_{cm}\right\}_{[b,c,d]} = 0$ (typed indices) (6.10c)

We can also lower the n indices to get an equivalent form for the Bianchi and triplet identities assuming that only barred and unbarred indices are used. We are, however, cautious about extending the results to slashed indices n, which do not have an obvious ability to be raised and lowered with a metric tensor like the other two.

Bianchi Identities: $\left\{R_{nabd\|c} + K^{m}{}_{bd}R_{nacm}\right\}_{[b,c,d]} = 0$ (typed indices) (6.10a)

Triplet R Symmetry: $\left\{R_{nbcd}\right\}_{[b,c,d]} = 0$ (typed indices) (6.10b)

Modified Bianchi Identities: To put these equalities into a slightly different form we can use the Equation (6.10b) (with a lowered first index), which says that the sum over all six signed permutations of the last three indices of R_{nabc} equals 0. By keeping one term on the left and putting the other 5 terms with opposite sign on the right, we can rewrite the Bianchi identities in Equations (6.10b) as (6.11a). (Note that the c and d index names have been switched here to make the relationship look simpler.)

Since we are antisymmetrizing with the indices (b, c, d), the first term on the right changes sign when we flip the cb indices, which means it can be added to the left side—keeping everything equal to zero. Similarly the second and fourth terms on the right simply double, and the third and fifth terms also double. Dividing everything by two, we end up with Equation (6.9b).

$$\left\{R_{eabc\|d}\right\}_{[b,c,d]} = \left\{R_{eacb\|d} + R_{ebac\|d} + R_{ecba\|d} - R_{ecab\|d} - R_{ebca\|d}\right\}_{(b,c,d)} = \left\{-K^{m}{}_{bd}R_{eacm}\right\}_{[b,c,d]} \quad (6.11a)$$

$$\left\{R_{ebac\|d} + R_{ecba\|d}\right\}_{[b,c,d]} = \left\{-K^{m}{}_{bd}R_{eacm}\right\}_{[b,c,d]} \text{ (typed indices)} \quad (6.11b)$$

Now we can use the antisymmetric property in the last two typed indices (Equation (6.1)) combined with the antisymmetry in the $[b,c,d]$ indices to make the two terms equal, resulting in Equation (6.12a) or equivalently in Equation (12b) and (12c), using the antisymmetric conditions in the first two and last two indices. This is exactly the same symmetry condition as in classical relativity.

$$\left\{R_{ebac\|d}\right\}_{[b,c,d]}=\left\{-K^{m}{}_{bd}R_{eacm}\right\}_{[b,c,d]} \text{ (typed indices)} \qquad (6.12a)$$

$$\left\{R_{ebca\|d}\right\}_{(b,c,d)}=\left\{K^{m}{}_{bd}R_{eacm}\right\}_{[b,c,d]} \text{ (typed indices)} \qquad (6.12b)$$

$$\left\{R_{beca\|d}\right\}_{(b,c,d)}=\left\{-K^{m}{}_{bd}R_{eacm}\right\}_{[b,c,d]} \text{ (typed indices)} \qquad (6.12c)$$

Note that if $[b,c,d]$ indices are all the same type, as we will require to derive the Einstein tensor below, then all the K terms vanish since the lower two indices are always of the same type. In particular, Equation (6.12a) will be the starting point for the derivation of the generalized Einstein tensor.

Einstein Tensor: With this background, we can now show the generalization of the zero divergence condition of the Einstein tensor, which will lay the basis for the field equations. The symmetries required for the field equations apply only to a particular set of barred and unbarred indices, namely $R_{\alpha\bar{\beta}\gamma\bar{\delta}}$. Since we need to be able to contract the tensors in the Einstein field equation, which requires both barred and unbarred indices to be present. First, however, we need to define some contracted tensors for clarity. With two barred and two unbarred indices in $R_{\alpha\bar{\beta}\gamma\bar{\delta}}$, we can contract to a two dimensional tensor in four different ways—all of which are needed to make the zero divergence Einstein tensor.

$$R_{\bar{\beta}\gamma}=-R_{\alpha\bar{\beta}\gamma\bar{\delta}}c^{\alpha\bar{\delta}} \qquad (6.13a)$$

$$R_{\gamma\bar{\beta}}=-R_{\gamma\bar{\delta}\alpha\bar{\beta}}c^{\alpha\bar{\delta}}=-R_{\bar{\delta}\gamma\bar{\beta}\alpha}c^{\alpha\bar{\delta}}=R_{\bar{\gamma}\beta}{}^{*} \qquad (6.13b)$$

$$R'_{\beta\bar{\gamma}}=R_{\alpha\bar{\delta}\beta\bar{\gamma}}c^{\alpha\bar{\delta}}=R_{\bar{\delta}\alpha\bar{\gamma}\beta}c^{\alpha\bar{\delta}}=R'_{\gamma\bar{\beta}}{}^{*} \qquad (6.13c)$$

$$R''_{\beta\bar{\gamma}}=R_{\beta\bar{\gamma}\alpha\bar{\delta}}c^{\alpha\bar{\delta}}=R_{\bar{\gamma}\beta\bar{\delta}\alpha}c^{\alpha\bar{\delta}}=R''_{\gamma\bar{\beta}}{}^{*} \qquad (6.13d)$$

Note that two of the double contractions are equal i.e.

$$R' = R'' \tag{6.14}$$

Now to get the generalization of the zero divergence Einstein tensor, begin with Equation (6.15a) (a modified Bianchi identity) and write out all the terms explicitly in Equation (6.15a). Note that the generalized Latin indices have been made specific typed indices here. Also note that here all the permuted terms are the same type, which means that the K term is zero, resulting in a very traditional and familiar result.

$$\left\{R_{\alpha\bar{\beta}\gamma\bar{\delta}\|\bar{\eta}}\right\} = R_{\alpha\bar{\beta}\gamma\bar{\delta}\|\bar{\eta}} - R_{\alpha\bar{\beta}\gamma\bar{\eta}\|\bar{\delta}} + R_{\alpha\bar{\eta}\gamma\bar{\beta}\|\bar{\delta}} - R_{\alpha\bar{\eta}\gamma\bar{\delta}\|\bar{\beta}} + R_{\alpha\bar{\delta}\gamma\bar{\eta}\|\bar{\beta}} - R_{\alpha\bar{\delta}\gamma\bar{\beta}\|\bar{\eta}} = 0 \tag{6.15a}$$

Now contract Equation (6.15a) with $c^{\alpha\underline{\delta}}$ and $c^{\gamma\underline{\eta}}$ to get (6.15b).

$$-R_{\bar{\beta}\gamma}{}^{\|\gamma} - R''{}_{\gamma\bar{\beta}}{}^{\|\gamma} - R_{\gamma\bar{\beta}}{}^{\|\gamma} + R_{\|\bar{\beta}} + R'{}_{\|\bar{\beta}} - R'{}_{\gamma\bar{\beta}}{}^{\|\gamma} \tag{6.15b}$$

This can be rewritten as follows when we remember that $R' = R''$.

$$\left(R_{\bar{\beta}\gamma} + R_{\gamma\bar{\beta}} - c_{\gamma\bar{\beta}}R\right)^{\|\gamma} + \left(R'{}_{\gamma\bar{\beta}} - \frac{1}{2}c_{\gamma\bar{\beta}}R'\right)^{\|\gamma} + \left(R''{}_{\gamma\bar{\beta}} - \frac{1}{2}c_{\gamma\bar{\beta}}R''\right)^{\|\gamma} = 0 \tag{6.15c}$$

This then will satisfy a zero divergence criterion if we define the generalized Einstein tensor as

$$2G_{\bar{\beta}\gamma} = \left(R_{\bar{\beta}\gamma} + R_{\gamma\bar{\beta}} - c_{\gamma\bar{\beta}}R\right) + \left(R'{}_{\gamma\bar{\beta}} - \frac{1}{2}c_{\gamma\bar{\beta}}R'\right)^{\|\gamma} + \left(R''{}_{\gamma\bar{\beta}} - \frac{1}{2}c_{\gamma\bar{\beta}}R''\right)^{\|\gamma} = 0 \tag{6.16a}$$

The generalized Einstein tensor can also be written in the more compact (and = familiar looking) form of Equation (6.16b) where the contracted curvature tensor R^s is a sum of the four different contracted forms as shown in Equation (6.16c).

$$G_{\bar{\beta}\gamma} = R^s{}_{\bar{\beta}\gamma} - \frac{1}{2}c_{\gamma\bar{\beta}}R^s \tag{6.16b}$$

where

$$2R^{s}{}_{\bar{\beta}\gamma} = R_{\bar{\beta}\gamma} + R_{\gamma\bar{\beta}} + R'{}_{\gamma\bar{\beta}} + R''{}_{\gamma\bar{\beta}} \tag{6.16c}$$

Since the sum $R_{\bar{\beta}\gamma} + R_{\gamma\bar{\beta}}$ is Hermitian from Equations (6.13a) and (6.13b), and since $R''{}_{\gamma\bar{\beta}}$ and $R''{}_{\gamma\bar{\beta}}$ are each Hermitian from Equations (6.13c) and (6.13d), the tensor $R^{s}{}_{\bar{\beta}\gamma}$ is also Hermitian. Since the metric tensor $c_{\gamma\underline{\beta}}$ is also Hermitian, the full Einstein tensor is thus Hermitian.

The final result of zero divergence is then summarized in Equation (6.17), which looks exactly like the familiar classical conservation equation except for the bar above the first index.

$$G_{\bar{\beta}\gamma}{}^{\|\gamma} = 0 \tag{6.17}$$

Overview of the Field Equations: In classical relativity, the Einstein tensor is set equal to a general stress- energy-momentum tensor that also satisfies the zero divergence condition. This was an ad hoc assumption extrapolated from known gravitational laws at the time, and its solutions have occupied a significant number of scientists for many different scenarios. Equation (6.18) is thus expected to be one of the field equations of the generalized Riemannian geometry. However, it does contain more than gravity. It contains four different types of field: gravity $g_{\mu\nu}$, electromagnetism $f_{\mu\nu}$ and two other third rank fields $S_{\mu\nu\rho}$ and $W_{\mu\nu\rho}$, which together form mass-energy and charge source terms. Thus this form of Riemannian geometry offers the possibility of a self-contained geometry including all mass-energy and charge sources.

7. SUMMARY

In this paper we have explored complex Riemannian geometry further. In particular, we have derived the Riemann curvature tensors (three different ones) and the symmetries that go with them. Using those results, we found the generalized Einstein tensor which satisfies the familiar zero divergence condition suitable for field equations. The next paper will propose and explore field equations for all four of the geometric fields, gravity, electromagnetism and two others that provide mass and charge sources.

8. CONCLUSION

We see that complex Riemannian geometry inherently has a metric with a real gravity field and an imaginary part resembling the electromagnetic field plus two wholly new fully symmetric third rank tensor fields $W_{\alpha\beta\gamma}$ and $S_{\alpha\beta\gamma}$ which represent additional potential interactions. The real symmetric fields W and S each have 24 degrees of freedom, and are PT symmetry breaking in that S and W reverse sign when $x^{\alpha} \Rightarrow -x^{\alpha}$. Assuming that CPT will remain a valid symmetry, one might expect that S and W will be involved in charge-current sources so that under all three inversions S and W would remain unchanged. The third and final paper of this series explores the Einstein tensor field Equation (6.18) above plus other field equations required for the electromagnetic field and the two third rank S and W fields. Only more work can tell if this complex Riemannian geometry offers a pathway into a fully unified theory.

REFERENCES

1. Weyl, H. (1918) Gravitation und Elektrizität. Sitzungsberichte der K öniglich Preussischen Akademie der Wissenschaften, 465.
2. Kaluza, T. (1921) Zum Unittsproblem in der Physik. 2nd Edition, Math. Phys, Berlin, 966-972.
3. Eddington, A.S. (1924) The Mathematical Theory of Relativity. 2nd Edition, Cambridge University Press, Cambridge.
4. Mie, G. (1912) Annals of Physics, 37, 511-534.
5. Reichenbächer, E. (1917) Annals of Physics, 52, 134-173.
6. Hutchin, R.A. (2015) Journal of Modern Physics, 6, 749-757.
7. Adler, R., Basin, M. and Schiffer, M. (1975) Introduction to General Relativity. McGraw Hill, New York.

CHAPTER 2

Noncommutative-Geometry Inspired Charged Wormholes with Low Tidal Forces

Peter K. F. Kuhfittig, Vance D. Gladney

Department of Mathematics, Milwaukee School of Engineering, Milwaukee, WI, USA

ABSTRACT

When Morris and Thorne first proposed that wormholes might be actual physical structures suitable for interstellar travel, they needed to pay close attention to certain traversability conditions such as low tidal forces, which placed severe constraints on the wormhole geometry. Even more problematical was the need for "exotic matter" resulting from the unavoidable violation of the null energy condition required to hold a wormhole open. The purpose of this paper is to overcome these problems by starting with the charged wormhole model of Kim and Lee and assuming a noncommutative-geometry background: the violation of the null energy condition can be attributed to the latter, while the electric charge allows the reduction of the tidal forces to acceptable levels without invoking the trivial zero-tidal-force assumption.

Keywords: Charged Wormholes, Noncommutative Geometry, Tidal Constraints

1. INTRODUCTION

Wormholes are handles or tunnels in spacetime that link widely separated regions of our Universe or different universes altogether. Morris and Thorne [1] proposed the following line element for the wormhole spacetime:

$$ds^2 = -e^{2\Phi(r)}dt^2 + \frac{dr^2}{1-b(r)/r} + r^2\left(d\theta^2 + \sin^2\theta d\phi^2\right), \tag{1}$$

using units in which $c = G = 1$. In this line element, $b = b(r)$ is called the shape function and $\Phi = \Phi(r)$ is called the redshift function, which must be everywhere finite to avoid an event horizon. For the shape function we must have $b(r_0) = r_0$, where $r = r_0$ is the radius of the throat of the wormhole. An important requirement is the flare-out condition at the throat: $b'(r_0) < 1$, while $b(r) < r$ near the throat. The flare-out condition can only be met by violating the null energy condition

$$T_{\alpha\beta}k^\alpha k^\beta \geq 0 \tag{2}$$

for all null vectors k^α , where $T_{\alpha\beta}$ is the stress-energy tensor. Matter that violates this condition is referred to as "exotic" in Ref. [1] . In particular, for the radial outgoing null vector $(1,1,0,0)$, the violation reads $T_{\alpha\beta}k^\alpha k^\beta = \rho + p_r < 0$. Here $T_t^t = -\rho(r)$, the energy density, and $T_r^r = p_r(r)$, the radial pressure. ($T_\theta^\theta = T_\phi^\phi = p_t(r)$, the lateral pressure.)

In Ref. [1] , Morris and Thorne also discussed the tidal forces that a traveler would be subjected to and subsequently put the following constraint on the redshift function: $\Phi'(r) \leq g_\oplus / \left(c^2\sqrt{1-b(r)/r}\right)$. The feasibility of this assumption will be discussed later.

The purpose of this paper is to revisit these requirements by starting with the modified charged wormhole model due to Kim and Lee [2] and assuming a non- commutative-geometry background.

Similar issues involving traversable wormholes in Lovelock gravity are discussed in Refs. [3] [4] [5] , with emphasis on the energy conditions. In particular, Ref. [5] considers higher-dimensional thin-shell wormholes.

Einstein-Gauss- Bonnet wormholes satisfying the weak energy condition are studied in Ref. [6] , while higher-dimensional evolving wormholes satisfying the null energy condition are discussed in Ref. [7] . Regarding the tidal constraints, it was noted in Ref. [8] that the zero-tidal-force assumption is incompatible with quantum field theory in classical general relativity.

2. NONCOMMUTATIVE GEOMETRY

An important outcome of string theory is the realization that coordinates may become noncommutative operators on a D-brane [9] [10] . The result is a fundamental discretization of spacetime due to the commutator $\left[x^{\mu}, x^{\nu}\right] = i\theta^{\mu\nu}$, where $\theta^{\mu\nu}$ is an antisymmetric matrix. According to Refs. [11] [12] [13] , noncommutativity replaces point-like objects by smeared objects, thereby eliminating the divergences that normally appear in general relativity.

Applications of noncommutative geometry are numerous and varied. One of the first was to wormholes in semi-classical gravity by Garattini and Lobo [14] . The same authors analyzed the stability of gravastars [15] . Gravitational lensing of wormholes in noncommutative geometry was studied in Ref. [16] , while the search for higher-dimensional wormholes was proposed in Ref. [17] .

It seems natural to model the smearing by means of a Gaussian distribution of minimal length $\sqrt{\alpha}$ instead of the Dirac delta function [12] [13] [17] [18] . A simpler but equally effective way is to assume that the energy density of the static and spherically symmetric and particle-like gravitational source has the form

$$\rho(r) = \frac{M\sqrt{\alpha}}{\pi^2\left(r^2+\alpha\right)^2}; \tag{3}$$

(see Refs. [19] and [20]). The point is that the mass M is diffused throughout the region of linear dimension $\sqrt{\alpha}$ due to the uncertainty. It should be noted that noncommutative geometry is an intrinsic property of spacetime and does not depend on particular features such as curvature.

Next, we note that the Einstein field equations $G_{\hat{\mu}\hat{\nu}} = 8\pi T_{\hat{\mu}\hat{\nu}}$ result in the following forms:

$$\rho(r) = \frac{b'}{8\pi r^2}, \tag{4}$$

$$p_r(r) = \frac{1}{8\pi}\left[-\frac{b}{r^3} + 2\left(1-\frac{b}{r}\right)\frac{\Phi'}{r}\right], \tag{5}$$

and

$$p_t(r) = \frac{1}{8\pi}\left(1-\frac{b}{r}\right)\left[\Phi'' - \frac{b'r-b}{2r(r-b)}\Phi' + (\Phi')^2 + \frac{\Phi'}{r} - \frac{b'r-b}{2r^2(r-b)}\right]. \tag{6}$$

Here $p_r(r)$ is the radial pressure and $p_t(r)$ is the lateral pressure. The conservation law $T^{\mu\nu}_{;\nu} = 0$ implies that only two of Equations (4)-(6) are independent. So Equation (6) can be obtained from the other two. (See Ref. [21] for further details.)

As a final comment, to make use of Equation (3), we can keep the standard form of the Einstein field equations in the sense that the Einstein tensor retains its original form, but the stress-energy tensor is modified [12] . It follows that the length scale used need not be restricted to the Planck scale.

3. THE MODIFIED KIM-LEE MODEL

Since black holes can carry an electric charge, it is natural to assume that wormholes can do likewise. So it was proposed by Kim and Lee [2] that for a wormhole with constant electric charge Q, the Einstein field equations take on the form

$$G^{(0)}_{\mu\nu} + G^{(1)}_{\mu\nu} = 8\pi\left[T^{(0)}_{\mu\nu} + T^{(1)}_{\mu\nu}\right]. \tag{7}$$

Given that the usual form is $G^{(0)}_{\mu\nu} = 8\pi T^{(0)}_{\mu\nu}$, the modified form in Equation (7) is obtained by adding the matter term $T^{(1)}_{\mu\nu}$ to the right side and the corresponding back reaction term $G_{\mu\nu}^{(1)}$ to the left side. The proposed metric is

$$ds^2 = -\left(1+\frac{Q^2}{r^2}\right)dt^2 + \left(1-\frac{b(r)}{r}+\frac{Q^2}{r^2}\right)^{-1} dr^2 + r^2\left(d\theta^2 + \sin^2\theta d\phi^2\right). \tag{8}$$

Observe that whenever $b \equiv 0$, the wormhole becomes a Reissner-Nordström black hole, and if $Q = 0$, the spacetime reverts to a Morris-Thorne wormhole, where the shape function $b = b(r)$ meets the usual requirements. Kim and Lee go on to show that the metric, Equation (8), is a self-consistent solution of the Einstein field equations. The shape function $b = b(r)$ of the Morris-Thorne wormhole is now replaced by the effective shape function

$$b_{\text{eff}}(r) = b(r) - \frac{Q^2}{r}, \tag{9}$$

to be discussed later.

To overcome certain difficulties in determining the tidal constraints, we will use the following modified metric already introduced in Refs. [22] and [23] :

$$ds^2 = -\left(1+S(r)+\frac{Q^2}{r^2}\right)dt^2 + \left(1-\frac{b(r)}{r}+\frac{Q^2}{r^2}\right)^{-1} dr^2 + r^2\left(d\theta^2 + \sin^2\theta d\phi^2\right). \tag{10}$$

here $S(r)$ is a differentiable function such that $S(r) > 0$ and $S'(r) > 0$. As noted in Ref. [22] , line element (10) remains a valid solution of the Einstein field equations.

4. THE MODIFIED KIM-LEE MODEL IN NONCOMMUTATIVE GEOMETRY

In the Kim-Lee model, Equation (8), and in the modified model, Equation (10),

$$\text{total charge} = \iiint_V \rho_q(r)\,\mathrm{d}V, \tag{11}$$

where $\rho_q(r)$ is the charge density. Adapting this to the noncommutative-geo- metry background, we assume that the charge density has the form

$$\rho_q(r) = \frac{Q^2\sqrt{\alpha}}{\pi^2\left(r^2+\alpha\right)^2}, \tag{12}$$

where Q^2 refers to the Kim-Lee model. To obtain the smeared charge $Q_\alpha^2(r)$, we evaluate

$$Q_\alpha^2(r) = \int_{r_0}^{r} 4\pi(r')^2 \frac{Q^2\sqrt{\alpha}}{\pi^2\left(r^2+\alpha\right)^2}\mathrm{d}r' = \frac{2Q^2\sqrt{\alpha}}{\pi}\left(\frac{1}{\sqrt{\alpha}}\tan^{-1}\frac{r}{\sqrt{\alpha}} - \frac{r}{r^2+\alpha}\right). \tag{13}$$

Observe that $Q_\alpha^2(r_0)=0$; also, seen from a large distance, $Q_\alpha^2(r)$ becomes Q^2.

Similarly, from Equation (4) and Equation (9), we get for the effective shape function,

$$b_{\text{eff}}(r) = \frac{4M\sqrt{\alpha}}{\pi}\left(\frac{1}{\sqrt{\alpha}}\tan^{-1}\frac{r}{\sqrt{\alpha}} - \frac{r}{r^2+\alpha}\right) - \frac{1}{r}\frac{2Q^2\sqrt{\alpha}}{\pi}\left(\frac{1}{\sqrt{\alpha}}\tan^{-1}\frac{r}{\sqrt{\alpha}} - \frac{r}{r^2+\alpha}\right) + C. \tag{14}$$

Since $Q_\alpha^2(r_0)=0$, we obtain from $b(r_0)=r_0$,

$$b_{\text{eff}}(r_0) = \frac{4M\sqrt{\alpha}}{\pi}\left(\frac{1}{\sqrt{\alpha}}\tan^{-1}\frac{r_0}{\sqrt{\alpha}} - \frac{r_0}{r_0^2+\alpha}\right) - 0 + C = r_0 \tag{15}$$

and hence the constant C. Thus

$$b_{\text{eff}}(r) = \frac{4M\sqrt{\alpha}}{\pi}\left(\frac{1}{\sqrt{\alpha}}\tan^{-1}\frac{r}{\sqrt{\alpha}} - \frac{r}{r^2+\alpha}\right) - \frac{1}{r}\frac{2Q^2\sqrt{\alpha}}{\pi}\left(\frac{1}{\sqrt{\alpha}}\tan^{-1}\frac{r}{\sqrt{\alpha}} - \frac{r}{r^2+\alpha}\right)$$
$$-\frac{4M\sqrt{\alpha}}{\pi}\left(\frac{1}{\sqrt{\alpha}}\tan^{-1}\frac{r_0}{\sqrt{\alpha}} - \frac{r_0}{r_0^2+\alpha}\right) + r_0. \tag{16}$$

For a wormhole solution, we require that $b'_{\text{eff}}(r_0) > 0$ and $b'_{\text{eff}}(r_0) < 1$, the flare-out condition mentioned earlier. To that end, we determine

$$b'_{\text{eff}}(r_0) = \frac{4M\sqrt{\alpha}}{\pi}\frac{2r_0^2}{\left(r_0^2+\alpha\right)^2} - \frac{1}{r_0}\frac{2Q^2\sqrt{\alpha}}{\pi}\frac{2r_0^2}{\left(r_0^2+\alpha\right)^2} = \frac{4r_0^2\sqrt{\alpha}}{\pi\left(r_0^2+\alpha\right)^2}\left(2M - \frac{1}{r_0}Q^2\right) \tag{17}$$

since $Q_\alpha^2(r_0) = 0$. Looking ahead to the traversability conditions which involve measurements in meters, we will assume that $r_0 > 1$ m. So to keep $b'_{\text{eff}}(r_0)$ positive, we must have

$$Q^2 \leq 2M. \tag{18}$$

Since $\sqrt{\alpha} \ll M$, it also follows that

$$b'_{\text{eff}}(r_0) < 1, \tag{19}$$

as required. Moreover, $b_{\text{eff}}(r) < r$ near the throat and

$$\lim_{r\to\infty}\frac{b_{\text{eff}}(r)}{r} = 0,$$

so that the wormhole spacetime is asymptotically flat.

From the exoticity condition in Ref. [1], $b'_{\text{eff}}(r_0) < 1$ is equivalent to the violation of the null energy condition, requiring the need for "exotic matter." According to Equation (17) and Equation (19), however, this violation is simply a consequence of the noncommutative-geometry background.

To study the effect that the tidal forces may have on the traveler, we return to the criterion

$$\Phi'(r) \leq \frac{g_{\oplus}}{c^2 \sqrt{1-b(r)/r}} \tag{20}$$

mentioned in Sec. 1. At the throat of the wormhole, this condition is trivially satisfied. It is subsequently proposed in Ref. [1] that the space stations be far enough away from the throat so that $b(r)/r$ becomes negligible. The result is the requirement

$$\Phi' \lesssim \left(9.2\times10^{15}\ \mathrm{m}\right)^{-1} \approx \left(10^{16}\ \mathrm{m}\right)^{-1}. \tag{21}$$

The modified Kim-Lee model, Equation (10),

$$\mathrm{e}^{2\Phi} = 1 + R(r) + \frac{Q^2}{r^2}$$

yields

$$\Phi(r) = \frac{1}{2}\ln\left[1 + R(r) + \frac{1}{r^2}\frac{2Q^2\sqrt{\alpha}}{\pi}\left(\frac{1}{\sqrt{\alpha}}\tan^{-1}\frac{r}{\sqrt{\alpha}} - \frac{r}{r^2+\alpha}\right)\right] \tag{22}$$

and

$$\Phi'(r) = \frac{\frac{1}{2}R'(r) - \frac{1}{r^3}\frac{2Q^2\sqrt{\alpha}}{\pi}\left(\frac{1}{\sqrt{\alpha}}\tan^{-1}\frac{r}{\sqrt{\alpha}} - \frac{r}{r^2+\alpha}\right) + \frac{2Q^2\sqrt{\alpha}}{\pi\left(r^2+\alpha\right)^2}}{1 + R(r) + \frac{1}{r^2}\frac{2Q^2\sqrt{\alpha}}{\pi}\left(\frac{1}{\alpha}\tan^{-1}\frac{r}{\sqrt{\alpha}} - \frac{r}{r^2+\alpha}\right)}. \tag{23}$$

As noted earlier, $R(r) > 0$ and $R'(r) > 0$, but it quickly becomes apparent that in view of Inequality (21), $R(r)$ has to be fine-tuned so that

$$\frac{1}{2}R'(r) - \frac{1}{r^3}\frac{2Q^2\sqrt{\alpha}}{\pi}\left(\frac{1}{\sqrt{\alpha}}\tan^{-1}\frac{r}{\sqrt{\alpha}} - \frac{r}{r^2+\alpha}\right) \tag{24}$$

becomes negligibly small near the proposed location of the station.

Remark: While the need for fine-tuning may not be desirable, it does not present a serious conceptual problem here: for large r, expression (24) approaches

$$\frac{1}{2}R'(r)-\frac{Q^2}{r^3}+\frac{2Q^2\sqrt{\alpha}}{\pi r^2\left(r^2+\alpha\right)}.$$

So $\frac{1}{2}R'\approx Q^2/r^3$ near the station where $R(r)$ has the approximate form $-Q^2/r^2+C$, $C>0$; C can be chosen so that the denominator in Equation (23) exceeds unity. As a result

$$\Phi'(r)\lesssim\frac{2Q^2\sqrt{\alpha}}{\pi\left(r^2+\alpha\right)^2}.$$

Equation (21) and Equation (23) now yield the condition

$$\Phi'(r)\lesssim\frac{2Q\sqrt{\alpha}}{\pi\left(r^2+\alpha\right)^2}<\left(10^{16}\text{ m}\right)^{-1}. \tag{25}$$

This requirement places the following constraint on the parameter α:

$$\sqrt{\alpha}<\frac{\left(r^2+\alpha\right)^2\times10^{-16}}{\left(2Q^2\right)/\pi}\text{ m}. \tag{26}$$

According to Ref. [12] , noncommutativity is not visible at presently accessible energies if $\sqrt{\alpha}<10^{-18}$ m. So according to Inequality (26), the constraint on α is much less severe and so would in principle be observable.

5. CONCLUSIONS

While Morris and Thorne [1] had demonstrated that wormholes could be macroscopic structures suitable for interstellar travel, they could not avoid certain problems with traversability. In particular, the structures would have to be designed to produce low tidal forces, while the unavoidable violation of the null energy condition called for the use of "exotic matter." The latter requirement is not only difficult to meet classically, but may even be unphysical. The former was met by assuming a constant redshift function, so that $\Phi' \equiv 0$, called the "zero-tidal-force solution".

To overcome these problems, this paper discusses the modified wormhole model due to Kim and Lee in conjunction with a noncommutative-geometry background. The violation of the null energy condition can be attributed to the latter, thereby avoiding the need for exotic matter. The presence of Q^2 in the modified Kim-Lee model allows the reduction of the tidal forces to acceptable levels without invoking the trivial zero-tidal-force assumption, again made pos- sible by the noncommutative-geometry background. Finally, the condition on Φ' places a constraint on the value of $\sqrt{\alpha}$, but one that is well above the limit of observability.

REFERENCES

1. Morris, M.S. and Thorne, K.S. (1988) Wormholes in Spacetime and Their Use for Interstellar Travel: A Tool for Teaching General Relativity. American Journal of Physics, 56, 395-412.
2. Kim, S.-W. and Lee, H. (2001) Exact Solutions of Charged Wormhole. Physical Review D, 63, Article ID: 064014.
3. Dehghani, M.H. and Dayyani, Z. (2009) Lorentzian Wormholes in Lovelock Gravity. Physical Revue D, 79, Article ID: 064010.
4. Zangeneh, M.K., Lobo, F.S.N. and Dehghani, M.H. (2015) Traversable Wormholes Satisfying the Weak Energy Condition in Third-Order Lovelock Gravity. Physical Review D, 92, Article ID: 124049.
5. Mehdizadeh, M.R., Zangeneh, M.K. and Lobo, F.S.N. (2015) Higher-Dimensional Thin-Shell Wormholes in Third-Order Lovelock Gravity. Physical Review D, 92, Article ID: 044022.
6. Mehdizadeh, M.R., Zangeneh, M.K. and Lobo, F.S.N. (2015) Einstein-Gauss-Bonnet Traversable Wormholes Satisfying the Weak Energy Condition. Physical Review D. 91, Article ID: 084004.
7. Zangeneh, M.K., Lobo, F.S.N. and Riazi, N. (2014) Higher-Dimensional Evolving Wormholes Satisfying the Null Energy Condition. Physical Review D, 90, Article ID: 024072.
8. Kuhfittig, P.K.F. (2015) Macroscopic Traversable Wormholes with Zero Tidal Forces Inspired by Noncommutative Geometry. International Journal of Modern Physics D, 24, Article ID: 1550023.
9. Witten, E. (1996) Bound States of Strings and p-Branes. Nuclear Physics B, 460, 335-350.

10. Seiberg, N. and Witten, E. (1999) String Theory and Noncommutative Geometry. Journal of High Energy Physics, 9909, Article ID: 032.
11. Smailagic, A. and Spallucci, E. (2003) Feynman Path Integral on the Non-Commutative Plane. Journal of Physics A, 36, L467-L471.
12. Nicolini, P., Smailagic, A. and Spallucci, E. (2006) Noncommutative Geometry Inspired Schwarzschild Black Hole. Physics Letters B, 632, 547-551.
13. Rinaldi, M. (2011) A New Approach to Non-Commutative Inflation. Classical and Quantum Gravity, 28, Article ID: 105022.
14. Garattini, R. and Lobo, F.S.N. (2009) Self-Sustained Traversable Wormholes in Noncommutative Geometry. Physics Letters B, 71, 146-152.
15. Lobo, F.S.N. and Garattini, R. (2013) Linearized Stability Analysis of Gravastars in Noncommutative Geometry. Journal of High Energy Physics, 1312, Article ID: 065.
16. Kuhfittig, P.K.F. (2016) Gravitational Lensing of Wormholes in Noncommutative Geometry. Scientific Voyage, 2, 1-8.
17. Rahaman, F., Kuhfittig, P.K.F., Ray, S. and Islam, S. (2012) Searching for Higher Dimensional Wormholes with Noncommutative Geometry. Physical Review D, 86, Article ID: 106101.
18. Kuhfittig, P.K.F. (2013) Macroscopic Wormholes in Noncommutative Geometry. International Journal of Pure and Applied Mathematics, 89, 401-408.
19. Liang, J. and Liu, B. (2012) Thermodynamics of Noncommutative Geometry Inspired BTZ Black Holes Based on Lorentzian Smeared Mass Distribution. Europhysics Letters, 100, Article ID: 30001.
20. Nozari, K. and Mehdipour, S.H. (2008) Hawking Radiation as Quantum Tunneling for a Noncommutative Schwarzschild Black Hole. Classical and Quantum Gravity, 25, Article ID: 175015.
21. Sushkov, S.V. (2005) Wormholes Supported by a Phantom Energy. Physical Review D, 71, Article ID: 043520.
22. Kuhfittig, P.K.F. (2011) On the Feasibility of Charged Wormholes. Central European Journal of Physics, 9, 1144-1150.
23. Kuhfittig, P.K.F. (2016) The Effect of Conformal Symmetry on Charged Wormholes. Journal of Applied Mathematics and Physics, 4, 2117-2125.

CHAPTER 3

The Geometry of the Mappings by General Dirichlet Series

Dorin Ghisa

York University, Glendon College, Toronto, Canada

ABSTRACT

We dealt in a series of previous publications with some geometric aspects of the mappings by functions obtained as analytic continuations to the whole complex plane of general Dirichlet series. Pictures illustrating those aspects contain a lot of other information which has been waiting for a rigorous proof. Such a task is partially fulfilled in this paper, where we succeeded among other things, to prove a theorem about general Dirichlet series having as corollary the Speiser's theorem. We have also proved that those functions do not possess multiple zeros of order higher than 2 and the double zeros have very particular locations. Moreover, their derivatives have only simple zeros. With these results at hand, we revisited GRH for a simplified proof.

Keywords: General Dirichlet Series, S_k Strips, Intertwining Curves, Fundamental Domains, Riemann Hypothesis

1. INTRODUCTION

The study of general Dirichlet series has its origins in the works [1] - [18] of E. Cahen, J. Hadamard, E. Landau, H. Bohr, G. H. Hardi, M. Riesz, T. Kojima, M. Kuniyeda, G. Valiron, etc. A lot of contemporary mathematicians created a diversified theory of general Dirichlet series, some insisting on the connection with the Laplace-Stieltjes transforms [19] - [24] as J. Yu, Y. Kong, S. Daochun, X. Luo, Y. Yan, C. Singhal, G. S. Srivastava, etc., others as P.K. Kamthan, S. K. Shing Gautam, L.H. Khoi, [25] [26] endowing them with some topological structures, extending to them the Nevanlinna theory (too many to be cited) or dealing with vector valued Dirichlet series [27] - [32] as A. Defant, M. Maestre, D. Perez-Garcia, J. Bonet, B.L. Srivastava, A. Sharma, G. S. Srivastava, etc.

We will use normalized series defined as follows. To any sequence $A=(1=a_1,a_2,\cdots)$ of complex numbers and any increasing sequence $\Lambda=(0=\lambda_1<\lambda_2<\cdots)$ such that $\lim_{n\to\infty}\lambda_n=\infty$ we associate the series

$$\zeta_{A,\Lambda}(s)=\sum_{n=1}^{\infty}a_n e^{-\lambda_n s} \tag{1}$$

Λ is called the type of the series (1) and the series defined by the same Λ will be called series of the same type. When $\lambda_n=\ln n$, we obtain the ordinary (proprement dites, [18]) Dirichlet series.

Suppose that A and Λ are such that the abscissa of convergence (see [14] [18]) of the series (1)

$$\sigma_c=\limsup_{n\to\infty}\frac{1}{\lambda_n}\log\left|\sum_{k=1}^{n}a_k\right| \text{ is finite.} \tag{2}$$

Then (see [1]) $z=\zeta_{A,\Lambda}(s)$ is an analytic function in the half plane $\Re(s)>\sigma_u$, where σ_u is the abscissa of uniform convergence of (1), and where σ_u is at most σ_c+D with

$$D=\limsup_{n\to\infty}\frac{\log n}{\lambda_n} \tag{3}$$

For the ordinary Dirichlet series $D=1$ and it is known that in the case of Dirichlet L-series defined by imprimitive Dirichlet characters $\sigma_c=0$ and $\sigma_u=1$, while in the case of primitive characters $\sigma_c=\sigma_u=0$.

In general, when $D=0$ then $\sigma_u=\sigma_c$, and the series (1) is an analytic function in the half plane $\Re(s)>\sigma_c$.

Suppose that this function can be continued analytically to the whole complex plane, except possible at $s=1$ which is a simple pole. We keep the notation $\zeta_{A,\Lambda}(s)$ for the function obtained by analytic continuation.

With the exception of a discrete set of points from the complex plane, this function is locally injective, i.e. it maps conformally and hence bijectively small neighbourhoods of every point onto some domains. Enlarging these neigh- bourhoods, the image domains get bigger. How big can they get? The answer is: they become the whole complex plane with some slits (see [33] [34] [35] [36]). A region with this property is called fundamental domain of $\zeta_{A,\Lambda}(s)$.

Our aim is to show that the complex plane can be divided into a countable number of sets whose interiors are fundamental domains of this function. We have done this previously for the Riemann Zeta function [37] [38] as well as for Dirichlet L-functions [35] . To do the same thing for functions obtained by analytic continuations of general Dirichlet series we made [36] the assumption that (1) satisfies a Riemann type of functional equation. We will show next that such a strong assumption is not necessary and we can obtain the same result by using just elementary properties of the conformal mapping.

It has been proved [36] that for normalized series (1) the limit $\lim_{\sigma\to+\infty}\zeta_{A,\Lambda}(\sigma+it)=1$ is uniform with respect to t. This means that for any $\epsilon>0$ there is $\sigma_\epsilon\geq\sigma_c$ such that $\sigma>\sigma_\epsilon$ implies $|\zeta_{A,\Lambda}(s)-1|<\epsilon$, therefore the whole half plane $\Re(s)>\sigma_\epsilon$ is mapped by $\zeta_{A,\Lambda}(s)$ into a disc centred at $z=1$ of radius ϵ.

An immediate consequence of this fact is that the abscissa of convergence of a normalized series is less than $+\infty$. Moreover, for $\epsilon<1$ there is no zero of $\zeta_{A,\Lambda}(s)$ in the half plane $\Re(s)>\sigma_\epsilon$. Also, if $\zeta_{A,\Lambda}(s)$ satisfies a Riemann type of functional equation, then there cannot be non trivial zeros of $\zeta_{A,\Lambda}(s)$ in the half plane $\Re(s)<-\sigma_\epsilon$ neither.

2. PRE-IMAGES OF LINES AND CIRCLES

Suppose that, for a point s_0 with $\Re(s_0) > \sigma_c$ the function $\zeta_{A,\Lambda}(s)$ has a real value $\zeta_{A,\Lambda}(s_0) > 1$. The continuation from s_0 along the interval $[1, +\infty)$, is a curve Γ'_k such that when $z = \zeta_{A,\Lambda}(s)$ tends to 1, we have that σ tends to $+\infty$ on Γ'_k, or there is a point $u_{k,j}$ such that $\zeta_{A,\Lambda}(u_{k,j}) = 1$ and the con- tinuation can be carried along the whole real axis giving rise to a curve $\Gamma_{k,j}$. We will show later that when $z = \zeta_{A,\Lambda}(s)$ tends to $+\infty$ on the real axis then σ tends to $-\infty$ on Γ'_k, or on $\Gamma_{k,j}$, in other words these curves cannot remain in a right half plane.

Let us notice first that $\zeta_{A,\Lambda}(s)$ are transcendental functions and $s = \infty$ is an essential singular point for them. The value $z = 0$ cannot be a lacunary value for $\zeta_{A,\Lambda}(s)$ since then $\frac{1}{\zeta_{A,\Lambda}(s)}$ would be a non polynomial integer function for which $s = \infty$ is an essential singular point and hence $s = 0$ would be also an essential singular point for $\zeta_{A,\Lambda}(s)$, which is not true. Then, by the Big Picard Theorem, $\zeta_{A,\Lambda}(s)$ has infinitely many zeros in every neigh- bourhood of $s = \infty$.

Given a bounded region of the plane, there is $r > 0$ such that the pre-image of the circle C_r centred at the origin and of radius r has only disjoint com- ponents, which are closed curves containing each one a unique zero of $\zeta_{A,\Lambda}(s)$ belonging to that region. When r increases those curves expand and they can touch one another at some points v_k (see Figure 1(b)). These are branch points of the function, since in every neighbourhood of v_k the function takes at least twice any value on the image circle. Therefore the derivative of $\zeta_{A,\Lambda}(s)$ cancels at v_k. It is obvious that any zero of the derivative, which is not a zero of the function itself, can be obtained in this way. Indeed, if v is such a zero, we can take $r = |\zeta_{A,\Lambda}(v)|$ and necessarily at least two components of the pre-image of C_r will pass through v.

What happens with those components of the pre-image of a circle C_r when $r = 1$? We have proved [36] , Theorem 1 that there is at least one unbounded component of the pre-image of the unit circle. That proof did not use the as- sumption of $\zeta_{A,\Lambda}(s)$ satisfying a Riemann type of functional equation and therefore it is true for any function $\zeta_{A,\Lambda}(s)$.

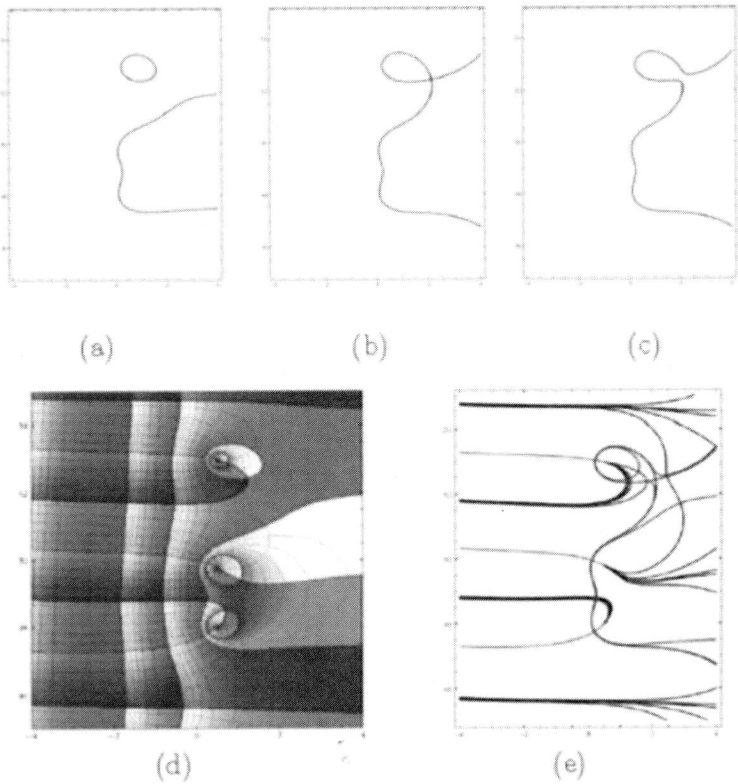

Figure 1. Pre-images of circles and the birth of a strip.

Let us notice that two curves Γ'_k and Γ'_l cannot intersect each other. Indeed, if s_0 would be a common point of these curves, then when $z = \zeta_{A,\Lambda}(s)$ moves on the interval I between $z_0 = \zeta_{A,\Lambda}(s_0)$ and 1 the point s describes an unbounded curve which bounds a domain mapped by $\zeta_{A,\Lambda}(s)$ onto the complex plane with a slit alongside the interval I. That domain should contain a pole of $\zeta_{A,\Lambda}(s)$ which is not true. Therefore an intersection point s_0 of the two curves cannot exist and consecutive curves Γ'_k and Γ'_{k+1}

bound infinite strips S_k. We suppose that s_0 is the strip containing the point $s = 1$ and for every integer k, the curve Γ'_{k+1} is situated above Γ'_k. Figure 1(e) illustrates the birth of a strip S_k when the pre-image of a ray making a small angle α with the positive real half axis is taken and then we let $\alpha \to 0$.

We have also proved [36] , Theorem 2 that every unbounded component of the pre-image by $\zeta_{A,\Lambda}(s)$ of the unit circle is contained between two consecutive curves Γ'_k and Γ'_{k+1} and vice-versa, if $k \neq 0$ between two consecutive curves Γ'_k and Γ'_{k+1} there is a unique unbounded component of the pre-image of the unit circle. It has been shown that the respective component does not intersect any one of these curves. For $k \neq 0$ every strip S_k contains also a unique curve $\Gamma_{k,0}$ which is mapped bijectively by $\zeta_{A,\Lambda}(s)$ onto the interval $(-\infty, 1)$ of the real axis, as well as a certain number of curves $\Gamma_{k,j}$, $j \neq 0$ which are mapped bijectively by $\zeta_{A,\Lambda}(s)$ onto the whole real axis. There are infinitely many strips S_k covering the whole complex plane [36] , Theorem 4. Therefore, there are infinitely many unbounded components of the pre-image by $\zeta_{A,\Lambda}(s)$ of the unit circle. Some strips S_k can contain also bounded components of the pre-image of the unit circle as well as bounded components of the pre-image of C_r with $r > 1$. Figure 1(d) portrays the strip S_5 of $\zeta(s)$ containing two components of the pre-image of the unit circle: the unbounded one containing two zeros and the other bounded, containing one zero.

The use of the pre-image of the real axis can be traced back to Speiser's work [39] on the Riemann Zeta function. After that the pre-image of the real axis does not appear any more in literature as a tool except for the paper of Arias-de-Reina [40] , who revisits Speiser's theorem. In the same year John Derbyshire uses both: the pre-image of the real axis and that of the imaginary axis in his popular book Prime Obsession and declares that they are at the heart of that book. The classification of the components of the pre-image of the real axis by the Riemann Zeta function appears for the first time in [37] , where the strips S_k are also introduced and a method is devised of partitioning them into fundamental domains. Later, such a classification has been extended to Dirichlet L-functions and finally to functions defined by general Dirichlet series.

A different approach, namely that of phase diagram, has been used by Elias Wegert for visual exploration of complex functions [41] [42] . Applied to the Riemann Zeta function, his phase plots revealed interesting patterns pertaining to the universality property of that function. It is known that such a property extends to more general Dirichlet series and probably it can complement our fundamental domains approach.

An even better way to visualize the complexity of conformal mappings by analytic functions of a complex variable is to use an orthogonal mesh in the z-plane formed with rays issuing from the origin and circles centred at the origin. Moreover, a spectre of colors can be superposed to the mesh as seen in Figures 2(a)-(c). By taking the pre-image of that mesh we obtain a coloured orthogonal mesh in the s-plane in which the color of every point coincides with that of its image. In this way we can locate the corresponding points in the two planes and have also a global view of the mapping. The pre-image of the real axis and the S_k-strips are still identifiable. Figures 2(d)-(f) above illustrates the mapping by the Riemann Zeta function in the rectangles $[-20,20]\times[-30,30]$, $[-1,5]\times[1,000;1,020]$ and $[0.4,1.4]\times[10^9,10^9+20]$. By comparing them, one can notice the increasing number of zeros in the strips with increasing t. The pattern we can see here is proper to any function $\zeta_{A,\Lambda}(s)$.

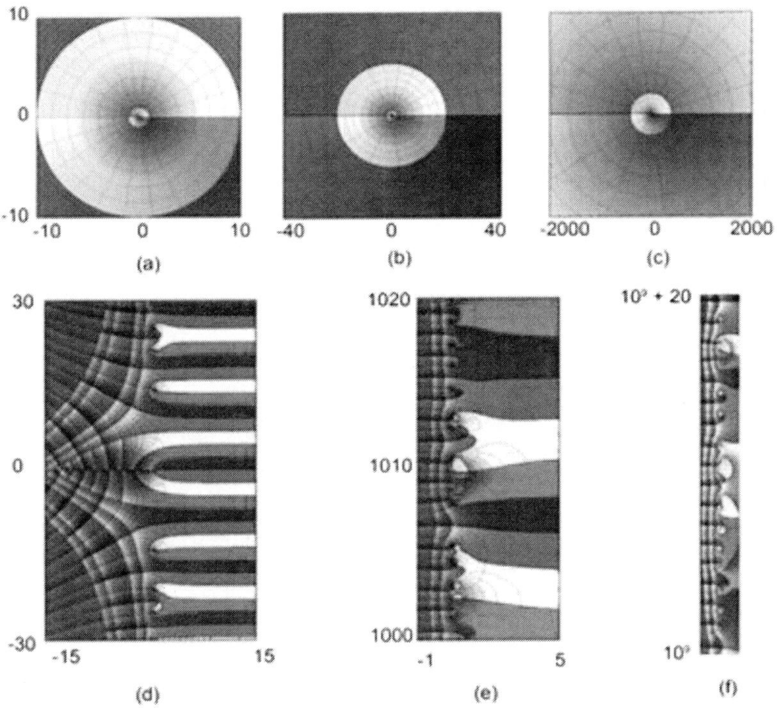

Figure 2. Colour-visualization of the conformal mapping by $\zeta(s)$.

Theorem 1 No zero of $\zeta_{A,\Lambda}(s)$ or of $\zeta'_{A,\Lambda}(s)$ can belong to a curve Γ'_k.

Proof. The affirmation of the theorem is obvious for the zeros of $\zeta_{A,\Lambda}(s)$ since 0 does not belong to the interval $[1,+\infty)$ of the real axis. A more intricate argument guarantees that the same is true for the zeros of the derivative of $\zeta_{A,\Lambda}(s)$. Indeed, even if $r>1$, no bounded component of the pre-image of C_r can reach Γ'_k, despite of the fact that C_r intersects the interval $[1,+\infty)$. Indeed, in the contrary case, the respective component should intersect Γ'_k at lest twice, or it should be tangent to it, fact which requires that C_r intersects the interval $[1,+\infty)$ the same number of times or to be tangent to it, which is not possible. The fact that C_r intersects the interval $[1,+\infty)$ has as effect the pre-image of C_r intersecting the curves $\Gamma_{k,j}$ (and not Γ'_k). It results that no two bounded components of the pre-image of C_r can meet on Γ'_k, neither can one of these components meet on Γ'_k an unbounded component of the pre-image of C_r into a zero of $\zeta'_{A,\Lambda}(s)$.

Remark: Theorem 1 does not imply that $\zeta'_{A,\Lambda}(s)$ cannot have zeros on some $\Gamma_{k,j}$. Such zeros appeared as possible for the Dirichlet L-function $L(5,2,s)$ as seen in Figure 3 when t has approximately the values 169.2 and 179.2. However, we suspected that this was due to the poor resolution of the picture and indeed, when we zoomed on the respective points, we obtained configu- rations which show clearly that $\zeta'_{A,\Lambda}(s)$ does not cancel there. However, as we will see later, the possibility of such zeros cannot be excluded.

We have seen that the unbounded components of the pre-image of the unit circle do not intersect any Γ'_k. On the other hand the bounded components of the pre-image of the unit circle intersect curves $\Gamma_{k,j}$ at points $u_{k,j}$ where $\zeta_{A,\Lambda}(u_{k,j})=\pm 1$. In the same way the bounded components of the pre-image of C_r with $r>1$ will intersect $\Gamma_{k,j}$ at points u where $\zeta_{A,\Lambda}(u)=\pm r$. However the story of the unbounded components of the pre-image of C_r is a little more complicated.

When increasing r past 1 all the unbounded components of the pre-image of C_1 fuse together into a unique unbounded curve γ_r intersecting every curve Γ'_k, hence they do not generate by this fusion zeros of $\zeta'_{A,\Lambda}(s)$.

Indeed, since the mapping of Γ'_k onto the interval $(1,+\infty)$ is bijective, there should be a unique point s_k on every Γ'_k such that $\zeta_{A,\Lambda}(s_k)=r$. The continuation over C_r from each one of these points can be made clockwise and counter clockwise into s_k, respectively S_{k-1}, for every k, giving rise to that unbounded curve. The final conclusion is that $\zeta'_{A,\Lambda}(s)$ does not cancel on any Γ'_k. ■

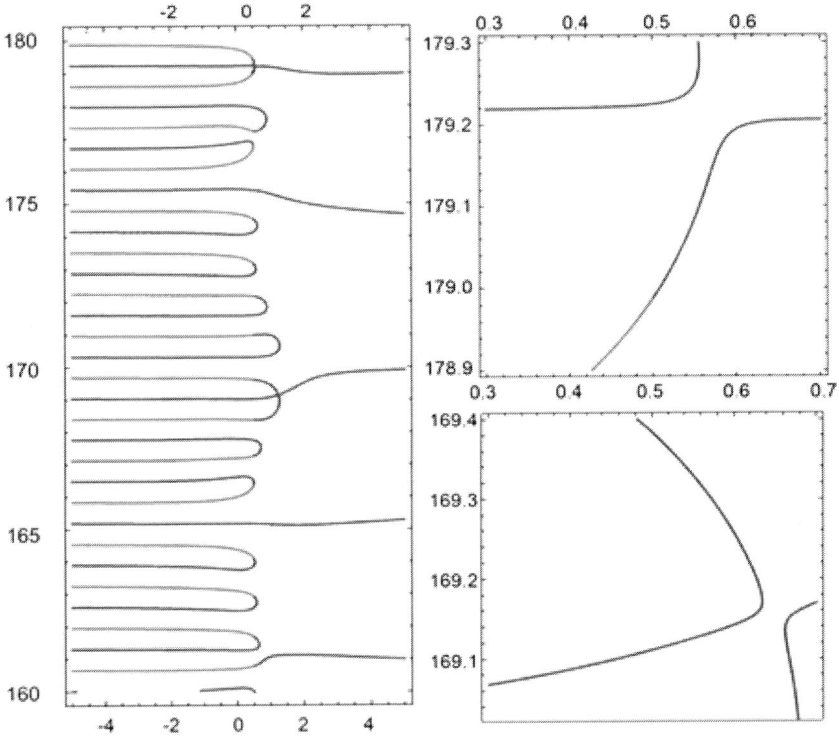

Figure 3. Possible zeros of $\zeta'_{A,\Lambda}(s)$ on $\Gamma_{k,J}$ which are not double zeros of $\zeta_{A,\Lambda}(s)$.

Given any bounded domain in the plane (s), we can take $r>1$ close enough to 1 such that γ_r does not touch any bounded component of the pre-image of C_r included in that domain. However, for bigger values of r the curve γ_r

comes into contact with bounded components which were turning around one or several zeros $s_{k,j}$ of $\zeta_{A,\Lambda}(s)$, fusing with them and getting to the left of those zeros, hence intersecting also the curves $\Gamma_{k,j}$ which contain the respective zeros. The curve γ_r is orthogonal to every component of the pre-image of the real axis if it intersects that component at a point where $\zeta'_{A,\Lambda}(s)$ does not cancel.

Since a point turning around the origin in the same direction on an arbitrary circle C_r centred at the origin will meet consecutively the positive and the negative real half axis, the components of the pre-image of C_r (including γ_r when $r > 1$) should meet consecutively the pre-image of the positive and the negative half axis (coloured differently). This is [33] [35] the so-called color alternating rule.

An immediate consequence of this rule is that in every strip S_k the first and the last curve $\Gamma_{k,j}$ should be such that the pre-image of the negative real half axis faces the corresponding Γ'_k. Then, consecutive $\Gamma_{k,j}$ have the same orien- tation, as long as they are on the same side of $\Gamma_{k,0}$. A zero s_0 of $\zeta'_{A,\Lambda}(s)$ being on $\Gamma_{k,j}$ and not being a zero of $\zeta_{A,\Lambda}(s)$ (where the colors are changing) implies that different components of the pre-image of a circle C_r fuse (for a value r_0 of r) on $\Gamma_{k,j}$. The respective components must obey the color alternating rule for every $r < r_0$ and we realize that after fusion the component can be such that the rule is still in force. Indeed, if $\Gamma_{k,j}$ and $\Gamma_{k,j'}$ have the same color at s_0 then this curve can continue to meet the respective color without affecting the color alternation by simply switching the branches of $\Gamma_{k,j}$ and $\Gamma_{k,j'}$ which come to s_0. However, as we will see next, the position of s_0 with respect to the two zeros of $\zeta_{A,\Lambda}(s)$ on these curves cannot be arbitrary.

Theorem 2 For k different of 0, there is no zero of the derivative of $\zeta'_{A,\Lambda}(s)$ at the left of the leftmost zero of $\zeta_{A,\Lambda}(s)$ in S_k

Proof. We can reproduce the proof of Theorem 1 from [43] for arbitrary functions $\zeta_{A,\Lambda}(s)$. Let $s_{k,j}$ be the leftmost zero of $\zeta_{A,\Lambda}(s)$ from S_k and suppose that the simple zero $v_{k,j}$ of $\zeta'_{A,\Lambda}(s)$ is a progenitor of $s_{k,j}$, i.e. a component of the pre-image of the circle C_r with $r=\left|\zeta_{A,\Lambda}\left(v_{k,j}\right)\right|$ contains the point $s_{k,j}$. It is obviously enough to deal with the case where $j>0$, i.e. where $\Gamma_{k,j}$ is above $\Gamma_{k,0}$. Then $\zeta_{A,\Lambda}\left(v_{k,j}\right)$ (indicated in Figure 4 as $L\left(v_{k,j}\right)$, since it has been partially computer generated by a Dirichlet L-function) belongs to the upper half plane and $0\leq\alpha<\pi$, where $\alpha=\arg\zeta_{A,\Lambda}\left(v_{k,j}\right)$.

The pre-image of the ray determined by $\zeta_{A,\Lambda}\left(v_{k,j}\right)$ contains two curves which are orthogonal at $v_{k,j}$. The angles at $v_{k,j}$ are doubled by $\zeta_{A,\Lambda}(s)$, hence the four arcs of the pre-image of that ray make angles of $\alpha/2$, $\pi/2+\alpha/2$, $\pi+\alpha/2$ and $3\pi/2+\alpha/2$ with a horizontal line whose image passes through $\zeta_{A,\Lambda}\left(v_{k,j}\right)$. The angle $\pi/2+\alpha/2$ made by the second arc (which ends in $s_{k,j}$) with this line is less than the angle β made by the tangent to the pre-image of the respective ray at any point between $s_{k,j}$ and $v_{k,j}$ with the same horizontal line. If $\Re\left(v_{k,j}\right)\leq\Re\left(s_{k,j}\right)$, then there must be a point on that arc for which $\beta=\pi/2$, therefore $\alpha<0$, which is absurd. ■

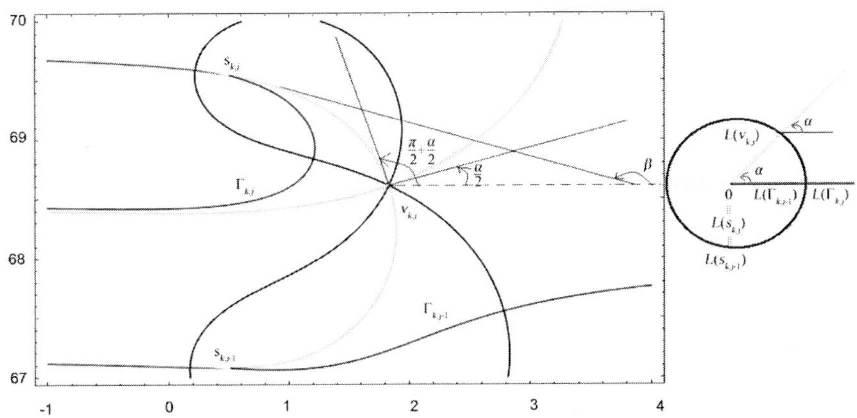

Figure 4. The location of the zeros of $\zeta'_{A,\Lambda}(s)$.

Corollary: If $\zeta_{A,\Lambda}(s)$ satisfies RH, then the zeros of $\zeta'_{A,\Lambda}(s)$ from every strip S_k are at the right side of the critical line. When $\zeta_{A,\Lambda}(s)$ is the Riemann Zeta function this corollary represents the Speiser's theorem [39] .

The existence of multiple zeros of functions obtained by analytic continuations of Dirichlet series has been documented (probably for the first time) in [44] , where double zeros of a linear combination of Dirichlet L-functions have been found (see Figure 5 below).

We have shown that all those double zeros are located on the critical line. In this example the function is $L(7,2,s)+0.34375L(7,4,s)$ and the double zero is obtained for the approximate value of s of $0.5+31.6i$. The double zeros we have found for all the functions of this type were located at the intersection of $\Gamma_{k,0}$ and $\Gamma_{k,1}$ or of $\Gamma_{k,0}$ and $\Gamma_{k,-1}$. We can make now a much more general affirmation about the multiple zeros of functions obtained by analytic con- tinuation of general Dirichlet series.

Theorem 3 In every strip S_k of a function $\zeta_{A,\Lambda}(s)$ this function has at most one double zero. Such a zero is found at the intersection of $\Gamma_{k,0}$ and $\Gamma_{k,1}$ or of $\Gamma_{k,0}$ and $\Gamma_{k,-1}$. There is no multiple zero of $\zeta'_{A,\Lambda}(s)$ in S_k and hence no zero of a higher order than two of $\zeta_{A,\Lambda}(s)$.

We need to postpone the proof of this theorem for a while.

3. INTERTWINING CURVES

When studying functions $z=\zeta_{A,\Lambda}(s)$ it is useful to consider besides the planes (s) and (z) also a plane (w), where $w=\zeta'_{A,\Lambda}(s)$. Sometimes the planes (z) and (w) will be identified in order to make more obvious certain relations between the configurations defined by the two functions in the respective planes. The configurations we have in view are pre-images by both $\zeta_{A,\Lambda}(s)$ and by

$\zeta'_{A,\Lambda}(s)$, of some curves or domains.

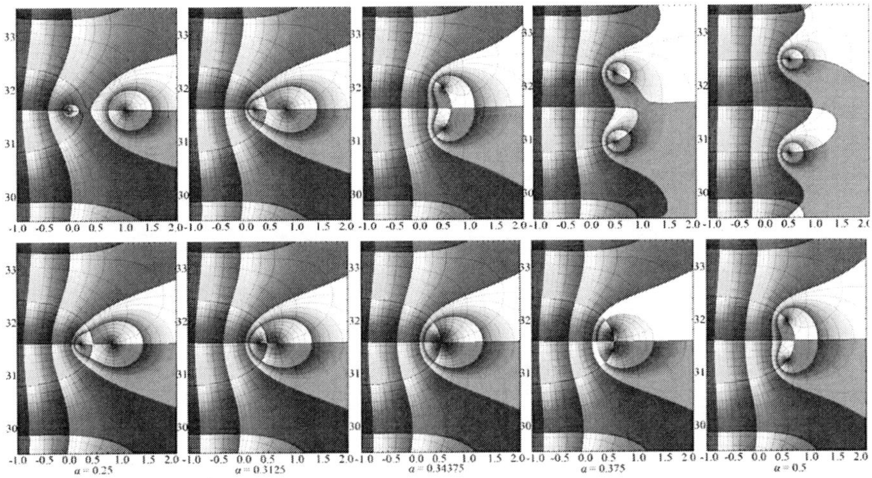

Figure 5. A double zero of a linear combination of dirichlet L-functions.

Regarding the pre-image by $\zeta'_{A,\Lambda}(s)$ and by $\zeta'_{A,\Lambda}(s)$ of the real axis it has been found [34] [36] that the components of these pre-images are paired in such a way that only the components of the same pair can intersect each other. The respective pairs form the so-called intertwining curves.

Three kinds of intertwining curves have been distinguished [33] [34] [35] [36] , namely:

1) Γ'_k and Υ'_k, $k \in \mathbb{Z}$, which are mapped bijectively by $\zeta_{A,\Lambda}(s)$ and by $\zeta'_{A,\Lambda}(s)$ onto the interval $(1,+\infty)$, respectively $(-\infty,0)$,

2) $\Gamma_{k,0}$ and $\Upsilon_{k,0}$, $k \in \mathbb{Z}$, which are mapped bijectively by $\zeta_{A,\Lambda}(s)$ and by $\zeta'_{A,\Lambda}(s)$ onto the interval $(-\infty,1)$, respectively $(0,+\infty)$

3) $\Gamma_{k,j}$ and $\Upsilon_{k,j}$, $j \neq 0$, $k \in \mathbb{Z}$, $j \in J_k \setminus \{0\} =$ (a finite set of integers), which are mapped bijectively by $\zeta_{A,\Lambda}(s)$, respectively by $\zeta'_{A,\Lambda}(s)$ onto the whole real axis.

Theorem 4 The intertwining curves touch each other at the points where the tangent to $\Gamma_{k,j}$, respectively Γ'_k is horizontal. Vice-versa, if at a point of such a curve the tangent is horizontal, then a component of the pre-image of the real axis by $\zeta'_{A,\Lambda}(s)$ passes also through that point.

Proof. Indeed, suppose that $s=s(x)$ is the equation of a curve Γ'_k or of a curve $\Gamma_{k,j}$ such that $\zeta_{A,\Lambda}(s(x))=x$. Then (see Figure 6 below)

$$\zeta'_{A,\Lambda}(s(x))s'(x)=1 \quad (4)$$

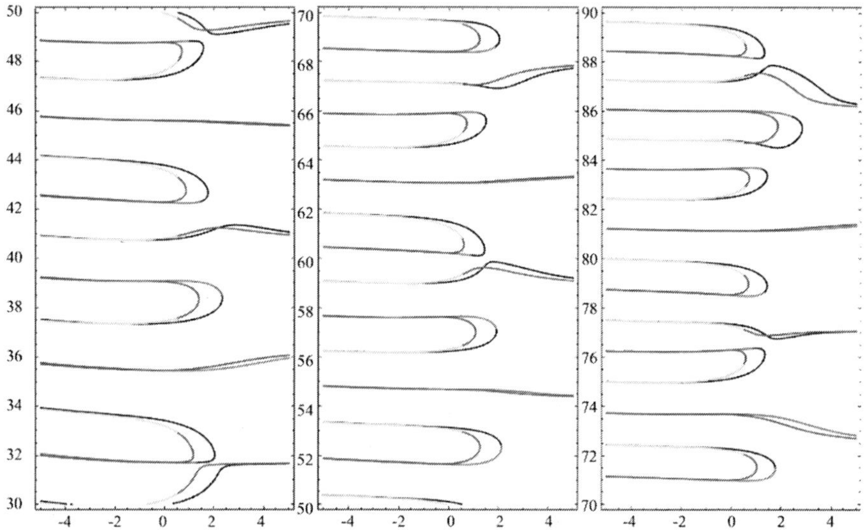

Figure 6. Intertwining curves of $\zeta(\sigma+it)$ for $t\in[0,100]$.

The Equation (4) shows that the argument of $\zeta'_{A,\Lambda}(s(x))$ is opposite to that of $s'(x)$, therefore they cancel simultaneously, and when one is π, the other should be $-\pi$. Yet these values of the argument of a point mean that the respective point is on the real axis, therefore $s(x)$ belongs to both pre-images of the real axis, by $\zeta_{A,\Lambda}(s)$ and $\zeta'_{A,\Lambda}(s)$ which completely proves the theorem.

Remark: The Theorem 4 is a corollary of a much more general property which says that if $z=f(s)$ is an analytic function in a domain D of the complex plane and γ is the image by $f(s)$ of a smooth curve $\Gamma:s=s(z)$, then denoting by Υ the pre-image by $f'(s)$ of γ, at every point s_0 where Γ and Υ intersect each other we have $\arg f'(s_0)+\arg s'(z_0)=0(\bmod 2\pi)$, where $s_0=s(z_0)$.

Indeed, suppose that the planes (z) and (w) are identified, where $w=f'(s)$ and write $S=S(w)$ for the curve Υ. Then at an intersection point s_0 of Γ and Υ we have $s_0=s(z_0)=S(f'(s_0))$ and $f(s(z))=z$ implies $f'(s(z))s'(z)=1$, hence $f'(s_0)s'(z_0)=1$, etc.

We can show now that:

Theorem 5 No strip S_k can be included in a right half plane.

Proof. The formula (4) written for Γ'_k tells us that when $s'(x)$ tends to ∞ then $\zeta'_{A,\Lambda}(s(x))$ must tend to 0. Yet, there is no zero of $\zeta'_{A,\Lambda}(s(x))$ on any Γ'_k. On the other hand, $\zeta'_{A,\Lambda}(s(x))$ is an unlimited continuation of $\zeta'_{A,\Lambda}(s)$ alongside Γ'_k, hence when $s(x)$ tends to ∞ we have that $\zeta'_{A,\Lambda}(s(x))$ tends to ∞ not to 0 and this is a contradiction. The conclusion is that the geometry of the pre-image of the real axis is in the whole complex plane similar to that we can see in a bounded region of the plane in Figure 2, Figure 3, Figure 6, Figure 7. ■

Summarizing these facts and having in view [36] we can say:

Theorem 6 The variable σ takes any real value on every curve Γ'_k. Consecutive curves Γ'_k form infinite strips S_k which are mapped (not necessarily bijectively) onto the whole complex plane with a slit alongside the interval $[1,+\infty)$ of the real axis. If S_k contains m_k zeros of the function $\zeta_{A,\Lambda}(s)$, then it will contain m_k-1 zeros of $\zeta'_{A,\Lambda}(s)$. For any given k, the function $\zeta_{A,\Lambda}(s)$ either has a finite number of zeros $s_{k,j}$ in S_k, or $\lim_{j\to\infty} s_{k,j}=\infty$. If $\zeta_{A,\Lambda}(s)$ satisfies a Riemann type of functional equation, every strip S_k, $k\neq 0$ contains a finite number of zeros of this function. The strip S_0 may contain infinitely many zeros of $\zeta_{A,\Lambda}(s)$.

Proof. We only need to justify the numbers m_k and m_k-1 to which the theorem makes reference. We have seen that every zero of $\zeta'_{A,\Lambda}(s)$ is obtained when two components of the pre-image of some circle C_r centred at the origin and of radius r come into contact. If we consider the zeros of $\zeta_{A,\Lambda}(s)$ as the leafs of a binary tree whose internal nodes are obtained in this way, that tree is a complete binary tree and it is known that it must have exactly $m_k - 1$ internal nodes. Figure 7 illustrates this situation for ten strips of the Riemann Zeta function. The location of the zeros of $\zeta(s)$ and those of its first two derivative are also clearly indicated. If there is a double zero of $\zeta_{A,\Lambda}(s)$ in S_k, then the respective leaf is counted twice as a leaf and once as an internal node, etc. When the number of zeros $s_{k,j}$ in S_k is infinite we must have $\lim_{j\to\infty} s_{k,j} = \infty$, otherwise $s_{k,j}$ would have an accumulation point in S_k and hence $\zeta_{A,\Lambda}(s)$ would be identically zero. ■

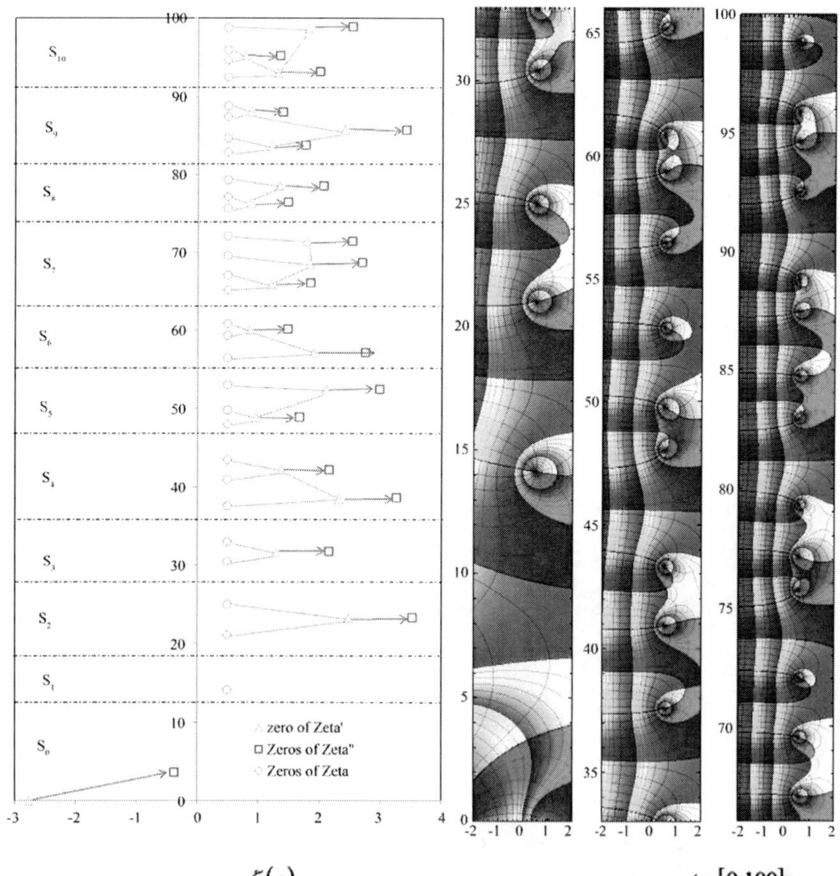

Figure 7. The zeros of $\zeta(s)$ and of its first two derivatives for $t \in [0,100]$.

We can prove now also the Theorem 3.

Proof. Exactly 2 curves $\Gamma_{k,j}$ and $\Gamma_{k,j'}$ should meet at a double zero s_0 of $\zeta_{A,\Lambda}(s)$ belonging to S_k. But s_0 is also a simple zero of $\zeta'_{A,\Lambda}(s)$ and therefore a curve, say $\Upsilon_{k,j}$ should pass through s_0. Yet, $\Gamma_{k,j'}$ intertwines with a curve $\Upsilon_{k,j'}$ and unless this curve is $\Upsilon_{k,0}$, which does not contain any zero of $\zeta'_{A,\Lambda}(s)$, the point s_0 would be a double zero of for $\zeta'_{A,\Lambda}(s)$, and this is impossible.

Hence , necessarily one of the curves $\Gamma_{k,j}$ passing through s_0 is $\Gamma_{k,0}$. This shows that there can be only one double zero in S_k and the other curve passing through s_0 in that case is either $\Gamma_{k,1}$, or $\Gamma_{k,-1}$.

We notice that the curve $\Upsilon_{k,0}$ does not contain any zero of $\zeta'_{A,\Lambda}(s)$, hence it cannot pass through a multiple zero of order m of $\zeta'_{A,\Lambda}(s)$. Since every curve passing through that zero has an intertwining curve defined by the second derivative of $\zeta_{A,\Lambda}(s)$, the respective point should be a multiple zero of order m of this second derivative, which is absurd. Therefore $\zeta'_{A,\Lambda}(s)$ has no multiple zero and then $\zeta_{A,\Lambda}(s)$ cannot have any zero of the higher order than 2. ■

4. FUNDAMENTAL DOMAINS

For $j \neq 0$, the curves $\Gamma_{k,j}$ as well as $\Upsilon_{k,j}$ are parabola like curves with branches extending to infinity as $\sigma \to -\infty$. Therefore we can distinguish between the interior and the exterior of such curves. They can be viewed as oriented curves, with the same orientation as the real axis whose components of the pre-image they are. By the same rule Γ'_k and Υ'_k, as well as $\Upsilon_{k,0}$ are also oriented, the positive orientation of Γ'_k and of $\Upsilon_{k,0}$ being from the right to the left, while that of Γ'_k, is from the left to the right. Some curves $\Gamma_{k,j}$, $j \neq 0$ can contain in interior some other curves of the same type (embraced curves) and by the color alternating rule the orientation of the embracing curve and that of the embraced curves must be different. We did not find any instance where an embraced curve is in turn embracing, yet there is no reason to believe that such a situation is impossible.

Theorem 7 If the strip S_k contains m_k zeros counted with multiplicities, then S_k can be partitioned into m_k sub-strips which are fundamental domains for $\zeta_{A,\Lambda}(s)$

Proof. Suppose that curves $\Gamma_{k,j}$ and $\Gamma_{k,j'}$ are containing the simple zeros $s_{k,j}$ and $s_{k,j'}$ and two components of the pre-image of a circle C_r which are going around each one of these zeros touch at a point $v_{k,j}$. This is a zero of $\zeta'_{A,\Lambda}(s)$. The pre-image of the segment of line connecting $z=1$ and $\zeta_{A,\Lambda}(v_{k,j})$ has as component an arc $\eta_{k,j}$ connecting the points on the two curves where $\zeta_{A,\Lambda}(s)=1$ and passing through $v_{k,j}$. If one of these curves is $\Gamma_{k,0}$ then the respective point is ∞ and $\eta_{k,j}$ is an unbounded curve. The strip $\Omega_{k,j}$ bounded by this curve and the branches of $\Gamma_{k,j}$ and $\Gamma_{k,j'}$ corresponding to the interval $[1,+\infty)$, turns out to be a fundamental domain of $\zeta_{A,\Lambda}(s)$. Indeed, $\Omega_{k,j}$ is mapped conformally by $\zeta_{A,\Lambda}(s)$ onto the whole complex plane with a slit alongside the interval $[1,+\infty)$ followed by a slit alongside the segment from $z=1$ to $\zeta_{A,\Lambda}(v_{k,j})$. If $\Gamma_{k,j}$ embraces $\Gamma_{k,j'}$ then $\Omega_{k,j}$ is bounded to the right.

Suppose now that one of the two zeros is a double zero of $\zeta_{A,\Lambda}(s)$. We know that $\Gamma_{k,0}$ must pass through that zero and the part of $\Gamma_{k,0}$ corresponding to the interval $[0,1]$ becomes part of the boundaries of adjacent fundamental domains. One of them will have as image the complex plane with a slit alongside the positive real half axis and for the other one a slit from $z=1$ to $z=\zeta_{A,\Lambda}(v_{k,j})$ should be added, where j is 1 or −1.

We know that $s_{k,j}$ cannot have a higher order of multiplicity and therefore the cases analysed exhaust all the possibilities. Having in view Theorem 6, we conclude that the strip S_k can be always divided into m_k fundamental domains. ∎

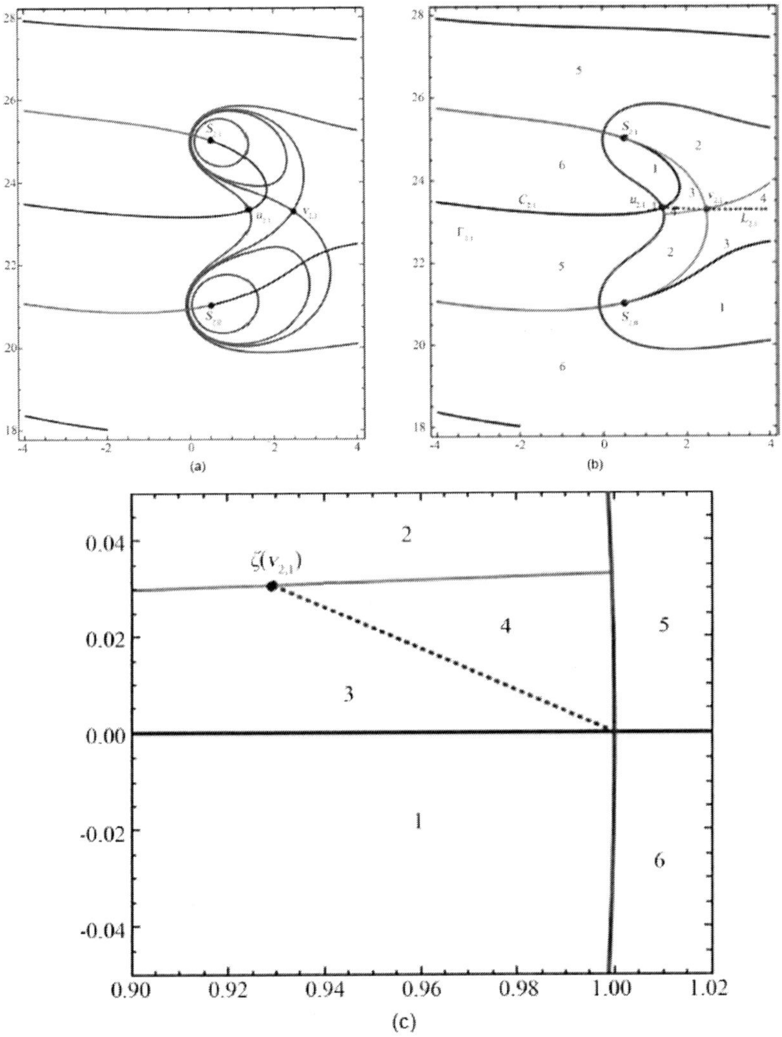

Figure 8. Two fundamental domains and their images.

Figure 8 below illustrates this theorem for the case of the Riemann Zeta function and the strip S_2.

As seen in the case of Dirichlet L-functions (including the Riemann Zeta function) the strip S_0 has infinitely many zeros, yet following the same technique we can divide it into infinitely many fundamental domains.

If four different colors are used, say color 1 and 2 for the pre-images by $\zeta_{A,\Lambda}(s)$ of the positive and of the negative real half axis and 3 and 4 for the pre-images by $\zeta'_{A,\Lambda}(s)$ of the same half axes, then two simple topological facts can be established [33] :

a) The color alternating rule, which states that as a point turns indefinitely in the same direction on a circle centred at the origin, the pre-images of this point by each one of the functions $\zeta_{A,\Lambda}(s)$ and $\zeta'_{A,\Lambda}(s)$ will meet alternatively the colors 1 and 2, respectively 3 and 4.

b) The color matching rule, which states that when intertwining curves meet each other, then if these are not $\Gamma_{k,0}$ and $\Upsilon_{k,0}$ color 1 will always meet color 4 and color 2 will always meet color 3. Only the curves $\Gamma_{k,0}$ and $\Upsilon_{k,0}$ can intersect each other at points where color 2 can meet color 4.

The series (1) which are Euler products display special important properties.

5. EULER PRODUCTS

It is known that the Dirichlet L-functions are meromorphic continuations of ordinary Dirichlet series defined by Dirichlet characters and these series can be expressed as Euler products. This property is a corollary of the fact that the Dirichlet characters are totally multiplicative functions (see for example [33] . Yet the property of being total multiplicative can be extended to general Dirichlet series, as done in [35] , and therefore some of the general Dirichlet series $\zeta_{A,\Lambda}(s)$ (see the details in [35]) can also be written as Euler products:

$$\zeta_{A,\Lambda}(s)=\sum_{n=1}^{\infty}a_n e^{-\lambda_n s}=\Pi_{p\in\mathbb{P}}\left(1-a_p e^{-\lambda_p s}\right)^{-1}, \tag{5}$$

where $\mathbb{P}$ is the set of prime numbers. This convention will be kept in the following for all the products and sums involving the subscript p. The product has the same abscissa of convergence as the series itself.

Looking for counterexamples to the Grand Riemann Hypothesis (GRH), some Dirichlet series satisfying a Riemann type of functional equation have been found, whose analytic continuation exhibit off critical line non trivial zeros, namely the Davenport and Heilbronn type of functions and linear combinations of L-functions satisfying the same functional equation.

Although these are not counterexamples to GRH, their study allowed us to draw interesting conclusions. We have seen in [36] that if $\zeta_{A,\Lambda}(s)$ does not satisfy the GRH, then for every two distinct non trivial zeros $s_1 = \sigma + it$ and $s_2 = 1 - \sigma + it$ there is τ_0, $0 < \tau_0 < 1$ such that $\zeta'_{A,\Lambda}(s(\tau_0)) = 0$, where $s(\tau) = (1-\tau)s_1 + \tau s_2$, i.e. the derivative of $\zeta_{A,\Lambda}(s)$ cancels at a point $s_0 = s(\tau_0)$ of the interval I deter- mined by s_1 and s_2. Moreover, $\Re(s_0) < 1/2$.

Let us rephrase and give a simplified proof to [36] , Theorem 3.

Theorem 8 Suppose that the function (5) satisfies a Riemann type of functional equation and the respective series has the abscissa of convergence $\sigma_c < 1/2$. Then for every non trivial zero $\sigma + it$ of $\zeta_{A,\Lambda}(s)$ we have $\sigma = 1/2$.

Proof. Suppose that there is a zero $\sigma + it$ of $\zeta_{A,\Lambda}(s)$ for which $\sigma > 1/2$. Then, due to the functional equation, $1 - \sigma + it$ is also a zero of $\zeta_{A,\Lambda}(s)$ (see Figure 9 below). There is $r > 0$ such that in the components D_1 and D_2 of the pre-image of the disc centred at the origin and of radius r containing the respective zeros the function $\zeta_{A,\Lambda}(s)$ is injective. Then we can define the function $\phi : D_1 \to D_2$ as follows:

$$\phi(s) = \zeta_{A,\Lambda}(s)\Big|_{D_2}^{-1} \circ \zeta_{A,\Lambda}(s) \tag{6}$$

The function ϕ can be continued as an analytic involution of the union $\Omega \cup \Omega'$ of the fundamental domains Ω and Ω' of $\zeta_{A,\Lambda}(s)$ containing the respective zeros. The boundaries of the domains Ω and Ω' have a common component L and the union $\Omega \cup \Omega' \cup L$ is a simply connected domain H. The function ϕ can be continued to an analytic involution of H having s_0 as a fixed point. We have $\phi(\phi(s)) = s$ in H, in particular $\phi(\sigma + it) = 1 - \sigma + it$ and $\phi(1 - \sigma + it) = \sigma + it$. Moreover, $\zeta_{A,\Lambda}(\phi(s)) = \zeta_{A,\Lambda}(s)$ in H. Let us define the function Φ by

$$\Phi(s) = \zeta_{A,\Lambda}(s) / \zeta_{A,\Lambda}(\phi(s)) \tag{7}$$

Since the numerator and the denominator of Φ are analytic functions in H and the denominator cancels only at $\sigma+it$ and $1-\sigma+it$, the function Φ is analytic in H except at these two points. Since $\zeta_{A,\Lambda}(s)=\zeta_{A,\Lambda}(\phi(s))$, we have that $\Phi(s)=1$ for s not equal to one of these points. Yet they are removable singularities, and we can set $\Phi(s)=1$ in H. By the formula (5) we have

$$\Phi(s)=\Pi_{p\in\mathbb{P}}\left(1-a_p e^{-\lambda_p\phi(s)}\right)/\left(1-a_p e^{-\lambda_p s}\right) \tag{8}$$

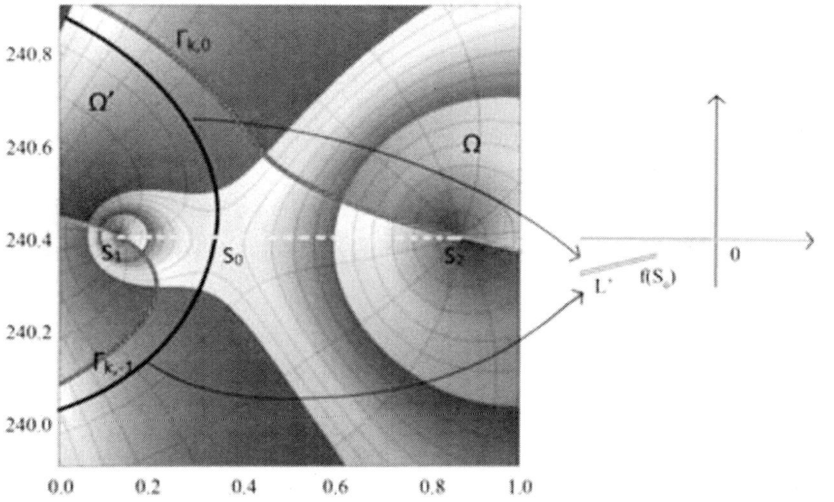

Figure 9. Symmetric zeros with respect to the critical line.

for sÎH, $\Re(s)$> σc. In particular,

$$\begin{aligned}1&=\Phi(\sigma+it)\\&=\Pi_{p\in\mathbb{P}}\left[1-a_p e^{-\lambda_p(1-\sigma+it)}\right]/\left[1-a_p e^{-\lambda_p(\sigma+it)}\right]\\&=\Pi_{p\in\mathbb{P}}\left[e^{\lambda_p ti}-a_p e^{-\lambda_p(1-\sigma)}\right]/\left[e^{\lambda_p ti}-a_p e^{-\lambda_p\sigma}\right]\end{aligned}$$

The arguments of these ratios represent the angles under which the segment between $a_p e^{-\lambda_p(1-\sigma)}$ and $a_p e^{-\lambda_p\sigma}$ is seen from the point $e^{\lambda_p ti}$ on the unit circle. If a Ramanujan type condition is fulfilled, namely A and Λ are such that for every $\epsilon > 0$ we have $\lim_{n\to\infty} |a_n| e^{-\lambda_n \epsilon} = 0$, then the respective angles tend to zero as $p \to \infty$. This appears to be a necessary condition for the convergence of the series.

$$\sum_{p\in\mathbb{P}} \arg\left[e^{\lambda_p ti} - a_p e^{-\lambda_p(1-\sigma)}\right] \Big/ \left[e^{\lambda_p ti} - a_p e^{-\lambda_p\sigma}\right]$$

However, the condition is implicitly satisfied since we know that $\Phi(s)$ is well defined in the domain H. The series is a continuous function of σ, yet it can take only integer multiple values of 2π, which is possible only if it is a constant. Since $\lim_{n\to\infty} a_n e^{-\lambda_n\delta} = 0,$ the series can remain constant only if there is p_0 such that for $p > p_0$ we have

$$\arg\left[e^{\lambda_p ti} - a_p e^{-\lambda_p(1-\sigma)}\right] \Big/ \left[e^{\lambda_p ti} - a_p e^{-\lambda_p\sigma}\right] = 0.$$

This can happen in two situations: either the three points $e^{\lambda_p ti}$, $a_p e^{-\lambda_p\sigma}$ and $a_p e^{-\lambda_p(1-\sigma)}$ are collinear, or $a_p e^{-\lambda_p(1-\sigma)} = a_p e^{-\lambda_p\sigma}$ and in this last case $1-\sigma=\sigma$, i.e. $\sigma = 1/2$. In the first case, a shift in t will not affect the real part of the zeros, yet it will destroy the collinearity of the three points and therefore this situation can be ignored. The final conclusion is that $\varsigma_{A,\Lambda}(s)$ cannot have any non trivial zero $\sigma + it$ with σ strictly greater than 1/2. The zero $1/2 + it$ can be either a simple zero and then we have only one fundamental domain Ω, or a double zero and then it is the fixed point of the involution $\phi(s)$. In both cases it is located on the critical line, which completely proves the

We need to point out the fact that Figure 9 above illustrates a situation in which the Equation (5) is not satisfied.

Remark: In all of our publications we understood by trivial zeros of an L-function those zeros which can be trivially computed. In this respect, the non trivial zeros of the alternating Zeta function $\zeta_a(s) = (1-2^{1-s})\zeta(s)$ are the same as those of $\zeta(s)$ and $\zeta_a(s)$ satisfies the conditions of Theorem 8, therefore its non trivial zeros are located on the critical line. Thus, RH is also true for $\zeta(s)$.

Similarly, the non trivial zeros of a Dirichlet L-function $L(q,1,s)=(1-q^{1-s})\zeta(s)$ defined by the principal character modulo q are the non trivial zeros of the Riemann Zeta function.

Also, the non trivial zeros of a Dirichlet L-function induced by an imprimitive character are the non trivial zeros of the function defined by the associated primitive character. Consequently, the RH for any Dirichlet L-function is fulfilled. With this understanding of the concept of non trivial zeros, Theorem 8 represents the proof of GRH for a wide class of functions.

ACKNOWLEDGEMENTS

The author is very much thankful to Florin Alan Muscutar for his contribution with computer generated graphics.

REFERENCES

1. Cahen, E. (1894) Sur la function ζ(s) de Riemann et sur des fonctions analogues, Ann. Ecole Norm., 3e serie, t. 11.
2. Cahen, E. (1918) Sur les séries de Dirichlet, Compte rendue, t. 166.
3. Hadamard, J. (1896) Sur la distribution des zéros de la function ζ(s) de Riemann et ses cons équences arithmé tiques, Bull. Soc. math., t. 24.
4. Hadamard, J. (1908) Sur les s éries de Dirichlet, Rendiconti di Palermo, t. 25.
5. Landau, E. (1907) über die Multiplication Dirichletscher Reihen, rendiconti di Palermo, t. 24.
6. Landau, E. (1909) über das Konvergenzproblem der Dirichletschen Reihen, Rendiconti di Palermo, t. 28.
7. Landau, E. (1917) Neuer Beweis eines Haptsatzes aus der Theorie der Dirichletschen Reihen, Leipziger Ber., t. 69.
8. Landau, E. (1921) über die gleichmassige Konvergenz Dirichletscher Reihen, Math. Zeitschrift, t. 11.
9. Bohr, H. (1913) über die gleichmassige Konvergenz Dirichletscher Reihen, J. fuer Math, t. 143.
10. Bohr, H. (1913) Einige Bemerkungen über das konvergenzproblem Dirichletscher Reihen, Rendicnti di Palermo, t. 37.
11. Bohr, H. (1913) Darstellung der gleichmassigen Konvergenzabszisse einer Dirichletschen reihe $\sum_{n=1}^{\infty} a_n/n^s$ als Funktionen der Koeffizienten der Reihe. Archiv der Mathematik und Physik, 21, 326-330.
12. Bohr, H. (1913) Zür Theorie der allgemeinen Dirichletschen Reihen. Mathematische Annalen, 79, 136-156.
13. Hardi, G.-H. (1911) The Multiplication of Dirichlet's Series. Vol. 10, London Mathematical Society, London.

14. Hardy, G.H. and Riesz, M. (1915) The General Theory of Dirichlet's Series. Cambridge University Press, Cambridge.
15. Kojima, T. (1914) On the Convergence Abscissa of General Dirichlet's Series. Tohoku Mathematical Journal, 6, 140-142.
16. Kojima, T. (1916) Note on the Convergence Abscissa of Dirichlet's Series. Tohoku Mathematical Journal, 9, 28-37.
17. Kuniyeda, M. (1916) Uniform Convergence Abscissa of General Dirichlet's Series. Tohoku J., t. 9, 7-27.
18. Valiron, G. (1926) Théorie générale des séries de Dirichlet. Mémorial des sciences mathématiques, fascicule, 17, 1-56.
19. Yu, J. (1963) Borel's Line of Entire Functions Represented by Laplace-Stieltjes Transformations. Acta Mathematica Sinica, 13, 471-484.
20. Kong, Y.Y. and Daochun, S. (2008) On Type-Function and Growth of Laplace-Stieltjes Transformations Convergent in the Right Half Plane, 2008. Advances in Mathematics, 37, 197-205.
21. Luo, X. and Kong, Y.Y. (2012) On the Order and Type of Laplace-Stieltjes Transforms of Slow Growth. Acta Mathematica Sinica, 32A, 601-607.
22. Kong, Y.Y. and Yang, Y. (2014) On the Growth Properties of the Laplace-Stieltjes Transform. Complex Variables and Elliptic Equations, 59, 553-563.
23. Singhal, C. and Srivastava, G.S. (2015) On the Approximation of an Analytic Function Represented by Laplace-Stieltjes Transformation. Analysis in Theory and Applications, 31, 407-420.
24. Srivastava, G.S. and Singhal, C. (2015) On the Generalized Order and the Generalized Type of Laplace-Stieltjes Transformation Convergent in the Right Half Plane. Global Journal of Pure and Applied Mathematics, 11, 469-477.
25. Kamthan, P.K. and Shing Gautam, S.K. (1975) Bases in a Certain Space of Functions Analytic in the Half Plane. Indian Journal of Pure and Applied Mathematics, 6, 1066-1075.
26. Khoi, L.H. (1998) Holomorphic Dirichlet Series in the Half Plane. Vietnam Journal of Mathematics, 26, 259-271.
27. Defant, A., Garcia, D., Maestre, M. and Perez-Garcia, D. (2008) Bohr's Strip for Vector Valued Dirichlet Series. Mathematische Annalen, 342, 533-555.
28. Bonet, J. (2015) Abscissas of Weak Convergence of Vector Valued Dirichlet Series. arXiv: 1502.00418v1.
29. Srivastava, B.L. (1983) A Study of Spaces of Certain Classes of Vector Valued Dirichlet Series. Thesis, Indian Institute of Technology Kanpur, Kalyanpur.
30. Srivastava, G.S. and Sharma, A. (2011) Bases in the Space of Vector Valued Analytic Dirichlet Series. International Journal of Pure and Applied Mathematics, 7, 993-1000.
31. Srivastava, G.S. and Sharma, A. (2012) Spaces of Entire Functions Represented by Vector Valued Dirichlet Series of Slow Growth. Acta Universitatis Sapientiae, Mathematica, 4, 154-167.
32. Sharma, A. and Srivastava, G.S. (2016) Coefficient Multipliers on Spaces of Vector Valued Entire Dirichlet Series.
33. Ghisa, D. (2013) Fundamental Domains and the Riemann Hypothesis. Lambert Academic Publishing, Saarbrücken.
34. Ghisa, D. (2014) On the Generalized Riemann Hypothesis, Complex Analysis and Potential Theory with Applications. Cambridge Scientific Publishers, Cambridge, 77-94.
35. Ghisa, D. (2016) On the Generalized Riemann Hypothesis II. International Journal of Scientific and Innovative Mathematical Research, 4, 46-55.
36. Ghisa, D. (2016) Fundamental Domains and Analytic Continuation of General Dirichlet Series. British Journal of Mathematics & Computer Science, 9, 94-111. [Citation Time(s):11]
37. Andreian-Cazacu, C. and Ghisa, D. (2009) Global Mapping Properties of Analytic Functions. Proceedings of 7th ISAAC Congress, London, 13-18 July 2009, 3-12.
38. Andreian-Cazacu, C. and Ghisa, D. (2011) Fundamental Domains of Gamma and Zeta Functions. International Journal of Mathematics and Mathematical Sciences, 2011, Article ID: 985323.

39. Speiser, A. (1934) Geometrisches zür Riemannschen Zetafunction. Mathematische Annalen, 110, 514-521.
40. Arias-de-Reina, J. (2003) X-Ray of Riemann's Zeta-Function. ArXiv: math/0309433, 42 p.
41. Wegert, E. (2012) Visual Complex Functions. Springer, Basel.
42. Wegert, E. (2016) Visual Exploration of Complex Functions. In: Qian, T. and Rodino, L.G., Eds., Mathematical Analysis, Probability and Applications—Plenary Lectures, Vol. 177, Springer International Publishing, 253-279.
43. Barza, I., Ghisa, D. and Muscutar, F.A. (2014) On the Location of the Zeros of the Derivatives of Dirichlet L-Functions. Annals of the University of Bucharest (Mathematics Series), 5, 21-31.
44. Cao-Huu, T., Ghisa, D. and Muscutar, F.A. (2016) Multiple Solutions of Riermann Type of Functional Equations. British Journal of Mathematics and Computer Science, 17, 1-12.

CHAPTER 4

Geometry of the Standard Model of Quantum Physics

Claude Daviau[1], Jacques Bertrand[2]

[1]*Le Moulin de la Lande, Pouillé-les-coteaux, France*

[2]*15 Avenue Danielle Casanova, Saint-Gratien, France*

ABSTRACT

General relativity links gravitation to the structure of our space-time. Nowadays physics knows four types of interactions: Gravitation, electromagnetism, weak interactions, strong interactions. The theory of everything (ToE) is the unification of these four domains. We study several necessary cornerstones for such a theory: geometry and mathematics, adapted manifolds on the real domain, Clifford algebras over tangent spaces of these manifolds, the real Lagrangian density in connection with the standard model of quantum physics. The geometry of the standard model of quantum physics uses three Clifford algebras. The algebra Cl_3 of the 3-dimensional physical space is sufficient to describe the wave of the electron. The algebra $Cl_{1,3}$ of space-time is sufficient to describe the wave of the pair electron-neutrino. A greater space-time with two additional dimensions of space generates the algebra $Cl_{1,5}$. It is sufficient to get the wave equation for all fermions, electron, its neutrino and quarks u and d of the first generation, and the wave equations for the two other generations. Values of these waves allow defining, in each point of space-time, geometric transformations from one intrinsic manifold of space-time into the usual manifold. The Lagrangian density is the scalar part of the wave equation.

Keywords: Geometry, Invariance Group, Dirac Equation, Electromagnetism, Weak Interactions, Strong Interactions, Clifford Algebras

1. INTRODUCTION

Geometry is equivalent to physics since Einstein has linked gravitation to the geometry of our Universe [1] [2]. The mathematical frame of this geometry is a 4-dimensional manifold where the dimension is this number of real parameters. Nowadays everyone in physics tries to produce a ToE (theory of everything), a theory including gravitation, electromagnetism, weak and strong interactions. String theory, supersymmetry, superstring theory, M-theory, loop quantum gravity, are most famous recent tentatives. This means that the problem is very hard. The aim of this paper is the analysis of these difficulties, preparing new trails to construct such a theory. The basic idea is: perhaps the solution is not so difficult, perhaps we have not be seeking in the right direction. After his discovery of general relativity, Einstein made numerous attempts [3] to get a theory unifying electromagnetism and gravitation. In the same time quantum mechanics was quickly growing, a relativistic wave equation was given by Dirac for the electron. It was then necessary to account not only for gravitation and electromagnetism, but also for the spin 1/2 of the electron and for the quantum wave. Difficulties here begin with the terrible difference between mathematical tools: the 4-dimensional real manifold of the general relativity versus quantum states using hermitical linear spaces over the complex field. Electromagnetism and weak interactions have been partially unified in the Weinberg-Salam model of electro-weak interactions [4], then strong interactions have been described by chromodynamics [5] and both are the main parts of the synthesis known as standard model of quantum physics.

All preceding attempts have supposed that the starting point was necessarily the standard model of quantum physics. The philosophical background of this choice is the believing in the necessity of the theoretical construction made from quantum mechanics. This comes historically from the fact that the first wave equation, found by E. Schrödinger, was able to account not only for one particle, but for a lot of them. Unhappily this wave equation is not relativistic. Consequently it is also unable to account for the spin 1/2, a common property for all fundamental fermions of relativistic quantum physics.

Since 1912 general relativity has made tremendous progress. Solutions were found justifying the behavior of stars with a very strong gravitational field, emitting gravitational waves. Calculations perfectly account for the measurements. On the contrary even if the quantum field theory claims still more precise results, such as the calculation of the Lamb effect or the calculation of the anomalous magnetic moment of the electron, the calculation is difficult: it is a process of successive approximations with the necessity to cut off not a small error but an infinite part of the result: this is not mathematically correct. In fact there are two quantum theories, the first and the second quantification. The first quantification uses for the electron a wave that is a function of space and time with value into the complex field or into a complex vector space. Calculations are exact in the case of the hydrogen atom and results are very precise. The second quantification transforms this wave into something very different, a field of operators operating on something, an object with no mathematical definition. Therefore on the mathematical point of view the departure of the ToE must be the general relativity and neither the second quantification nor any theory using the second quantification.

On the physical point of view it is the same from time properties. Time in physics is not space: in space we can go in any direction, in time we go only from past to future and there is strictly no way to go backwards. Gravitation propagates with gravitational waves that are emitted by stars and the move of these stars is changing with time. All laws of quantum physics are obtained from Lagrangian densities that give the wave equations from a principle of stationarity. This principle was inherited from classical mechanics where the time is not oriented if we forget frictions. Since the true time is the oriented time of thermodynamics, not the time of classical mechanics, only general relativity, allowing the use of an oriented time, is able to be the starting point of the ToE. General relativity and standard model are both obtained from a principle of stationarity, this is certainly another cornerstone to build the ToE (private discussion with D. Girardot).

2. CLIFFORD ALGEBRA OF THE 3-DIMENSIONAL SPACE

Clifford algebras on the real field are a necessary tool because particles with spin 1/2 exist. We use here all notations of "New insights in the standard model of quantum physics in Clifford Algebra" [6]. Its first chapter study at an elementary level the three Clifford algebras used in the present work. The algebra Cl_3 of the physical space is sufficient to describe the physical space-time because

$$x = x^\mu \sigma_\mu = \begin{pmatrix} x^0 + x^3 & x^1 - ix^2 \\ x^1 + ix^2 & x^0 - x^3 \end{pmatrix} \tag{1.1}$$

forms the auto-adjoint subspace of the Clifford algebra Cl_3. The wave of an electron is a function of this part of Cl_3 into the full algebra: $x \mapsto \phi(x)$. The Dirac wave equation reads in this frame

$$0 = \nabla \hat{\phi}\sigma_{21} + qA\hat{\phi} + m\phi \tag{1.2}$$

$$\phi = \sqrt{2}(\xi - i\sigma_2\eta^*) = \sqrt{2}\begin{pmatrix} \xi_1 & -\eta_2^* \\ \xi_2 & \eta_1^* \end{pmatrix}; \hat{\phi} = \sqrt{2}(\eta - i\sigma_2\xi^*) = \sqrt{2}\begin{pmatrix} \eta_1 & -\xi_2^* \\ \eta_2 & \xi_1^* \end{pmatrix}$$

$$\nabla = \sigma^\mu \partial_\mu ; A = \sigma^\mu A_\mu ; q = \frac{e}{\hbar c};\ m = \frac{m_0 c}{\hbar}, \sigma_{21} = \sigma_2\sigma_1 = -i\sigma_3. \tag{1.3}$$

where ξ and η are respectively right and left Weyl spinors, m_0 is the proper mass and e is the negative charge of the electron[1]. The Dirac equation is the linear approximation of our non-linear homogeneous wave equation

$$\overline{\phi}(\nabla\hat{\phi})\sigma_{21} + \overline{\phi}qA\hat{\phi} + m\rho = 0 \tag{1.4}$$

$$\overline{\phi} = \hat{\phi}^\dagger ; \overline{\phi}\phi = \phi\overline{\phi} = \det(\phi) = \rho e^{i\beta} \tag{1.5}$$

where β is the Yvon-Takabayasi angle. The non-linear wave equation has the Dirac equation as linear approximation if this angle is null or very small.

I recall now how the wave equation is said form invariant: If M is any element in Cl_3 and if R is the transformation from the space-time into itself, which to any x associates x' such as

$$x = x^\mu \sigma_\mu ;\ x' = x'^\mu \sigma_\mu = R(x) = MxM^\dagger \tag{1.6}$$

we note, if $\det(M) \neq 0$:

$$\det(M) = re^{i\theta},\ r = |\det(M)|. \tag{1.7}$$

We then get:

$$\det(x') = \det(MxM^\dagger)\,(x'^0)^2 - (x'^1)^2 - (x'^2)^2 - (x'^3)^2 =$$

$$= re^{i\theta}\det(x)re^{-i\theta} = r^2[(x^0)^2 - (x^1)^2 - (x^2)^2 - (x^3)^2] \tag{1.8}$$

R multiplies then by r any space-time distance and is called "Lorentz dilation with ratio r". If we let, with the usual convention summing the up and down indices:

$$x'^\mu = R^\mu_\nu x^\nu \tag{1.9}$$

we get for $M = \begin{pmatrix} a & b \\ c & d \end{pmatrix} \neq 0$:

$$2R^0_0 = |a|^2 + |b|^2 + |c|^2 + |d|^2 > 0 \tag{1.10}$$

x'^0 has then the same sign as x^0 at the origin: R conserves the arrow of the time. Even more we get (the calculation is in [7] A.2.1 and A.2.4):

$$\nabla = \overline{M}\nabla'\hat{M}\,;\nabla' = \sigma^\mu\partial'_\mu \tag{1.11}$$

$$\det(R^\mu_\nu) = r^4 \tag{1.12}$$

R conserves therefore the orientation of space-time and as it conserves the time orientation it conserves also the space orientation. Relativistic quantum mechanics lets

$$\phi'(x') = M\phi(x) \tag{1.13}$$

and non relativistic quantum mechanics uses the same relation with M only in $SU(2)$. All properties of particles with spin $\frac{1}{2}$ such as Clebsch-Gordon coefficients are deduced from (1.13) and from the properties of the matrix representations of $SU(2)$. The wave Equation (1.4) is form invariant because

$$\bar{\phi}(\nabla\hat{\phi})\sigma_{21} = \bar{\phi}'(\nabla'\hat{\phi}')\sigma_{21} \tag{1.14}$$

$$\bar{\phi}qA\hat{\phi} = \bar{\phi}'q'A'\hat{\phi}' \tag{1.15}$$

$$m\rho = m'\rho'. \tag{1.16}$$

Here (1.14) comes from (1.11) and (1.13). The relation (1.15) is necessary to get the gauge invariance of the electromagnetism and comes from

$$A' = \sigma^{\mu} A'_{\mu};\, qA = \overline{M} q' A' \hat{M}. \tag{1.17}$$

The relation (1.16) comes from (1.5), (1.7), (1.13) that give

$$\rho' = r\rho \tag{1.18}$$

and (1.16) is then equivalent to

$$m = rm' \tag{1.19}$$

The set $Cl_3^* = GL(2,\mathbb{C})$ of the M matrices is a 8-dimensional real Lie group. The application $M \mapsto R$ is an homomorphism of Cl_3^* into the 7-dimensional Lie group D of Lorentz dilations, each dilation being the product, in any order, of an element of the restricted Lorentz group $\mathcal{L}_+^\uparrow$ and of a homothety with ratio r. The kernel of this homomorphism is 1-dimensional, it is the $U(1)$ set of the $M = e^{i\theta}$ where $i = \sigma_1\sigma_2\sigma_3$ gives the orientation of the physical space. We consider the Cl_3^* group as the true group of invariance of the Dirac wave and of the electromagnetism[2] [6]-[8]. This group is greater than the $SL(2,\mathbb{C})$ group of relativistic quantum mechanics, then there are less invariant terms: now ρ and m are not separately invariant, (1.16) means that only the $m\rho$ product remains invariant under Cl_3^*.

Our wave equation is gauge invariant under two gauges. The first one is well known since it is the electric gauge

$$\phi \mapsto \phi' = \phi e^{ia\sigma_3} \; ; \; A_\mu \mapsto A_\mu' = A_\mu - \frac{1}{q}\partial_\mu a \tag{1.20}$$

The second one is only a global gauge, it is the chiral gauge

$$\phi \mapsto \phi' = \phi e^{ia} \; ; \; \partial_\mu a = 0. \tag{1.21}$$

Each gauge is linked to a conserved current. The current linked to the electric gauge is the current of probability

$$D_0 = \phi\phi^\dagger \tag{1.22}$$

The second conserved current is the D_3 current satisfying

$$D_3 = \phi\sigma_3\phi^\dagger \tag{1.23}$$

Components D_0^μ and D_3^μ of these space-time vectors are 8 of the 36 components of tensorial densities available from the 8 real parameters of the wave of the electron (the Dirac theory knows only 16 of them). At each point x the value of the wave $\phi(x)$ has exactly the same form as the M matrix, then we can associate to this value the geometric transformation D, with ratio ρ :

$$D : y \mapsto x = \phi y \phi^\dagger. \tag{1.24}$$

And the components D_μ^ν of the four D_μ are the 16 terms of the matrix of the dilation because

$$x = x^\mu \sigma_\mu = \phi y^\nu \sigma_\nu \phi^\dagger = y^\nu \phi \sigma_\nu \phi^\dagger = y^\nu D_\nu = y^\nu D_\nu^\mu \sigma_\mu ; x^\mu = D_\nu^\mu y^\nu. \tag{1.25}$$

There is no difference between the matrix product $M'M$ which gives the composition of dilations $R' \circ R$ and the product $M\phi$ which gives the transformation of the wave under a dilation, and this induces then a composition of dilations $R \circ D$:

$$x' = MxM^{\dagger} = M\phi y\phi^{\dagger}M^{\dagger} = (M\phi)y(M\phi)^{\dagger} = \phi' y\phi'^{\dagger}. \tag{1.26}$$

This means that the y introduced into (1.24) does not change, either seen by the observer of x or by the observer of x'. It is independent on the observer, intrinsic to the wave.

And since ϕ is function of x, the dilation is also a function of x, and varies from a point to another in space-time: y is not an element of the global space-time, only of the local space-time. So we must see y as the general element of the tangent space-time, at x, to a space-time manifold which depends only on the wave, not on the observer, and that we will name "intrinsic manifold". On the contrary the dilation depends on the observer, the observer of x sees D, the one of x' sees $D' = R \circ D$.

The geometric meaning of the Dirac wave is as follows: At each point of the space-time we have, not only one space-time manifold, but two space-time manifolds: the manifold of the x and x', for which each relativistic observer is associated to a Lorentzian tangent space-time, and the intrinsic manifold, this of the y which is a manifold with torsion [7] [8]. Moreover each tangent space-time has an orthonormal basis allowing to construct a Clifford algebra. The fiber of each of these two manifolds is then a same Clifford algebra.

2. WAVE OF ELECTRON + NEUTRINO IN $CL_{1,3}$

The electric gauge is the reason of the electric interaction of the electron. Similarly the second gauge of the electron is linked to the weak interaction of the electron. The Weinberg-Salam model of electro-weak interactions [4] considers the right ξ_e and left η_e waves of the electron and the left wave η_n of the electronic neutrino. If a right ξ_n wave exists it does not interact neither electrically nor weakly. Therefore the standard model supposes $\xi_n = 0$.

Relativistic quantum mechanics supposes also a link between the wave of the particle and the wave of its anti-particle. Noting now ϕ_e the wave of the electron, this link reads in space algebra (see [6] 2.4.2)

$$\hat{\phi}_{\bar{e}} = \hat{\phi}_e\sigma_1\ ;\ \hat{\phi}_{\bar{n}} = \hat{\phi}_n\sigma_1 \tag{2.1}$$

Then the wave of electron + positron + neutrino + antineutrino is a function of x with value in a 12-dimen- sional subspace of $Cl_{1,3}$. The form invariance of the wave implies that ϕ_e and ϕ_n must transform similarly. The lepton wave Ψ_l is a function of x into the Clifford algebra of space-time $Cl_{1,3}$ [9]:

$$\phi_e = \sqrt{2}\begin{pmatrix}\xi_e & -i\sigma_2\eta_e^*\end{pmatrix};\hat{\phi}_e = \sqrt{2}\begin{pmatrix}\eta_e & -i\sigma_2\xi_e^*\end{pmatrix} \tag{2.2}$$

$$\phi_n = \sqrt{2}\begin{pmatrix}0 & -i\sigma_2\eta_n^*\end{pmatrix};\hat{\phi}_n = \sqrt{2}\begin{pmatrix}\eta_n & 0\end{pmatrix} \tag{2.3}$$

$$\Psi_l = \begin{pmatrix}\phi_e & \phi_n \\ \hat{\phi}_{\bar{n}}\sigma_1 & \hat{\phi}_{\bar{e}}\sigma_1\end{pmatrix} = \begin{pmatrix}\phi_e & \phi_n \\ \hat{\phi}_n & \hat{\phi}_e\end{pmatrix}. \tag{2.4}$$

The link between Cl_3 and $Cl_{1,3}$ is made by the Pauli matrices σ_j and the matrix representation[3]

$$\gamma_0 = \gamma^0 = \begin{pmatrix}0 & I \\ I & 0\end{pmatrix};\gamma_j = -\gamma^j = \begin{pmatrix}0 & \sigma_j \\ -\sigma_j & 0\end{pmatrix};I = \sigma_0 = \sigma^0 = \begin{pmatrix}1 & 0 \\ 0 & 1\end{pmatrix} \tag{2.5}$$

With this matrix representation the reverse of Ψ_l is

$$\tilde{\Psi}_l = \begin{pmatrix}\overline{\phi}_e & \phi_n^\dagger \\ \overline{\phi}_n & \phi_e^\dagger\end{pmatrix}. \tag{2.6}$$

The invariant form of the wave equation of electron + neutrino reads [10]

$$\tilde{\Psi}_l(\mathbf{D}\Psi_l)\gamma_{012} + m\rho\tilde{\Psi}_l\chi_l = 0 \tag{2.7}$$

where we have to define ρ, χ_l and $\mathbf{D}$. The covariant derivative $\mathbf{D}$ satisfies

$$\mathbf{D} = \partial + \frac{g_1}{2}\mathbf{B}P_0 + \frac{g_2}{2}(\mathbf{W}^1P_1 + \mathbf{W}^2P_2 + \mathbf{W}^3P_3) \tag{2.8}$$

$$\mathbf{D} = \gamma^\mu D_\mu \; ; \; \partial = \gamma^\mu\partial_\mu \; ; \; \mathbf{B} = \gamma^\mu B_\mu \; ; \; \mathbf{W}^j = \gamma^\mu W_\mu^j \tag{2.9}$$

The P_μ are four operators so defined:

$$P_\pm(\Psi) = \frac{1}{2}(\Psi \pm \mathbf{i}\Psi\gamma_{21}); \ \mathbf{i} = \gamma_0\gamma_1\gamma_2\gamma_3 ; \ \gamma_{21} = \gamma_2\gamma_1 \tag{2.10}$$

$$P_0(\Psi) = \Psi\gamma_{21} + P_-(\Psi)\mathbf{i} \tag{2.11}$$

$$P_1(\Psi) = P_+(\Psi)\gamma_3\mathbf{i} \tag{2.12}$$

$$P_2(\Psi) = P_+(\Psi)\gamma_3 \tag{2.13}$$

$$P_3(\Psi) = P_+(\Psi)(-\mathbf{i}) \tag{2.14}$$

The ρ term generalizes the ρ of the electron:

$$a_1 = \det(\phi_e) = \phi_e\overline{\phi}_e = 2(\xi_{1e}\eta^*_{1e} + \xi_{2e}\eta^*_{2e}) = \rho e^{i\beta} \tag{2.15}$$

$$a_2 = 2(\xi_{1\bar{e}}\eta^*_{1n} + \xi_{2\bar{e}}\eta^*_{2n}) = 2(\eta^*_{2e}\eta^*_{1n} - \eta^*_{1e}\eta^*_{2n}) \tag{2.16}$$

$$a_3 = 2(\xi_{1e}\eta^*_{1n} + \xi_{2e}\eta^*_{2n}) \tag{2.17}$$

$$\rho = \sqrt{a_1a_1^* + a_2a_2^* + a_3a_3^*} \tag{2.18}$$

and χ_l satisfies

$$\chi_l = \frac{1}{\rho^2}\begin{pmatrix} a_1^*\phi_e + a_2^*\phi_n\sigma_1 + a_3^*\phi_n & -a_2^*\phi_{eL}\sigma_1 + a_3^*\phi_{eR} \\ a_2\hat{\phi}_{eL}\sigma_1 + a_3\hat{\phi}_{eR} & a_1\hat{\phi}_e - a_2\hat{\phi}_n\sigma_1 + a_3\hat{\phi}_n \end{pmatrix} \tag{2.19}$$

$$\phi_{eR} = \phi_e\frac{1+\sigma_3}{2}; \ \phi_{eL} = \phi_e\frac{1-\sigma_3}{2} \tag{2.20}$$

The invariant wave Equation (2.7) has a real scalar part [10] that is simply

$$\mathcal{L}=0 \tag{2.21}$$

where $\mathcal{L}$ is the Lagrangian density that gives the whole Equation (2.7) by Lagrange equations. Another of the real equations equivalent to (2.7) is

$$\partial_\mu(D_0^\mu + D_n^\mu)=0;\ D_0=\phi_e\phi_e^\dagger;\ D_n=\phi_n\phi_n^\dagger \tag{2.22}$$

Then there is only one conservative current, it is the sum of the current of the electron and the same current for the neutrino. The wave equation is gauge invariant [10] under the $U(1)\times SU(2)$ gauge group generated by operators P_μ. The wave equation is form invariant under the dilation R generated by any M matrix satisfying (1.6), because, if

$$N=\begin{pmatrix} M & 0 \\ 0 & \hat{M} \end{pmatrix};g:M\mapsto N \tag{2.23}$$

g is a group isomorphism, the transformation of the wave satisfies

$$\Psi_l'=N\Psi_l;\ \chi_l'=\tilde{N}^{-1}\chi_l;\ \rho'=r\rho \tag{2.24}$$

which gives, with (1.19)

$$\tilde{\Psi}_l(\mathbf{D}\Psi_l)\gamma_{012}+m\rho\tilde{\Psi}_l\chi_l=\tilde{\Psi}_l'(\mathbf{D}'\Psi_l')\gamma_{012}+m'\rho'\tilde{\Psi}_l'\chi_l'=0. \tag{2.25}$$

In the space-time algebra we use

$$\mathbf{x}=x^\mu\gamma_\mu=\begin{pmatrix} 0 & x \\ \hat{x} & 0 \end{pmatrix};\mathbf{y}=y^\mu\gamma_\mu=\begin{pmatrix} 0 & y \\ \hat{y} & 0 \end{pmatrix} \tag{2.26}$$

We note $\langle M\rangle_n$ the n-vector part of the multivector M. The geometric transformation D in (1.24) linked to the wave reads now

$$\mathbf{D}:\mathbf{y}\mapsto\mathbf{x}=<\Psi_l\mathbf{y}\tilde{\Psi}_l>_1=\frac{1}{2}(\Psi_l\mathbf{y}\tilde{\Psi}_l+\mathbf{i}\Psi_l\mathbf{y}\tilde{\Psi}_l\mathbf{i}). \tag{2.27}$$

Similarly to the electron case $\mathbf{y}$ is independent from the observer, intrinsic to the wave since under the dilation R we get

$$\Psi'_l=N\Psi_l\,;\;\tilde{\Psi}'_l=\Psi_l\tilde{N} \tag{2.28}$$

$$\mathbf{x}'=N\mathbf{x}\tilde{N}=N\frac{1}{2}(\Psi_l\mathbf{y}\tilde{\Psi}_l+\mathbf{i}\Psi_l\mathbf{y}\tilde{\Psi}_l\mathbf{i})\tilde{N}=\frac{1}{2}(\Psi'_l\mathbf{y}\tilde{\Psi}'_l+\mathbf{i}\Psi'_l\mathbf{y}\tilde{\Psi}'_l\mathbf{i})=<\Psi'_l\mathbf{y}\tilde{\Psi}'_l>_1 \tag{2.29}$$

Relation (2.27) is equivalent to

$$x=\phi_e y\phi_e^\dagger+\phi_n\hat{y}\phi_n^\dagger \tag{2.30}$$

$$x^\mu=(D_0^\mu+D_n^\mu)y^0+D_1^\mu y^1+D_2^\mu y^2+(D_3^\mu+D_n^\mu)y^3. \tag{2.31}$$

Then D is an affine transformation where indices 1 and 2 play a very different role. This is evidently linked to the fact that σ_{21} is present in the wave equation of the electron and γ_{012} is present in the wave of the electron + neutrino. This anisotropy is the reason of the existence of three kinds of leptons.

We know three generations of leptons and quarks and the standard model study separately these three generations. The reason is simply that our physical space is three dimensional, and we get the wave equation of leptons three times. One of the three is (2.7) that reads:

$$0=\tilde{\Psi}_3(\mathbf{D}_3\Psi_3)\gamma_{012}+m_3\rho\tilde{\Psi}_3\chi_3;\mathbf{D}_3=\mathbf{D};\Psi_3=\Psi_l;\chi_3=\chi_l\,;\;m_3=m \tag{2.32}$$

$$0=\tilde{\Psi}_1(\mathbf{D}_1\Psi_1)\gamma_{023}+m_1\rho\tilde{\Psi}_1\chi_1 \tag{2.33}$$

$$0=\tilde{\Psi}_2(\mathbf{D}_2\Psi_3)\gamma_{031}+m_2\rho\tilde{\Psi}_2\chi_2 \tag{2.34}$$

To go from one generation to another one is simple: we permute indices 1, 2, 3 of σ_j everywhere in all preceding formulas with the circular permutation p or p^2:

$$p:1\mapsto 2\mapsto 3\mapsto 1;\ p^2:1\mapsto 3\mapsto 2\mapsto 1. \qquad (2.35)$$

We do not know if the muon is obtained by p or by p^2. If it is p, the wave of the pair muon-muonic neutrino follows (2.33) and this explains why a muon is like an electron, generally. But the covariant derivative is different, because in the place of (2.10) to (2.14) we must use

$$P_\pm^1(\Psi)=\frac{1}{2}(\Psi\pm\mathbf{i}\Psi\gamma_{32}) \qquad (2.36)$$

$$P_0^1(\Psi)=\Psi\gamma_{32}+P_-(\Psi)\mathbf{i} \qquad (2.37)$$

$$P_1^1(\Psi)=P_+(\Psi)\gamma_1\mathbf{i} \qquad (2.38)$$

$$P_2^1(\Psi)=P_+(\Psi)\gamma_1 \qquad (2.39)$$

$$P_3^1(\Psi)=P_+(\Psi)(-\mathbf{i}) \qquad (2.40)$$

We must also change the link (2.1) between the wave of the particle and the wave of the antiparticle, link using a σ_1 for the first generation. The wave of the anti-muon must satisfy:

$$\hat{\phi}_{\bar{\mu}}=\hat{\phi}_\mu\sigma_2 \qquad (2.41)$$

and we shall have a 3 index in the case of the last generation. We must also change the definition of left and right wave. For the second generation this becomes

$$\phi_{\mu R}=\phi_\mu\frac{1}{2}(1+\sigma_1);\ \phi_{\mu L}=\phi_\mu\frac{1}{2}(1-\sigma_1) \qquad (2.42)$$

and me must also permute the indexes in the geometric transformation (2.31). We can then understand why the Lagrangian density, which comes from the scalar part of the invariant equation, must be calculated separately for the pair electron-electronic neutrino and for the pair muon-muonic neutrino or tau-tauic neutrino.

If there are only three objects like σ_{21}, there is one other term with square −1 in Cl_3, $i = \sigma_1\sigma_2\sigma_3$. this fourth term allows a fourth neutrino [11].

3. STANDARD MODEL IN CL1,5

The standard model adds to the leptons (electron and its neutrino) in the first "generation" two quarks u and d with three states each. Weak interactions acting only on left waves of quarks (and right waves of antiquarks) the wave of all fermions of the first generation satisfies:

$$\Psi = \begin{pmatrix} \Psi_l & \Psi_r \\ \Psi_g & \Psi_b \end{pmatrix}; \ \Psi_r = \begin{pmatrix} \phi_{dr} & \phi_{ur} \\ \hat{\phi}_{ur} & \hat{\phi}_{dr} \end{pmatrix} = \begin{pmatrix} \phi_{dr} & \varphi_{ur} \\ \hat{\phi}_{\overline{ur}}\sigma_1 & \hat{\phi}_{\overline{dr}}\sigma_1 \end{pmatrix} \tag{3.1}$$

$$\Psi_g = \begin{pmatrix} \phi_{dg} & \phi_{ug} \\ \hat{\phi}_{ug} & \hat{\phi}_{dg} \end{pmatrix} = \begin{pmatrix} \phi_{dg} & \phi_{ug} \\ \hat{\phi}_{\overline{ug}}\sigma_1 & \hat{\phi}_{\overline{dg}}\sigma_1 \end{pmatrix} \tag{3.2}$$

$$\Psi_b = \begin{pmatrix} \phi_{db} & \phi_{ub} \\ \hat{\phi}_{ub} & \hat{\phi}_{db} \end{pmatrix} = \begin{pmatrix} \phi_{db} & \phi_{ub} \\ \hat{\phi}_{\overline{ub}}\sigma_1 & \hat{\phi}_{\overline{db}}\sigma_1 \end{pmatrix} \tag{3.3}$$

The Ψ wave is now a function of space and time with value into $Cl_{1,5}$ which is a sub-algebra (on the real field) of $M_8(\mathbb{C})$ 4. The covariant derivative (2.8) becomes

$$\underline{D} = \underline{\partial} + \frac{g_1}{2}\underline{B}\,\underline{P}_0 + \frac{g_2}{2}\underline{W}^j\,\underline{P}_j + \frac{g_3}{2}\underline{G}^k\,\mathbf{i}\Gamma_k \tag{3.4}$$

$$\underline{D} = \sum_{\mu=0}^{3} L^\mu D_\mu \ ; \ \underline{\partial} = \sum_{\mu=0}^{3} L^\mu \partial_\mu \ ; \ \underline{B} = \sum_{\mu=0}^{3} L^\mu B_\mu \tag{3.5}$$

$$\underline{W}^j = \sum_{\mu=0}^{3} L^\mu W_\mu^j,\ j = 1,2,3 \tag{3.6}$$

$$\underline{G}^k = \sum_{\mu=0}^{3} L^\mu G_\mu^k, k = 1,2,\ldots,8 \tag{3.7}$$

$$L_\mu = \begin{pmatrix} 0 & \gamma_\mu \\ \gamma_\mu & 0 \end{pmatrix}; L_4 = \begin{pmatrix} 0 & -I_4 \\ I_4 & 0 \end{pmatrix}; L_5 = \begin{pmatrix} 0 & \mathbf{i} \\ \mathbf{i} & 0 \end{pmatrix}. \tag{3.8}$$

We use two projectors satisfying

$$\underline{P}_\pm(\Psi) = \frac{1}{2}(\Psi \pm \underline{\mathbf{i}}\Psi L_{21});\ \underline{\mathbf{i}} = L_{0123} \tag{3.9}$$

Three operators act on quarks like on leptons:

$$\underline{P}_1(\Psi) = \underline{P}_+(\Psi)L_{35}; \underline{P}_2(\Psi) = \underline{P}_+(\Psi)L_{5012} \tag{3.10}$$

$$\underline{P}_3(\Psi) = \underline{P}_+(\Psi)(-\underline{\mathbf{i}}). \tag{3.11}$$

The fourth operator acts differently on the lepton and on the quark sector:

$$\underline{P}_0(\Psi) = \begin{pmatrix} P_0(\Psi_l) & P_0'(\Psi_r) \\ P_0'(\Psi_g) & P_0'(\Psi_b) \end{pmatrix};$$
$$P_0(\Psi_l) = \Psi_l\gamma_{21} + P_-(\Psi_l)\mathbf{i} \tag{3.12}$$

$$P_0'(\Psi_c) = -\frac{1}{3}\Psi_c\gamma_{21} + P_-(\Psi_c)\mathbf{i},\ c = r,g,b. \tag{3.13}$$

These definitions are absolutely all that you have to change to go from the lepton case into the quark case, to get the gauge group of electro-weak interactions. We proved in [6] 6.3 that this gives:

$$D_\mu \xi_d = \partial_\mu \xi_d - i(-\frac{1}{3})g_1 B_\mu \xi_d \tag{3.14}$$

$$D_\mu \xi_u = \partial_\mu \xi_u - i(+\frac{2}{3})g_1 B_\mu \xi_u \tag{3.15}$$

$$D_\mu \eta_{\bar{d}} = \partial_\mu \eta_{\bar{d}} - i(+\frac{1}{3})g_1 B_\mu \eta_{\bar{d}} \tag{3.16}$$

$$D_\mu \eta_{\bar{u}} = \partial_\mu \eta_{\bar{u}} - i(-\frac{2}{3})g_1 B_\mu \eta_{\bar{u}} \tag{3.17}$$

This means that changing the coefficient 1 of $\Psi\gamma_{21}$ into $-1/3$ is sufficient to get the correct charges of u and d quarks, the correct charges of antiquarks $\bar{u}$ and $\bar{d}$. Moreover we get a doublet of left waves for the quarks and a doublet of right waves for the antiquarks:

$$\psi_L = \begin{pmatrix} \eta_u \\ \eta_d \end{pmatrix}; \ \psi_R = \begin{pmatrix} \xi_{\bar{u}} \\ \xi_{\bar{d}} \end{pmatrix}, \tag{3.18}$$

which gives

$$D_\mu \psi_L = \partial_\mu \psi_L - i\frac{g_1}{6} B_\mu \psi_L - i\frac{g_2}{2}(W_\mu^1 \tau_1 + W_\mu^2 \tau_2 + W_\mu^3 \tau_3)\psi_L \tag{3.19}$$

$$D_\mu \psi_R = \partial_\mu \psi_R + i\frac{g_1}{6} B_\mu \psi_R - i\frac{g_2}{2}(W_\mu^1 \tau_1 - W_\mu^2 \tau_2 + W_\mu^3 \tau_3)\psi_R \tag{3.20}$$

$$\tau_1 = \gamma_0 \ ; \ \tau_2 = \gamma_{123} \ ; \ \tau_3 = \gamma_5 . \tag{3.21}$$

Then all features of electro-weak interactions of leptons and quarks are simply obtained from the structure of the wave, itself linked to the structure of the space-time

The generators of the $SU(3)$ gauge group of chromodynamics use two projectors:

$$P^{+}=\frac{1}{2}(I_8+L_{012345})=\begin{pmatrix} I_4 & 0 \\ 0 & 0 \end{pmatrix};\ P^{-}=\frac{1}{2}(I_8-L_{012345})=\begin{pmatrix} 0 & 0 \\ 0 & I_4 \end{pmatrix} \tag{3.22}$$

and eight operators Γ_k, $k=1,2,\ldots,8$ so defined (with c instead Ψ_c):

$$\Gamma_1(\Psi)=\frac{1}{2}(L_4\Psi L_4+L_{01235}\Psi L_{01235})=\begin{pmatrix} 0 & g \\ r & 0 \end{pmatrix} \tag{3.23}$$

$$\Gamma_2(\Psi)=\frac{1}{2}(L_5\Psi L_4-L_{01234}\Psi L_{01235})=\begin{pmatrix} 0 & -\mathbf{i}g \\ \mathbf{i}r & 0 \end{pmatrix} \tag{3.24}$$

$$\Gamma_3(\Psi)=P^{+}\Psi P^{-}-P^{-}\Psi P^{+}=\begin{pmatrix} 0 & r \\ -g & 0 \end{pmatrix} \tag{3.25}$$

$$\Gamma_4(\Psi)=L_{01253}\Psi P^{-}=\begin{pmatrix} 0 & b \\ 0 & r \end{pmatrix};\Gamma_5(\Psi)=L_{01234}\Psi P^{-}=\begin{pmatrix} 0 & -\mathbf{i}b \\ 0 & \mathbf{i}r \end{pmatrix} \tag{3.26}$$

$$\Gamma_6(\Psi)=P^{-}\Psi L_{01253}=\begin{pmatrix} 0 & 0 \\ b & g \end{pmatrix};\Gamma_7(\Psi)=-\underline{\mathbf{i}}P^{-}\Psi L_4=\begin{pmatrix} 0 & 0 \\ -\mathbf{i}b & \mathbf{i}g \end{pmatrix} \tag{3.27}$$

$$\Gamma_8(\Psi)=\frac{1}{\sqrt{3}}(P^{-}\Psi L_{012345}+L_{012345}\Psi P^{-})=\frac{1}{\sqrt{3}}\begin{pmatrix} 0 & r \\ g & -2b \end{pmatrix}. \tag{3.28}$$

We explained in [9] how these operators are equivalent to the eight generators λ_k of $SU(3)$. Everywhere in (3.23) to (3.28) the eight matrices $\Gamma_k(\Psi)$ have a zero left up term, therefore all Γ_k project the wave on its quark sector. The physical translation is: leptons do not interact by strong interactions, this comes from the structure itself of the quantum wave. We consider this result as a great success of real Clifford algebras. Now with

$$\underline{D} = \sum_{\mu=0}^{3} L^{\mu} D_{\mu} \; ; \; \underline{\partial} = \sum_{\mu=0}^{3} L^{\mu} \partial_{\mu} \; ; \; \underline{B} = \sum_{\mu=0}^{3} L^{\mu} B_{\mu}$$

$$\underline{W}^{j} = \sum_{\mu=0}^{3} L^{\mu} W_{\mu}^{j} \; ; \; \underline{G}^{k} = \sum_{\mu=0}^{3} L^{\mu} G_{\mu}^{k} \tag{3.29}$$

where the eight $\underline{G}^k$ are part of what is named "gluons", the covariant derivative reads

$$\underline{D} = \underline{\partial} + \frac{g_1}{2} \underline{B}\, \underline{P}_0 + \frac{g_2}{2} \sum_{j=1}^{3} \underline{W}^{j} \underline{P}_j + \frac{g_3}{2} \sum_{k=1}^{8} \underline{G}^{k} \underline{\mathbf{i}}\, \Gamma_k . \tag{3.30}$$

The gauge group is obtained by exponentiation. We use four numbers a^{μ} and eight numbers b_k. We let

$$S = S_0 + S_1 + S_2 \tag{3.31}$$

$$S_0 = a^0 \underline{P}_0 \; ; \; S_1 = \sum_{j=1}^{3} a^j \underline{P}_j \; ; \; S_2 = \sum_{k=1}^{8} b^k \underline{\mathbf{i}}\, \Gamma_k . \tag{3.32}$$

We get

$$\begin{aligned} \exp(S) &= \exp(S_0)\exp(S_1)\exp(S_2) \\ &= \exp(S_0)\exp(S_2)\exp(S_1) \\ &= \exp(S_2)\exp(S_1)\exp(S_0) = \ldots \end{aligned} \tag{3.33}$$

in any order, because:

$$\underline{P}_0 \underline{P}_j = \underline{P}_j \underline{P}_0 \, , \; j = 1,2,3 \tag{3.34}$$

$$\underline{P}_{\mu} \underline{\mathbf{i}} \Gamma_k = \underline{\mathbf{i}} \Gamma_k \underline{P}_{\mu} \, , \; \mu = 0,1,2,3, \; k = 1,2,\ldots,8 \tag{3.35}$$

Therefore the set

$$G = \{\exp(S)\} \tag{3.36}$$

is a $U(1)\times SU(2)\times SU(3)$ Lie group. The gauge transformation reads

$$\begin{aligned}&\Psi' = [\exp(S)](\Psi);\\ &\underline{D} = L^{\mu}\,\underline{D}_{\mu}\ ;\ \underline{D}' = L^{\mu}\,\underline{D}'_{\mu}\end{aligned} \tag{3.37}$$

$$\underline{D}'\Psi' = [\exp(S)]\underline{D}(\Psi) \tag{3.38}$$

$$B'_{\mu} = B_{\mu} - \frac{2}{g_1}\partial_{\mu}a^{0} \tag{3.39}$$

$$W'^{j}_{\mu}\,\underline{P}_j = [\exp(S_1)W^{j}_{\mu}\,\underline{P}_j - \frac{2}{g_2}\partial_{\mu}[\exp(S_1)]]\exp(-S_1) \tag{3.40}$$

$$G'^{k}_{\mu}\,\underline{\mathbf{i}\Gamma}_k = [\exp(S_2)G^{k}_{\mu}\,\underline{\mathbf{i}\Gamma}_k - \frac{2}{g_3}\partial_{\mu}[\exp(S_2)]]\exp(-S_2) \tag{3.41}$$

The $SU(3)$ group of $\exp(S_2)$ operators, generated by projectors on the quark sector, acts only on this sector of the wave:

$$P^{+}[\exp(S_2)](\Psi)P^{+} = P^{+}\Psi P^{+} = \begin{pmatrix}\Psi_l & 0\\ 0 & 0\end{pmatrix}. \tag{3.42}$$

We then get the gauge group of the standard model, automatically, and not another group. It is impossible to get operators exchanging Ψ_l and $\Psi_c, c = r, g, b$ like Γ_1 exchanging Ψ_r and Ψ_g because $\Psi_l(x)$ is 12-dimensional on $\mathbb{R}$ while $\Psi_r(x)$ is 8-dimensional on $\mathbb{R}$. Then we cannot get a greater group than the preceding $U(1)\times SU(2)\times SU(3)$ gauge group. This limitation is linked to the geometric structure of the wave. The physical consequence is the experimental fact that the proton cannot disintegrate. This is said in the language of quantum field theory by saying that the baryonic quantum number is conservative.

We have still supposed nothing on ϕ_{ur}, ϕ_{ug}, ϕ_{ub}, ϕ_{dr}, ϕ_{dg}, ϕ_{db}. But the standard model uses only left waves for the particles in the case of electro-weak interactions (and right waves for the antiparticles). Why ? I think possible to give a mathematical reason to this physical hypothesis. If u and d quarks have only left wave, this means η waves, we have:

$$\Psi_l = \sqrt{2}\begin{pmatrix} \xi_1 & -\eta_2^* & 0 & -\eta_{2n}^* \\ \xi_2 & \eta_1^* & 0 & \eta_{1n}^* \\ \eta_{1n} & 0 & \eta_1 & -\xi_2^* \\ \eta_{2n} & 0 & \eta_2 & \xi_1^* \end{pmatrix};\ \Psi_r = \sqrt{2}\begin{pmatrix} 0 & -\eta_{2dr}^* & 0 & -\eta_{2ur}^* \\ 0 & \eta_{1dr}^* & 0 & \eta_{1ur}^* \\ \eta_{1ur} & 0 & \eta_{1dr} & 0 \\ \eta_{2ur} & 0 & \eta_{2dr} & 0 \end{pmatrix} \tag{3.43}$$

and two similar equalities for colors g and b. Now we define two matrices M_1 and M_2:

$$M_1 = \sqrt{2}\begin{pmatrix} \eta_1 & \eta_{1n} & \eta_{1dr} & \eta_{1ur} \\ \eta_2 & \eta_{2n} & \eta_{2dr} & \eta_{2ur} \\ \eta_{1dg} & \eta_{1ug} & \eta_{1db} & \eta_{1ub} \\ \eta_{2dg} & \eta_{2ug} & \eta_{2db} & \eta_{2ub} \end{pmatrix};\ M_2 = \sqrt{2}\begin{pmatrix} \eta_1 & -\xi_2^* & \eta_{1dr} & \eta_{1ur} \\ \eta_2 & \xi_1^* & \eta_{2dr} & \eta_{2ur} \\ \eta_{1dg} & 0 & \eta_{1db} & \eta_{1ub} \\ \eta_{2dg} & 0 & \eta_{2db} & \eta_{2ub} \end{pmatrix} \tag{3.44}$$

and we got the remarkable identity [6]

$$\det(\Psi) = |\det(M_1)|^2 + |\det(M_2)|^2 . \tag{3.45}$$

Usually $\det(\Psi) \neq 0$ therefore $\Psi(x)$ is invertible. We can see the wave Ψ, which implies by its structure itself the gauge group of the standard model, as having the maximum number (36) of degrees of freedom compatible with the existence of an inverse wave. And we need the existence of the inverse to allow the construction of the wave of systems of fermions (See [12] and [9] 4.4.1).

To add two quarks with three colors each in the second and in the third generation we need

$$P_0^1(\Psi_r) = -\frac{1}{3}\Psi_r\gamma_{32} + P_-^1(\Psi_r)\mathbf{i} \tag{3.46}$$

and two similar definitions for colors g and b. Now since the Γ_k generators of the $SU(3)$ group of chromodynamics are unchanged by the circular permutation p used to pass from one generation to another, strong interactions are unperturbed by the change of generation. This allows physical quarks composing particles to mix the generations. For instance the physical quark **d** present in protons and neutrons is thought as a mixing of the d of the first generation and the s that is the equivalent of d in the second generation. Even if the wave of an antiquark is linked to the wave of the corresponding quark, the mixing of waves of different generations, and the difference between what we call "left" and "right" part in each generation, induce the wave of physical quarks to have both a left and a right wave.

The link between the reverse in $Cl_{1,5}$ and the reverse in $Cl_{1,3}$ is not trivial and is similar to ([6] A.2). This link reads

$$\tilde{\Psi} = \begin{pmatrix} \tilde{\Psi}_b & \tilde{\Psi}_r \\ \tilde{\Psi}_g & \tilde{\Psi}_l \end{pmatrix} \tag{3.47}$$

This link does not exist in $Cl_{2,3} = M_4(\mathbb{C})$ nor in $Cl_{3,4} = M_8(\mathbb{C})$. We can see the link as a consequence both of the one-dimensionality of the time and of (2.1) that reduces the dimension of the Clifford algebra. It has then also a geometric meaning. If this link (2.1) between the wave of the particle and the wave of the antiparticle is not used we need one dimension more, we loose the link between the different kinds of reverse, then we loose the geometric transformation linked to the wave. We have then good reasons to think that $Cl_{1,5}$ is the true frame necessary and sufficient to get the gauge group required by experimental results. And this is so more interesting than we can consider the space-time manifold with signature + - - - as embedded in a space-time manifold with signature + - - - - -. The wave equation for all objects of the first generation reads [13]

$$0 = (\underline{D}\Psi)L_{012} + \underline{\mathbf{M}}; \underline{\mathbf{M}} = \begin{pmatrix} m_2\rho_2\chi_b & m_2\rho_2\chi_g \\ m_2\rho_2\chi_r & m_1\rho\chi_l \end{pmatrix} \tag{3.48}$$

where we use the scalar densities s_j and χ terms satisfying

$$\rho_1^2 = \rho^2 = a_1 a_1^* + a_2 a_2^* + a_3 a_3^* \; ; \; \rho_2^2 = \sum_{j=1}^{j=15} s_j s_j^* \tag{3.49}$$

$$s_1 = 2(\eta_{2ur}^* \eta_{1ug}^* - \eta_{1ur}^* \eta_{2ug}^*); s_2 = 2(\eta_{2ug}^* \eta_{1ub}^* - \eta_{1ug}^* \eta_{2ub}^*)$$

$$s_3 = 2(\eta_{2ub}^* \eta_{1ur}^* - \eta_{1ub}^* \eta_{2ur}^*); s_4 = 2(\eta_{2dr}^* \eta_{1dg}^* - \eta_{1dr}^* \eta_{2dg}^*)$$

$$s_5 = 2(\eta_{2dg}^* \eta_{1db}^* - \eta_{1dg}^* \eta_{2db}^*); s_6 = 2(\eta_{2db}^* \eta_{1dr}^* - \eta_{1db}^* \eta_{2dr}^*)$$

$$s_7 = 2(\eta_{2ur}^* \eta_{1dr}^* - \eta_{1ur}^* \eta_{2dr}^*); s_8 = 2(\eta_{2ug}^* \eta_{1dg}^* - \eta_{1ug}^* \eta_{2dg}^*)$$

$$s_9 = 2(\eta_{2ub}^* \eta_{1db}^* - \eta_{1ub}^* \eta_{2db}^*); s_{10} = 2(\eta_{2ur}^* \eta_{1dg}^* - \eta_{1ur}^* \eta_{2dg}^*)$$

$$s_{11} = 2(\eta_{2ug}^* \eta_{1db}^* - \eta_{1ug}^* \eta_{2db}^*); s_{12} = 2(\eta_{2ub}^* \eta_{1dr}^* - \eta_{1ub}^* \eta_{2dr}^*)$$

$$s_{13} = 2(\eta_{2ur}^* \eta_{1db}^* - \eta_{1ur}^* \eta_{2db}^*); s_{14} = 2(\eta_{2ug}^* \eta_{1dr}^* - \eta_{1ug}^* \eta_{2dr}^*)$$

$$s_{15} = 2(\eta_{2ub}^* \eta_{1dg}^* - \eta_{1ub}^* \eta_{2dg}^*). \tag{3.50}$$

We use

$$\rho_2^2 \chi_r = \begin{pmatrix} \begin{pmatrix} s_4^* \phi_{dg} - s_6^* \phi_{db} - s_7^* \phi_{ur} \\ - s_{12}^* \phi_{ub} - s_{14}^* \phi_{ug} \end{pmatrix} \sigma_1 & \begin{pmatrix} s_1^* \varphi_{ug} - s_3^* \phi_{ub} + s_7^* \phi_{dr} \\ + s_{10}^* \phi_{dg} + s_{13}^* \phi_{db} \end{pmatrix} \sigma_1 \\ \begin{pmatrix} - s_1 \hat{\phi}_{ug} + s_3 \hat{\phi}_{ub} - s_7 \hat{\phi}_{dr} \\ - s_{10} \hat{\phi}_{dg} - s_{13} \hat{\phi}_{db} \end{pmatrix} \sigma_1 & \begin{pmatrix} - s_4 \widehat{\varphi}_{dg} + s_6 \widehat{\varphi}_{db} + s_7 \widehat{\varphi}_{ur} \\ + s_{12} \hat{\phi}_{ub} + s_{14} \hat{\phi}_{ug} \end{pmatrix} \sigma_1 \end{pmatrix} \tag{3.51}$$

$$\rho_2^2\chi_g = \begin{pmatrix} \begin{pmatrix} s_5^*\phi_{db} - s_4^*\phi_{dr} - s_8^*\phi_{ug} \\ -s_{10}^*\phi_{ur} - s_{15}^*\phi_{ub} \end{pmatrix}\sigma_1 & \begin{pmatrix} s_2^*\phi_{ub} - s_1^*\phi_{ur} + s_8^*\phi_{dg} \\ +s_{11}^*\phi_{db} + s_{14}^*\phi_{dr} \end{pmatrix}\sigma_1 \\ \begin{pmatrix} -s_2\hat{\phi}_{ub} + s_1\hat{\phi}_{ur} - s_8\hat{\phi}_{dg} \\ -s_{11}\hat{\phi}_{db} - s_{14}\hat{\phi}_{dr} \end{pmatrix}\sigma_1 & \begin{pmatrix} -s_5\hat{\phi}_{db} + s_4\hat{\phi}_{dr} + s_8\hat{\phi}_{ug} \\ +s_{10}\hat{\phi}_{ur} + s_{15}\hat{\phi}_{ub} \end{pmatrix}\sigma_1 \end{pmatrix} \qquad (3.52)$$

$$\rho_2^2\chi_b = \begin{pmatrix} \begin{pmatrix} s_6^*\phi_{dr} - s_5^*\phi_{dg} - s_9^*\phi_{ub} \\ -s_{11}^*\phi_{ug} - s_{13}^*\phi_{ur} \end{pmatrix}\sigma_1 & \begin{pmatrix} s_3^*\phi_{ur} - s_2^*\phi_{ug} + s_9^*\phi_{db} \\ +s_{12}^*\phi_{dr} + s_{15}^*\phi_{dg} \end{pmatrix}\sigma_1 \\ \begin{pmatrix} -s_3\hat{\phi}_{ur} + s_2\hat{\phi}_{ug} - s_9\hat{\phi}_{db} \\ -s_{12}\hat{\phi}_{dr} - s_{15}\hat{\phi}_{dg} \end{pmatrix}\sigma_1 & \begin{pmatrix} -s_6\hat{\phi}_{dr} + s_5\hat{\phi}_{dg} + s_9\hat{\phi}_{ub} \\ +s_{11}\hat{\phi}_{ug} + s_{13}\hat{\phi}_{ur} \end{pmatrix}\sigma_1 \end{pmatrix} \qquad (3.53)$$

The wave Equation (3.48) is gauge invariant under the gauge transformation (3.37) to (3.41) [13]. It is form invariant under the transformation R in (1.6) because with

$$\underline{N} = \begin{pmatrix} N & 0 \\ 0 & N \end{pmatrix};\ \Psi' = \underline{N}\Psi \qquad (3.54)$$

we get [13]

$$0 = (\underline{D}\Psi)L_{012} + \underline{\mathbf{M}} = \underline{\tilde{N}}[(\underline{D}'\Psi')L_{012} + \underline{\mathbf{M}}'] \qquad (3.55)$$

From (3.54) which is equivalent to

$$\Psi'_c = N\Psi_c\ ,\ c = l, r, g, b \qquad (3.56)$$

and from the link $\hat{\phi}_{\bar{z}} = \hat{\phi}_z\sigma_1$ between the wave of the particle and the wave of the antiparticle (3.54) is equivalent to

$$\phi'_z = M\phi_z\ ;\ \hat{\phi}'_z = \hat{M}\hat{\phi}_z \qquad (3.57)$$

for any index z. It happens that $Cl_{1,5} = Cl_{5,1}$ because the n-vectors of $Cl_{1,5}$ are the (6-n)-vectors of $Cl_{5,1}$. The natural generalization of $\mathbf{x}' = N\mathbf{x}\tilde{N}$ uses vectors and pseudovectors of $Cl_{1,5}$. Noting with a 5 index all terms that are components of 5-vectors we let

$$\underline{x} = x^n L_n + x_5^n L_n L_{012345};\ \mathbf{x} = x^\mu \gamma_\mu;\ \mathbf{x}_5 = x_5^\mu \gamma_\mu \tag{3.58}$$

$$\underline{x}' = x'^n L_n + x_5'^n L_n L_{012345};\ \mathbf{x}' = x'^\mu \gamma_\mu;\ \mathbf{x}_5' = x_5'^\mu \gamma_\mu \tag{3.59}$$

$$\underline{x}' = \underline{N}\,\underline{x}\,\underline{\tilde{N}} \tag{3.60}$$

And since we can separate the different multivector parts, (3.60) is equivalent to the system:

$$x' = Nx\tilde{N} \tag{3.61}$$

$$x_5' = Nx_5\tilde{N} \tag{3.62}$$

$$x'^4 + ix_5'^5 = N(x^4 + ix_5^5)\tilde{N} \tag{3.63}$$

$$x_5'^4 + ix'^5 = N(x_5^4 + ix^5)\tilde{N}. \tag{3.64}$$

With (1.7) (3.63) and (3.64) read

$$x'^4 + ix_5'^5 = re^{i\theta}(x^4 + ix_5^5) \tag{3.65}$$

$$x_5'^4 + ix'^5 = re^{i\theta}(x_5^4 + ix^5) \tag{3.66}$$

This separation between the different components of the global space-time explains why we usually see only the real components of the 4-dimensional space-time vector x. Only the usual space-time has real components. Relations (3.65) and (3.66) indicates both that these two supplementary real dimensions give one complex dimension and that they separate completely the usual space-time in the global space-time. A space-time with one or two supplementary conditions has been used early [3]. The problem was always to explain why classical physics do not see these supplementary dimensions. Here this problem is automatically solved by the difference coming from the invariance group of physical laws.

The link (3.49) between the reversion into $Cl_{1,3}$ and the reversion into $Cl_{1,5}$ allows to extend the transformation linked to the wave. We let

$$\underline{x} = x^n L_n + x_5^n L_n L_{012345} \; ; \; \underline{y} = y^n L_n + y_5^n L_n L_{012345} \tag{3.67}$$

$$< M >_v = < M >_1 + < M >_5 \tag{3.68}$$

$$\underline{x} = < \Psi \, \underline{y} \tilde{\Psi} >_v \tag{3.69}$$

The transformation linked to the wave reads

$$f : y_v \mapsto x_v = 2 < \Psi y_v \tilde{\Psi} >_v \tag{3.70}$$

$$\begin{aligned} f : y_v \mapsto x'_v &= 2 < \Psi' y_v \tilde{\Psi}' >_v = 2 < \underline{N} \Psi y_v \tilde{\Psi} \underline{\tilde{N}} >_v \\ &= 2\underline{N} < \Psi y_v \tilde{\Psi} >_v \underline{\tilde{N}} = \underline{N} x_v \underline{\tilde{N}}. \end{aligned} \tag{3.71}$$

Similarly to what we said from (1.26), y_v is independent on the observer, intrinsic to the wave. Contrarily to the dilation R where the real space-time is well separated with (3.61), the transformation (3.70) is a transformation from the subspace of the $< M >_v$ into the same subspace of the relative manifold. We get from (3.1), (3.47) and (3.70)

$$x_v = \begin{pmatrix} 0 & x_2 \\ x_1 & 0 \end{pmatrix} ; y_v = \begin{pmatrix} 0 & y_2 \\ y_1 & 0 \end{pmatrix} \tag{3.72}$$

$$x_1 = 2(\Psi_b y_1 \tilde{\Psi}_b + \Psi_g y_2 \tilde{\Psi}_g) \tag{3.73}$$

$$x_2 = 2(\Psi_r y_1 \tilde{\Psi}_r + \Psi_l y_2 \tilde{\Psi}_l) \tag{3.74}$$

Next we let

$$x_1 = \mathbf{x} + \mathbf{x}_5 + x^4 + x_5^4 + (x^5 + x_5^5)\mathbf{i} \tag{3.75}$$

$$x_2 = \mathbf{x} - \mathbf{x}_5 - x^4 + x_5^4 + (x^5 - x_5^5)\mathbf{i} \tag{3.76}$$

$$y_1 = \mathbf{y} + \mathbf{y}_5 + y^4 + y_5^4 + (y^5 + y_5^5)\mathbf{i} \tag{3.77}$$

$$y_2 = \mathbf{y} - \mathbf{y}_5 - y^4 + y_5^4 + (y^5 - y_5^5)\mathbf{i} \tag{3.78}$$

and we get

$$\begin{aligned}
\mathbf{x} = {} & \phi_e(\mathbf{y} - \mathbf{y}_5)\phi_e^\dagger + \phi_n(\hat{\mathbf{y}} - \hat{\mathbf{y}}_5)\phi_n^\dagger + 2\Re\{\phi_e\phi_n^\dagger[-y^4 + y_5^4 + (y^5 - y_5^5)i]\} \\
& + \phi_{dr}(\mathbf{y} + \mathbf{y}_5)\phi_{dr}^\dagger + \phi_{ur}(\hat{\mathbf{y}} + \hat{\mathbf{y}}_5)\phi_{ur}^\dagger + 2\Re\{\phi_{dr}\phi_{ur}^\dagger[y^4 + y_5^4 + (y^5 + y_5^5)i]\} \\
& + \phi_{dg}(\mathbf{y} - \mathbf{y}_5)\phi_{dg}^\dagger + \phi_{ug}(\hat{\mathbf{y}} - \hat{\mathbf{y}}_5)\phi_{ug}^\dagger + 2\Re\{\phi_{dg}\phi_{ug}^\dagger[-y^4 + y_5^4 + (y^5 - y_5^5)i]\} \\
& + \phi_{db}(\mathbf{y} + \mathbf{y}_5)\phi_{db}^\dagger + \phi_{ub}(\hat{\mathbf{y}} + \hat{\mathbf{y}}_5)\phi_{ub}^\dagger + 2\Re\{\phi_{db}\phi_{ub}^\dagger[y^4 + y_5^4 + (y^5 + y_5^5)i]\}.
\end{aligned} \tag{3.79}$$

We consider now the geometric transformation $\mathbf{p} : \underline{y} \mapsto \mathbf{x}$. It is much more complicated than (2.30), because it is not possible to separate 1-vector and 5-vector. This comes from the identity between $Cl_{1,5}$ and $Cl_{5,1}$. The $\mathbf{p}$ transformation applies the linear space $\mathbb{R}_{1,5}$ onto $\mathbb{R}_{1,3}$.

4. CONCLUDING REMARKS

A quantum wave equation generalizing the Dirac wave equation has been proposed to explain the experimental results of the standard model of quantum physics. It does not require the second quantification nor the very tedious calculations brought by its method of approximation. The wave is a true function of space and time with value into well-defined Clifford algebras. It will then be easier to bring together this frame and the solid similar frame of general relativity.

It is very important to notice that these two frames use a same time. This time is our oriented time and the invariance group of the wave conserves this orientation. When the variational calculus is used to go from a Lagrangian density to the corresponding wave equations an integration by parts is necessary, and the assumption that it is possible to cancel solutions at infinity. This hypothesis seems physically equivalent to suppress the propagation. The result is that quantum fields theory is fully satisfactory for stationary states but is structurally unable to account for any irreversible process. Our study explains why Lagrangian densities take place everywhere in quantum physics: they are only the scalar part of the invariant wave equations of fermions. They are not at all fundamental, they are only a mathematical tool.

Fermions are useful and fundamental, but our Universe has not only fermions but also boson waves. We suspect that the boson part of our universe, that contains the massless photon, does not come from a Lagrangian density. For instance the theory of the photon made by L. de Broglie [14] [15] gives not one, but two Lagrangian densities. This is normal since his photon comes from two Dirac waves, and each one brings its own Lagrangian density.

Another important point is emphasized by T. Socroun [16]: in order to get the unification between gravitation and electromagnetism, it is necessary to incorporate charges into potentials terms. This is equivalent, in fact, to the prescription made by Einstein itself that all laws of physics are covariant. With Cl_3^* as fundamental group of invariance the difference between contravariant and covariant terms is a physical one: a contravariant vector in space-time transforms as x in (1.6) while a covariant vector transforms as ∇ in (1.11). These two transformations are not equivalent as soon as $\det(M)\neq 1$. It is easy to see that the incorporation of the charges g_1, g_2, g_3 into the potentials can be made inside each of our preceding equations. The electromagnetic potential gives instead (1.2), (1.3) and (1.15)

$$\nabla\hat{\phi}\sigma_{21}+\mathbf{a}\hat{\phi}+m\phi=0;\ \mathbf{a}=qA;\ \overline{\phi}\mathbf{a}\hat{\phi}=\overline{\phi}'\mathbf{a}'\hat{\phi}' \tag{4.1}$$

So the three charges disappear in the calculations. For electro-weak interactions and strong interactions we let

$$\underline{b}=\frac{g_1}{2}\underline{B};\ \underline{w}^j=\frac{g_2}{2}\underline{W}^j;\ \underline{g}^k=\frac{g_3}{2}\underline{G}^k. \tag{4.2}$$

Then the covariant derivative (3.4) is simply

$$\underline{D} = \underline{\partial} + \underline{b}\,\underline{P}_0 + \underline{w}^j\,\underline{P}_j + \underline{g}^k\,\mathbf{i}\underline{\Gamma}_k. \tag{4.3}$$

Why this has never been made? The reason is probably that theories of great unification consider the charges as slowly variable with the scale of energy and hope that when these charges become equal the structure of the gauge group is enlarged. This is strictly impossible when charges are integrated into potential terms. But this is a new scope as soon as we have linked the generators of the gauge group to the structure of the wave in (2.37) to (2.40), (3.10) to (3.13), (3.23) to (3.28). Since all attempts made to get the unification have failed, the incorporation of charges into potentials is a cornerstone for the ToE. To get the ToE perhaps will it be sufficient to link the Lagrangian density of the general relativity to the Lagrangian density coming from the standard model? (D. Girardot, private discussion) This density is the real scalar part of the invariant wave Equation (3.48)

$$\mathcal{L} = \langle (\underline{D}\Psi)L_{012} + \underline{\mathbf{M}} \rangle \tag{4.4}$$

This density is easily generalized to three generations of fundamental fermions:

$$\mathcal{L} = \langle (\underline{D}_3\Psi_3)L_{012} + \underline{\mathbf{M}}_3 \rangle + \langle (\underline{D}_1\Psi_1)L_{023} + \underline{\mathbf{M}}_1 \rangle + \langle (\underline{D}_2\Psi_2)L_{031} + \underline{\mathbf{M}}_2 \rangle \tag{4.5}$$

It is indeed very sophisticated but on the mathematical point of view it is not so different from the Lagrangian density of general relativity, they are both real functions of space and time.

Another cornerstone of a future ToE is the fact that both theories use a covariant derivative. We must understand how these two covariant derivative coexists. A simple hypothesis coming from the double manifold encountered in (1.24) is that each one can be the covariant derivative on one of the two manifolds. Is one of these two manifolds simply embedded into the other?

REFERENCES

1. Einstein, A. (1912) Lichtgeschwindigkeit und Statik des Gravitationsfeldes. Annalen der Physik (ser. 4), 38, 355-369.
2. Einstein, A. (1912) Theorie des statischen gravitationsfeldes. Annalen der Physik (ser. 4), 38, 443-458.
3. Tonnelat, M.A. (1965) Les théories unitaires de l'électromagnétisme et de la gravitation. Gauthier-Villars, Paris.
4. Weinberg, S. (1967) A Model of Leptons. Phys. Rev. Lett., 19, 1264-1266.

5. Matveev, V.A. and Tavkhelidze, A.N. (2006) The Quantum Number Color, Colored Quarks and QCD. In: RAS INR, Ed., 99th Session of the JINR Scientific Council, Dubna, Moscow.
6. Daviau, C. and Bertrand, J. (2014) New Insights in the Standard Model of Quantum Physics in Clifford Algebra. Je Publie, Pouillé-les-coteaux. http://hal.archives-ouvertes.fr/hal-00907848
7. Daviau, C. (2011) L'espace-temps Double. JePublie, Pouillé-les-coteaux.
8. Daviau, C. (2012) Double Space-Time and More. JePublie, Pouillé-les-coteaux.
9. Daviau, C. (2013) Invariant Quantum Wave Equations and Double Space-Time. Adv. in Imaging and Electron Physics, 179, Chapter 1, 1-137.
10. Daviau, C. and Bertrand, J. (2014) Relativistic Gauge Invariant Wave Equation of the Electron-Neutrino. Journal of Modern Physics, 5, 1001-1022.
11. Daviau, C. and Bertrand, J. (2013) A Lepton Dirac Equation with Additional Mass Term and a Wave Equation for a Fourth Neutrino. Ann. Fond. Louis de Broglie, 38.
12. Daviau, C. (2012) Invariance of the Dirac Equation and of Electromagnetism. Adv. Appl. Clifford Algebras, 22, 611-623.
13. Daviau, C. and Bertrand, J. (2014) A Wave Equation Including Leptons and Quarks for the Standard Model of Quantum Physics in Clifford Algebra.
14. de Broglie, L. (1940) La mécanique du photon, Une nouvelle théorie de la lumière: Tome 1 La lumière dans le vide. Hermann, Paris.
15. de Broglie, L. (1942) tome 2 Les interactions entre les photons et la matière. Hermann, Paris.
16. Socroun, T. (2014) Clifford to Unify General Relativity and Elec-tromagnetism. ICCA10, Tartu (Estonia), Submitted.

NOTES

[1]It seems that we are still using complex linear spaces. But actually anything in (1.2) is usual in classical physics: $(\sigma_1,\sigma_2,\sigma_3)$ is a direct orthonormal basis of the physical space, i is the name of the product $\sigma_{123}=\sigma_1\sigma_2\sigma_3$. For any vector $\vec{u}$, $\vec{v}$ we have $\vec{u}\vec{v}=\vec{u}\cdot\vec{v}+i(\vec{u}\times\vec{v})$ where $\vec{u}\cdot\vec{v}$ is the usual scalar product and $\vec{u}\times\vec{v}$ is the usual cross product.

[2]Relativistic quantum mechanics enlarges the invariance under $Sl(2,\mathbb{C})$ by using P and T transformations. We never use these transformations: they are purely theoretical since absolutely no way exists in the Nature to change the orientation of the time, no way exists to change the orientation of the space.

[3]The use for complex matrices here is also pure commodity. Matrix representations were used to get by induction on the dimension the main theorem of existence and unicity, up to an isomorphism, of all real Clifford algebras. All calculations may be done by using $(\gamma_0,\gamma_1,\gamma_2,\gamma_3)$ as a basis of space-time. The link with Cl_3 is made by $\sigma_j=\gamma_j\gamma_0$: this is equivalent to identify Cl_3 to the even subalgebra $Cl_{1,3}^{+}$ of the space-time algebra.

[4]As before me remark that we are always in a frame of classical physics on the real field. $(L_0, L_1, L_2, L_3, L_4, L_5)$ is an orthonormal basis in $\mathbb{R}_{1,5}$: $L_0^2 = 1$, $L_1^2 = -1$, $L_2^2 = -1$, $L_3^2 = -1$, $L_4^2 = -1$, $L_5^2 = -1$, $L_n \cdot L_p = 0, n \neq p$. We use $L^0 = L_0$, $L^n = -L_n, n > 0$. We can consider (L_0, L_1, L_2, L_3) as an orthonormal basis of $\mathbb{R}_{1,3}$. The use for complex matrices is convenient but not compulsory.

CHAPTER 5

The Relativistic Geometry of the Complex Matter Space

Reza Ahangar

Department of Mathematics, Texas A&M University, Kingsville, TX, USA

ABSTRACT

Duality behavior of photons in wave-particle property has posed challenges and opportunities to discover other frontiers of fundamental particles leading to the relativistic and quantum description of matter. The speed of particles faster than the speed of light could not be recognized, and matter was always described as a real number. A new fundamental view on matter as a complex value has been introduced by many authors who present a paradigm that is shifted from real or pure imaginary particles to Complex Matter Space. A new assumption will be imposed that matter has two intrinsic components: i) mass, and ii) charge. The mass will be measured by real number systems and charged by an imaginary unit. The relativistic concept of Complex Matter Space on energy and momentum is investigated and we can conclude that the new Complex Matter Space (CMS) theory will help get one step closer to a better understanding toward: 1) Un-Euclidean description of Minkowski Geometry in the context of the Complex Matter Space, 2) transformation from Euclidean to Minkowski space and its relativistic interpretation. Finally, geometrical foundations are essential to have a real picture of space, matter, and the universe.

Keywords: Relativistic Geometry, Mass and Charge, Complex Matter Space, Un-Euclidean, Minkowski's Geometry

1. INTRODUCTION

Several thousand years of human experiences in geometry, which was structured by Pythagoras and Euclid, has undergone challenges during the past few centuries. Among all were the attempts to prove Euclid's fifth postulate independently from other postulates and distinguish all undefined terms like point, line, plane, from terms that can be redefined like distances and parallel lines.

In seeking the truth behind the parallel postulate in 19th century, some alternatives and geometric models to Euclid's parallel postulate were presented.

In 1898, one of the great mathematicians of all time, David Hilbert presented a rigorous axiomatic approach to Euclidean Geometry (see Hilbert, 1902 [1]). Many others like Riemann, Bolyai, and Lobachevski created Un-Euclidean Geometry.

Thus, Hyperbolic Geometry was born and many geometrical models were developed (see Milnor, 1996 [2]).

Many were skeptical of the validity of this bizarre geometry which was axiomatically consistent and agreed with the models developed. But all these theories were not experimented physically to be consistent with the natural universe.

Certainly, in a small velocity for a shorter distance, Euclidean Geometry is a simplest model for our physical universe. But it is not the proper model for the entire universe particularly when the velocity of the object is near or approaches to the speed of light.

Einstein's General Relativity predicted that gravity alters the velocity of light and this causes the light ray to bend. Or, it can be interpreted that light travels in a geodesic line. This is a revolutionary conclusion that the physical space with the existing mass and gravity is not Euclidean (see Taylor et al., 2004 [3] , and Feynman, 1963 [4]).

NASA Experiment, May 4, 2011: Einstein was right again. There is a space-time vortex around Earth, and its shape precisely matches the predictions of Einstein's theory of gravity. Researchers confirmed these points at a press conference today at NASA headquarters where they announced the long-awaited results of Gravity Probe B (GP-B). "The space-time around Earth appears to be distorted just as general relativity predicts", says Stanford University physicist Francis Everitt, principal investigator of the Gravity Probe B mission (NASA, 2011 [5]).

A shortest path in traveling between two points can be defined by an arc-length or geodesic line. It may not be considered as straight lines because acceleration creates distortion. Due to high speed of the particle or a constant curvature there may two possibilities. The yard stick or rulers can get longer or shrink as they move away from the stationary point (Lavenda, 2009 [6]).

A massive object in space curves the space and bends the light, when a photon moves past a star; it's the star's gravity that causes it to deflect, regardless of whether we consider photon a particle or wave.

All evidence indicating the existence of dark matter and dark energy suggests that we may need to redefine matter. As a result we present the following two postulates:

Complex Matter Space (CMS) (see Ahangar, 2014 [7] [8] [9])

- Postulate (I): mass and charge are two intrinsic components of matter and they can be measured and determined in a complex plane. The real part is equal to the mass and the imaginary part is the charge of the particle. That is $M = m + iq$.
- Postulate (II): In a one dimensional space, position and velocity are assumed to be real numbers and the rate of change of the position of a particle can be assumed to be faster than the speed of light.

According to de Broglie's duality of particles used to describe matter—wave characteristic of photons (see Mayer, 2007 [10] and Gauthier, 2009 [11]). It is well known that the complex plane can help us to explain the duality of waves and explain a wave with two components in the complex plane with real and imaginary components:

$$\psi(r,t) = \psi_{\text{real}} + i\psi_{\text{imag}} \tag{1}$$

We will refer the value $|\psi|^2$ as the intensity of the wave function where the magnitude of the complex number is denoted by $|\psi| = \sqrt{|\psi_{\text{real}}|^2 + |\psi_{\text{imag}}|^2}$ [3] .

Traveling Beyond the Speed of Light:

According to the "Special Theory of Relativity", it is not allowed to cross the boundary of the speed of light c, in the real universe. But if we can pretend from this assumption that we exist in the imaginary world then it may be allowed to go faster than light, since our conjugate space will be the real world.

In section two we present an introduction to the geometric structure of Complex Matter Space. We will show its relation with Minkowski Geometry and pseudo-Euclidean space.

In section three, we use a bilinear transformation in hyperbolic geometry. In section four, we investigate the hyperbolic rotation which is equivalent to Lorentz transformation. In section five, we will show that the relativistic view of geometry is consistent with hyperbolic geometry. As a conclusion, the law of cosines is redeveloped in a hyperbolic triangle.

2. GEOMETRICAL STRUCTURE OF THE CMS

Assume V is a vector space and T represents a bilinear transformation from $V \times V$ into the R. Let us select vectors $u = \langle u^0, u^1, u^2, \cdots u^n \rangle$ and $v = \langle v^0, v^1, v^2, \cdots, v^n \rangle$ in the vector space V, where all of the first ($n+1$) components are real

$$u \text{ and } v \in V$$

define the transformation T from $V \times V \to R$ by

$$T(u,v) = \sum_{k=0}^{k=n} u^k v^k = u^0 v^0 + u^1 v^1 + u^2 v^2 + \cdots + u^n v^n \tag{2}$$

In the following theorem, we will show that the transformation T defined in (2) is a bilinear transformation.

Theorem (1): The operator $T : V \times V \to R$ defined by (2) is a bilinear operator meaning that for any real numbers a and b

$$T(au_1 + bu_2, v) = aT(u_1, v) + bT(u_2, v) \tag{3}$$

Proof: Notice that the superscripts are not exponents, thus

$$\begin{aligned} T(au_1 + bu_2, v) &= \sum_{k=0}^{k=n} \left[a\left(u_1^k\right) + b\left(u_2^k\right) \right] v^k \\ &= a\sum_{k=0}^{k=n} \left(u_1^k\right) v^k + b\sum_{k=0}^{k=n} \left(u_2^k\right) v^k \\ &= aT(u_1, v) + bT(u_2, v) \end{aligned}$$

which proves that T is a bilinear operator that can be used to define the Mikowski norm.

Also notice that the operator T mapping a vector u to itself will produce a value in R:

$$T(u,u) = \sum_{k=1}^{k=n} \left(u^k\right)^2 = \left(u^0\right)^2 + \left(u^1\right)^2 + \left(u^2\right)^2 + \cdots + \left(u^n\right)^2 \tag{4}$$

If the vector space V is selected from real valued vectors then

$$\forall k \in N, u^k \in R, \left(u^k\right)^2 \geq 0 \tag{5}$$

implies that the relation (4), is a positive definite inner product, that is for nonzero vector u, the transformation $T(u,u) > 0$.

But when the vectors are selected from the complex field then the relation (4) can be one of the following cases:

1) $T(u,u) > 0$, the transformation T is positive definite, when the vector u is in the region which is called space-like.

2) $T(u,u) < 0$ the transformation T is negative definite, when u is in the time-like, otherwise

3) $T(u,u) = 0$ and the transformation is called indefinite, when the vector u belongs in the light-like.

Many questions exist to justify the operator T for objects in complex space as well as explanation for the properties of the transformation T. The most important one is when T is positive or negative definite and the operator T from $R^{n+1} \times R^{n+1} --> R$ satisfies the properties of a metric space (see Naber, 1992 [12]).

To answer this question we need the definition of Metric Space first.

Metric Space: A set X, whose elements we shall call points, is said to be a Metric Space if with any two points P and Q of X there is associated a real number d(P,Q), called distance from P to Q such that

a) $\mathrm{d}(P,Q) > 0$ if $P \neq Q, \mathrm{d}(P,P) = 0$,

b) $\mathrm{d}(P,Q) = \mathrm{d}(Q,P)$, and c) $\mathrm{d}(P,Q) \leq \mathrm{d}(P,R) + \mathrm{d}(R,Q)$ for any R in X

One can verify that the transformation T defined in (2), (3), and (4) fails the postulate (a) of metric space, due to the case where $T(u,v)$ can be negative definite.

Thus as a conclusion T is not a Metric Space.

II) Special Complex Geometrical Structure: The standard Euclidean inner product of two vectors in (2) and (3) can be defined by the following matrix form

$$\boldsymbol{u}\cdot\boldsymbol{v}=\boldsymbol{u}^{t}I\boldsymbol{v} \tag{6}$$

where, I is the identity matrix, $\boldsymbol{u}^{t}=\langle u^{0},u^{1},u^{2},\cdots u^{n}\rangle,\boldsymbol{v}^{t}=\langle v^{0},v^{1},v^{2},\cdots,v^{n}\rangle$, and superscript t represents the transpose of these vectors.

To develop Minkowski space in (2) for k = 0, consider special coordinate for

$$u^{0}=v^{0}=\mathrm{i}ct,$$

where the symbol i is the imaginary number $\mathrm{i}=\sqrt{-1}$. The value c is assumed to be a constant real number representing the speed of light.

Thus the operator $T:V\times V\rightarrow R$ defined by the following

$$T(u,v)=\sum_{k=1}^{k=n}u^{k}v^{k}+(\mathrm{i}ct)(\mathrm{i}ct)=\sum_{k=1}^{k=n}u^{k}v^{k}-(c^{2}t^{2}) \tag{7}$$

is also a bilinear operator. The relation (7) for n = 3 is a four dimensional Minkowski space for events u and v and the transformation can be described by

$$T(u,v)=\sum_{k=1}^{k=3}u^{k}v^{k}+(\mathrm{i}ct)(\mathrm{i}ct)=u^{1}v^{1}+u^{2}v^{2}+u^{3}v^{3}-c^{2}t^{2} \tag{8}$$

3. PROPERTIES OF THE BILINEAR OPERATOR

i) The transformation $T:V\times V\rightarrow R$ generates Minkowski inner product space and

$$\|\boldsymbol{u}\|^2 = T(u,u) = \sum_{k=1}^{k=n} \left(u^k\right)^2 - (ct)^2 \tag{9}$$

Proof: The transformation is indefinite if it is neither positive nor negative definite.

Notice that the relations (2), (3), and (4) will demonstrate the four dimensional Minkowski space.

We can use the relation (8) to establish Minkowski Norm as follows:

$$\|\boldsymbol{u}\|^2 = T(u,u) = \sum_{k=1}^{k=3} \left(u^k\right)^2 - (ct)^2 = \left(u^1\right)^2 + \left(u^2\right)^2 + \left(u^3\right)^2 - (ct)^2 \tag{10}$$

where the Euclidean norm in 3D real event: $r^2 = \left(u^1\right)^2 + \left(u^2\right)^2 + \left(u^2\right)^2$ can change (10) into the following form of

$$T(u,u) = r^2 - (ct)^2 = \left\langle \mathrm{i}ct, u^1, u^2, u^3 \right\rangle I_{4\times4} \left\langle \mathrm{i}ct, u^1, u^2, u^3 \right\rangle \tag{11}$$

where I represents the identity matrix, as the identity matrix used in (6)

$$I_{4\times4} = \begin{bmatrix} 1 & 0 & 0 & 0 \\ 0 & 1 & 0 & 0 \\ 0 & 0 & 1 & 0 \\ 0 & 0 & 0 & 1 \end{bmatrix}.$$

The Minkowski inner product by the transformation $T(u,u)$ in (11) applied for event vector u may be positive definite, negative definite, or indefinite.

Orthogonal Subspace:

When the transformation $T(u,v) = 0$, then two vectors u and v are said to be orthogonal.

Assume a subspace $S \subset V$, such that

$$S^{\perp} = \left\{ v\varepsilon V : T(v,w) = 0 \ \text{ for all } w\varepsilon S \right\} \tag{12}$$

It can be concluded that the set $S^{\perp} \subset S$ as a subset of S is also a subspace of S?

In the next step we will explore the consistency of a two dimensional analogy of Minkowski-Hyperbolic system with Complex Matter space.

Proposition (1): Let $M(x,y)$ be a point on a hyperbola

$$x^2 - y^2 = 1 \tag{13}$$

For positive values of x there exists a unique value of u such that

$$x = \cosh(u) \text{ and } y = \sinh(u) \tag{14}$$

Proof: Let us take the transformation, $u = \sinh^{-1}(y)$. The result can be demonstrated by the geometric property of the hyperbolic identity

$$x^2 - y^2 = \cosh^2 u - \sinh^2(u) = 1. \tag{15}$$

Notice that the relation (13) for (x, y) is a two dimensional example in a complex plane when the transformation y = ct is described by

$$x^2 + (ict)^2 = 1 \Leftrightarrow \quad x^2 - (c^2t^2) = 1.$$

Pseudo-Euclidean Space: In our introduction it has been verified that the CMS transformation is not a metric space. We will show that this is a special pseudo Euclidean Space.

We define a general form of pseudo-Euclidean space as a finite dimensional n-space together with a quadratic form q. The quadratic form defined for any vector with a suitable choice of basis $(e_1, e_2, \cdots, e_n)$ can be applied to a vector u giving (with $1 \leq k < n$)

$$u = \sum_{k=1}^{n} u^k e_k = u^1 e_1 + u^2 e_2 + \cdots + u^n e_n \tag{16}$$

for a quadratic form

$$q(u)=(u^1)^2+(u^2)^2+\cdots+(u^k)^2-\left[(u^{k+1})^2+\cdots+(u^n)^2\right] \tag{17}$$

is called the magnitude of the vector u. When $k=1\cdots n$, then it is Euclidean Space. Notice that the superscripts k, for convenience, selected to be from 1 to n. This is slightly different from the transformation (2), (3), and (4). We will have the general Minkowski Space when k = 0.

In a pseudo-Euclidean space, unlike in a Euclidean space, there exist vectors with negative magnitude.

$$q(u)=\|u\|^2=(u^1)^2+(u^2)^2+\cdots+(u^n)^2-(u^0)^2 \tag{18}$$

for all positive integer k where, $0<k\leq n.$

A very important pseudo-Euclidean space is Minkowski's Space which is the mathematical setting in which Albert Einstein's theory of special relativity is conveniently formulated. The following is another example of pseudo-Euclidean space using Minkowski space, n = 4 and k = 3 so that

$$q(u)=u_1^2+u_2^2+u_3^2+u_4^2 \tag{19}$$

where assuming $\|u\|=\sqrt{q(u)}$ and $u_4=\mathrm{i}ct$ the above relation will be

$$\|u\|=\sqrt{u_1^2+u_2^2+u_3^2+(\mathrm{i}ct)^2}=\sqrt{u_1^2+u_2^2+u_3^2-(ct)^2} \tag{20}$$

The geometry associated with this pseudo-metric was investigated by Poincare and the rotation group is called Lorentz group.

This magnitude is used in Lorentzian manifold which represents the flat space-time in R^4 uses the coordinates (x, y, z, t) and the metric

$$\mathrm{d}s^2=\begin{bmatrix}\mathrm{icd}t & \mathrm{d}x & \mathrm{d}x & \mathrm{d}y\end{bmatrix}\begin{bmatrix}1&0&0&0\\0&1&0&0\\0&0&1&0\\0&0&0&1\end{bmatrix}\begin{bmatrix}\mathrm{icd}t\\ \mathrm{d}x\\ \mathrm{d}y\\ \mathrm{d}z\end{bmatrix}$$

$$ds^2 = dx^{\mu} 1_{4\times4} dx_{\nu}$$

$$ds^2 = (icdt)^2 + dx^2 + dy^2 + dz^2$$

4. FROM EUCLIDEAN ROTATION TO LORENTZ TRANSFORMATION

When we replace x^2 symbolically for the sum of the squares of all real components and $u^0 = ct$ in the relation (18) we will obtain the quadratic form

$$q(u) = x^2 - (ct)^2$$

Definition of Lorentz transformation Matrix: We also use the relations (14) to define the following transformation:

$$T(u) = \begin{bmatrix} \cosh(u) & \sinh(u) \\ \sinh(u) & \cosh(u) \end{bmatrix} \tag{21}$$

where $x = \cosh(u)$ and $y = \sinh(u)$ imply that b $\tanh(u) = \frac{y}{x}$.

The transformation T in (21) acting on an event vector $\begin{bmatrix} \zeta \\ \iota \end{bmatrix}$ on a hyperbola $\zeta^2 - t^2 = k$ such that

$$\begin{bmatrix} t \\ z \end{bmatrix} = \begin{bmatrix} \cosh(u) & \sinh(u) \\ \sinh(u) & \cosh(u) \end{bmatrix} \begin{bmatrix} \tau \\ \zeta \end{bmatrix}$$

will be converted into a new hyperbola $z^2 - t^2 = k$ and preserves Minkowski's norm.

The concept of space and time has been evolved and became paradoxical from the time of Descarts to Minkowskiin twentieth century.

We are introducing the space-time coordinate (x, y, z, t) such that it will transform into a new position point (x', y', z', t').

The units on all components x, y, z can be considered real numbers but we assume that the unit for time as an imaginary unit "ic" such that $i = \sqrt{-1}$ and c the speed of light c. Thus the component for t can be demonstrated symbolically by

$$\tau = ict \tag{22}$$

To keep the concepts simple, assume y and z stay fixed and the coordinate $(x \& \tau)$ under a rotation angle theta $= \theta$, then the new coordinate will be in the following form:

$$\begin{aligned} x &= x' \cos(\theta) - \tau' \sin(\theta) \\ \tau &= \tau' \cos(\theta) + x' \sin(\theta) \end{aligned}$$

or in the matrix form

$$\begin{bmatrix} x \\ \tau \end{bmatrix} = \begin{bmatrix} \cos(\theta) & -\sin(\theta) \\ \sin(\theta) & \cos(\theta) \end{bmatrix} \begin{bmatrix} x' \\ \tau' \end{bmatrix} \tag{23}$$

Let us call the initial reference frame $S = \{(\tau, x) \in C \times R\}$ and $S' = \{(\tau', x') \in C \times R\}$ after rotation with angle theta. In this reference angle, the evolution on time axis is very interesting. The tau axis in the S will be evolved into $(tau)'$ in the S' system. Any point on the tau axis will be transformed into a new point $(tau)'$ in the equation when the angle theta is 90°.

In other words let us replace $x' = 0$ in the system (23) then

$$\begin{aligned} x &= -\tau' \sin(\theta) \\ \tau &= \tau' \cos(\theta) \end{aligned} \tag{24}$$

This is showing the transform of all of the points on the time axis. By dividing these two relations it will give us

$$\tan(\theta) = -\frac{x}{\tau} \tag{25}$$

Replace tau from the relation (4.1) we will obtain

$$\tan(\theta) = -\frac{x}{\mathrm{i}ct} = \frac{\mathrm{i}v}{c} \tag{26}$$

According to the relativity postulates, the time evolution is a function of the velocity. Thus we can find sine and cosine values of the angle theta using (26):

$$\cos(\theta) = \frac{1}{\sqrt{1+\tan^2(\theta)}} = \frac{1}{\sqrt{1+\left(\frac{\mathrm{i}v}{c}\right)^2}} = \frac{1}{\sqrt{1-\left(\frac{v}{c}\right)^2}}$$

$$\sin(\theta) = \cos(\theta)\tan(\theta) = \frac{\frac{\mathrm{i}v}{c}}{\sqrt{1-\left(\frac{v}{c}\right)^2}}$$

Substitute in the system (23)

$$\begin{bmatrix} x \\ \tau \end{bmatrix} = \gamma \begin{bmatrix} 1 & -\frac{\mathrm{i}v}{c} \\ \frac{\mathrm{i}v}{c} & 1 \end{bmatrix} \begin{bmatrix} x' \\ \tau' \end{bmatrix} \tag{27}$$

where gamma is equal to: $\gamma = \frac{1}{\sqrt{1-\left(\frac{v}{c}\right)^2}}$. The relation (27) is the Lorentz transformation. In addition to these sine and cosine functional transformations, one can develop the additively property of rotations and velocity:

$$\tan(\theta+\phi) = \frac{\tan(\theta)+\tan(\phi)}{1-\tan(\theta)\cdot\tan(\phi)} = \mathrm{i}\frac{v+w}{1-v\cdot w}$$

Minkowski investigated the Lorentz transformation as an Euclidean rotation. This can be achieved by finding (x',τ') from the system (23) and replacing (22) for tau.

$$\begin{bmatrix} x' \\ ict' \end{bmatrix} = \begin{bmatrix} \cos(\theta) & ict\cdot\sin(\theta) \\ -\sin(\theta) & ict\cdot\cos(\theta) \end{bmatrix} \begin{bmatrix} x \\ ict \end{bmatrix} \tag{28}$$

We also replace the angle theta by pure imaginary angle (iw) and using the following identities

$$\cos(\theta) = \cos(\mathrm{i}w) = \cosh(w) \text{ and } \sin(\theta) = \sin(\mathrm{i}w) = \mathrm{i}\sinh(w) \tag{29}$$

Using these identities the hyperbolic version of the relation (28) will be

$\begin{cases} ct' = ct\cdot\cosh(w) - x\sinh(w) \\ x' = -ct\cdot\sinh(w) + x\cosh(w) \end{cases}$ and the inverse

$$\begin{cases} ct = ct'\cdot\cosh(w) + x'\cdot\sinh(w) \\ x = ct'\cdot\sinh(w) + x'\cdot\cosh(w) \end{cases} \tag{30}$$

These relations can be described by the following matrix form:

$$\begin{bmatrix} x' \\ ct' \end{bmatrix} = \begin{bmatrix} \cosh(w) & -\sinh(w) \\ -\sinh(w) & \cosh(w) \end{bmatrix} \begin{bmatrix} x \\ ct \end{bmatrix}$$

and

$$\begin{bmatrix} x \\ ct \end{bmatrix} = \begin{bmatrix} \cosh(w) & \sinh(w) \\ \sinh(w) & \cosh(w) \end{bmatrix} \begin{bmatrix} x' \\ ct' \end{bmatrix}$$

The hyperbolic systems in (30) explains the CMS with hyperbolic angle of rotation w and rapidity

$$\tanh(w) = \frac{v}{c} \tag{31}$$

Comparing the two tangent relationships of (26) and (31) we can discover many interesting connections between two spaces i) Minkowski Space and ii) CMS.

One is in the real world and another in the imaginary space.

In the first we are dealing with Euclidean trigonometry and in the second with the hyperbolic trigonometry.

It can be concluded that

$$\cosh(w) = \frac{1}{\sqrt{1-\left(\tanh(w)^2\right)}} = \frac{1}{\sqrt{1-\left(\frac{v}{c}\right)^2}} = \gamma$$

$$\sinh(w) = \tanh(w)\cdot\cosh(w) = \frac{\frac{v}{c}}{\sqrt{1-\left(\frac{v}{c}\right)^2}}$$

Geometry of Complex Matter in Complex Plane:

Assume a complex point z represents a particle $M = m + iq$ such that

$$|z| = m = \mathrm{Re}(M) \quad \text{and} \quad \arg(z) = \frac{q}{m} = \frac{\mathrm{Im}(z)}{|z|} = \arg(z) = \frac{\mathrm{Im}(z)}{\mathrm{Re}(z)} \tag{32}$$

In this case we need to consider two different cases in which that can happen:

When $|x| > |y|$ and $|x| < |y|$

Case (I): If $|x| > |y|$ then

$$r = |z| = \pm\sqrt{x^2 - y^2}$$

As a result:

$$\left(\frac{x}{r}\right)^2 - \left(\frac{y}{r}\right)^2 = \frac{x^2 - y^2}{x^2 - y^2} = 1$$

This statement describes the existence of an angle in Minkowskian Geometry such that

$$\cosh(\varnothing) = \frac{x}{r} = \frac{\mathrm{Re}(z)}{|z|} = \frac{m}{\sqrt{m^2+q^2}}$$

$$\sinh(\varnothing) = \frac{y}{r} = \frac{\mathrm{Im}(z)}{|z|} = \frac{q}{\sqrt{m^2+q^2}}$$

$$\tanh(\varnothing) = \frac{\sinh(\varnothing)}{\cosh(\varnothing)} = \frac{y}{x} = \frac{\mathrm{Im}(z)}{\mathrm{Re}(z)} = \frac{q}{m}$$

case (II): If $|x| < |y|$ then

$$r = |z| = \pm\sqrt{y^2 - x^2}$$

$$\left(\frac{y}{r}\right)^2 - \left(\frac{x}{r}\right)^2 = \frac{y^2 - x^2}{y^2 - x^2} = 1$$

Consequently, there exists a unique number phi such that:

$$\sinh(\varnothing) = \frac{x}{r} = \frac{\mathrm{Re}(z)}{|z|} = \frac{m}{\sqrt{m^2+q^2}}$$

$$\cosh(\varnothing) = \frac{y}{r} = \frac{\mathrm{Im}(z)}{|z|} = \frac{q}{\sqrt{m^2+q^2}}$$

$$\tanh(\varnothing) = \frac{\sinh(\varnothing)}{\cosh(\varnothing)} = \frac{x}{y} = \frac{\mathrm{Re}(z)}{\mathrm{Im}(z)}$$

As a result of this demonstration, we conclude that the complex number z in CMS can be described symbolically by

$$\begin{cases} z = r(\cosh(\varnothing) + \mathrm{i}\sinh(\varnothing)) = r^*\mathrm{CISH}(\varnothing) & \text{if } |x| > |y| \\ z = r(\sinh(\varnothing) + \mathrm{i}\cosh(\varnothing)) = r^*\mathrm{SICH}(\varnothing) & \text{if } |x| < |y| \end{cases} \tag{33}$$

5. RELATIVISTIC AND MINKOWSKI GEOMETRY IN CMS

The relations between Relativistic and Minkowski Geometry have been investigated extensively by many in the twentieth century. In this paper we would like to clarify the geometric concept and consistency of the Complex Matter Space with these two mathematical landmarks of the century.

By the relation (9) the Minkowski norm can be used to define an event vector

$$\boldsymbol{p} = [p_1, p_2, \cdots, p_n]$$

in the space-time $R^n \times C$ which is defined by

$$E(\boldsymbol{p}, t) = [p_1, p_2, \cdots, p_n, \mathrm{i}t] \tag{34}$$

where R^n is real Euclidean space and C is complex space, for $\mathrm{i} = \sqrt{-1}$.

Definition of CMS inner-product: For two event vectors, $E_1 = E(\boldsymbol{p}) = (\boldsymbol{p}, t_1)$ and $E_2 = E(\boldsymbol{q}) = (\boldsymbol{q}, t_2)$ we define the CMS inner product by:

$$E_1 \cdot E_2 = \sum_{k=1}^{n} p_k q_k + (\mathrm{i}t_1)(\mathrm{i}t_2) \tag{35}$$

As we defined in (18) and (19), the CNS pseudo-norm can be rearranged by

$$\|E\|^2 = (p^1)^2 + (p^2)^2 + \cdots + (p^n)^2 - (t)^2 = \|p\|^2 - t^2 \tag{36}$$

This value can be positive, negative, or indefinite. Consequently, the vector event will be inside the time-like, light-like, or space-like.

Lemma 5.1: Assume two vector events E_1 and E_2, the inner product for CMS

$$E_1 \cdot E_2 = (E_1)^t JE_2 \tag{37}$$

where J is the $(n+1)\times(n+1)$ identity matrix.

Proof: By the definition:

$$E_1 = T(\boldsymbol{p}, t_1) = \langle \boldsymbol{p}, \mathrm{i}t_1 \rangle$$

$$E_2 = E(\boldsymbol{q}, t_2) = \langle \boldsymbol{q}, \mathrm{i}t_2 \rangle$$

Using matrix multiplication, we can define

$$E_1 \cdot E_2 = (E_1)^t JE_2 = [p_1, p_2, \cdots, p_n, \mathrm{i}t_1] \begin{pmatrix} 1 & \cdots & 0 \\ \vdots & \ddots & \vdots \\ 0 & \cdots & 1 \end{pmatrix} \begin{bmatrix} q_1 \\ q_2 \\ \vdots \\ q_n \\ \mathrm{i}t_2 \end{bmatrix} = (\boldsymbol{p}, \mathrm{i}t_1)^t J (\boldsymbol{q}, \mathrm{i}t_2) \tag{38}$$

where the superscript t represents the transpose of the matrix.

In the next step, we will use the development that used for the future events described in "The Geometry of Spacetime ..." by James Callahan (2000) [13] .

We are targeting to apply for complex vector events which we need in the Complex Matter Space theory.

Event Transformation in CMS: Affine transformation includes translation and rotation that have been studied both in Euclidean and Minkowski geometries.

Let us denote the transformation of the, $\text{vector event} = \boldsymbol{r}$ by

$$T_\theta(\boldsymbol{r}) = A\boldsymbol{r}$$

where A is a matrix:

$$A = \begin{bmatrix} \cosh(\theta) & \sinh(\theta) \\ \sinh(\theta) & \cosh(\theta) \end{bmatrix} \tag{39}$$

With a little algebra one can verify the following inverse transform:

$$A^{-1} = \begin{bmatrix} \cosh(\theta) & -\sinh(\theta) \\ -\sinh(\theta) & \cosh(\theta) \end{bmatrix} = \left(T_\theta(\boldsymbol{r})\right)^{-1} \tag{40}$$

The angle θ of hyperbolic rotation in this transformation of the future vector $\boldsymbol{r}$ can be measured by hyperbolic radians.

Lemma 5.1: The hyperbolic transformation T has additive property on the rotation angle. That is

$$T_\theta(T_\alpha) \equiv T_{\theta+\alpha} \tag{41}$$

Proof: Assume a point P is the initial position on a unit hyperbola representing the event $E = \boldsymbol{OP} = \begin{bmatrix} \cosh(\alpha) \\ \sinh(\alpha) \end{bmatrix}$, such that the vector $P = (\cosh(\alpha), \sinh(\alpha))$

Thus using the definition,

$$T_\theta(E) = T_\theta \begin{bmatrix} \cosh(\alpha) \\ \sinh(\alpha) \end{bmatrix} = \begin{bmatrix} \cosh(\theta) & \sinh(\theta) \\ \sinh(\theta) & \cosh(\theta) \end{bmatrix} \begin{bmatrix} \cosh(\alpha) \\ \sinh(\alpha) \end{bmatrix}$$

By using algebra of multiplication of two matrices, we conclude that

$$T_\theta(E) = \begin{bmatrix} \cosh(\theta+\alpha) \\ \sinh(\theta+\alpha) \end{bmatrix} \tag{42}$$

The new position is a new point E' on the hyperbola with the angle rotation alpha and the initial hyperbolic angle theta. That is $T_\theta(E) = E'$.

Proposition 5.1: The Hyperbolic transformation in (39) preserves the CMS inner-product (40).

Proof: Assume two vector events $E_1 = [P_1, t_1]'$ and $E_2 = [P_2, t_2]'$. Using definition (37)

$$T_\theta(E_1)\cdot T_\theta(E_2)=\left[T_\theta(E_1)\right]^t JT_\theta(E_2)$$
$$=\left[E_1^t\cdot T_\theta^t\right]JT_\theta(E_2)$$
$$=E_1^t\cdot\begin{bmatrix}\cosh(\theta) & \sinh(\theta)\\ \sinh(\theta) & \cosh(\theta)\end{bmatrix}\begin{bmatrix}1 & 0\\ 0 & -1\end{bmatrix}\begin{bmatrix}\cosh(\theta) & \sinh(\theta)\\ \sinh(\theta) & \cosh(\theta)\end{bmatrix}\cdot E_2$$
$$=E_1^t JE_2=E_1\cdot E_2$$

Proposition 5.2: Assume two vector events $E_1^t=E(p,t_1)=[p_1,p_2,\cdots,p_n,\mathrm{i}\cdot t_1]$, and $E_2^t=E(q,t_2)=[q_1,q_2,\cdots,q_n,\mathrm{i}\cdot t_2]$ with a hyperbolic angle between the two event vectors $< E_1E_2=\beta$ then

$$\boldsymbol{E}_1\cdot\boldsymbol{E}_2=\|\boldsymbol{E}_1\|\|\boldsymbol{E}_2\|\cosh(\beta) \tag{43}$$

or equivalently; the angle between two hyperbolic events can be described by:

$$\beta=\cosh^{-1}\left(\frac{\boldsymbol{E}_1\cdot\boldsymbol{E}_2}{\|\boldsymbol{E}_1\|\|\boldsymbol{E}_2\|}\right) \tag{44}$$

Proof: Let us assume the following two unit vectors for only two events:

$$\boldsymbol{U}_i=\frac{\boldsymbol{E}_i}{\|\boldsymbol{E}_i\|}\quad\text{for}\quad i=1,2$$

where the angle between the two unit vectors have the same direction as the event vectors

$$\sqcup\ \boldsymbol{U}_1\boldsymbol{U}_2=\sqcup\ E_1E_2=\beta$$

Select the transformation T with initial hyperbolic angle theta such that

$$T_\theta(\boldsymbol{U}_1)=\begin{bmatrix}1\\0\end{bmatrix}$$

to calculate the transformation of the second unit vector using the conclusion of (42):

$$T_\theta(U_2) = \begin{bmatrix} \cosh(\theta+\beta) \\ \sinh(\theta+\beta) \end{bmatrix} = \begin{bmatrix} \cosh(\beta) \\ \sinh(\beta) \end{bmatrix}$$

Since the unit vector is in the direction of the horizontal axis when theta is zero we will find the second unit vector such that the hyperbolic rotation preserves the Minkowski inner product. That is

$$T_\theta(U_1) \cdot T_\theta(U_2) = U_1 \cdot U_2$$

$$\begin{bmatrix} 1 \\ 0 \end{bmatrix}^t \begin{bmatrix} \cosh(\beta) \\ \sinh(\beta) \end{bmatrix} = U_1 \cdot U_2$$

$$\cosh(\beta) = U_1 \cdot U_2 \Rightarrow \cosh(\beta) = \frac{E_1}{\|E_1\|} \cdot \frac{E_2}{\|E_2\|}$$

Consequently we can have the final result:

$$E_1 \cdot E_2 = \|E_1\| \|E_2\| \cosh(\beta) \tag{45}$$

Proposition 5.3: Hyperbolic Triangular Inequality:

In every triangle OE_1E_2 when E_1 and E_2 are two event vectors with the angle between two vectors is $\angle E_1E_2 = \beta$ then

$$\|E_1 - E_2\|^2 = \|E_1\|^2 + \|E_2\|^2 - 2\|E_1\|\|E_2\|\cosh(\beta) \tag{46}$$

The arc length connecting two points E_1 and E_2:

$$\|E_1E_2\|^2 = \|E_1\|^2 + \|E_2\|^2 - 2\|E_1\|\|E_2\|\cosh(\beta) \tag{47}$$

Proof: (i) To prove the first relation (i), it would be helpful to see event vectors $E_1 - E_2$ as a complex valued and their complex conjugates $E_1 - E_2$ or $(E_1 - E_2)^*$

$$\begin{aligned}\|E_1 - E_2\|^2 &= (E_1 - E_2)\cdot(E_1 - E_2)^* \\ &= (E_1 - E_2)\cdot(E_1^* - E_2^*) \\ &= E_1E_1^* - E_1 \cdot E_2^* - E_2 \cdot E_1^* + E_2E_2^* \\ &= \|E_1\|^2 + \|E_2\|^2 - 2E_1E_2^*\end{aligned}$$

Substitute the conclusion in (45)

$$\|E_1\|^2 + \|E_2\|^2 - 2E_1 \cdot E_2 = \|E_1\|^2 + \|E_2\|^2 - 2\|E_1\|\|E_2\|\cosh(\beta).$$

ii) For the proof of the second part, consider $\|E_1E_2\|$ is the hyperbolic distance between two event vectors E_1 and E_2. The relation (47) is a direct conclusion of the

$$\|E_1 - E_2\| = \|E_1E_2\|.$$

6. CONCLUDING REMARKS

The Beltrami-Klein Model: It is helpful, first to try redefining the primitive terms like “point” “line”, “lies on”, and “between”.

The Beltrami-Klein space is the set of all interior points of a circle which is called point in hyperbolic plane. Two points A and B inside a circle and a chord passing through these two points are imagined. The line is defined by an open chord joining these two points, which is called a hyperbolic line.

There exists an isomorphism between the Klein and Poincare models. That means there is a one-to-one correspondence can be set up between points and lines in one model and those in the others which preserve incidence, congruence, and betweenness.

By accepting the fifth postulate in Euclidean geometry, we can prove existence of a rectangle (quadrilateral with four right angles). Denying or having an alternative to the parallel postulate, particularly in hyperbolic geometry, one can prove that a triangle with angle sum equivalent to 2π radians or quadrilateral with four right angles does not exist (see Figures 1-3).

Thus a quadrilateral with three right angles is called a Lambert Quadrilateral (see Figure 4).

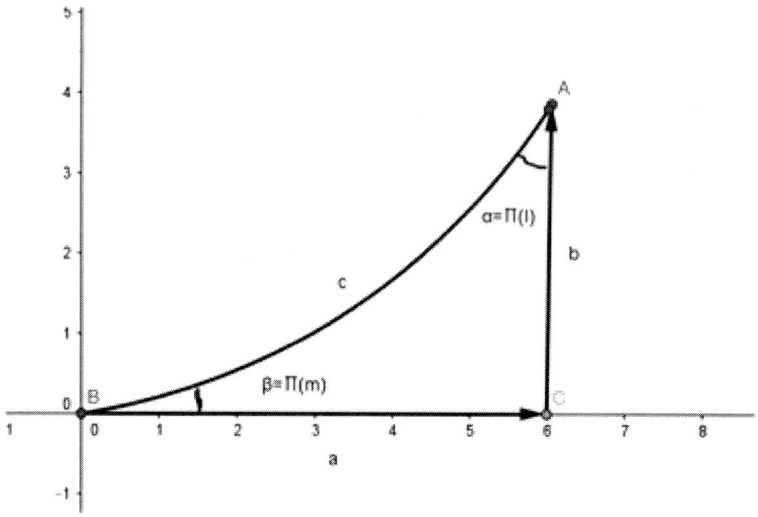

Figure 1. Hyperbolic right triangle with limiting angle of parallelism.

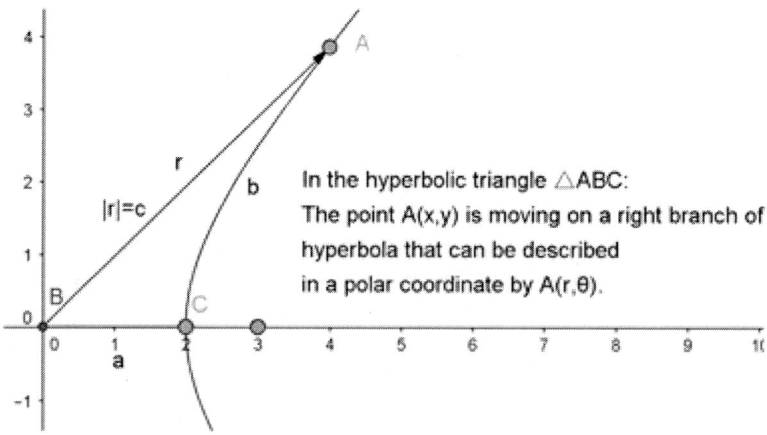

Figure 2. A point $A(x,y)$ is moving on a right branch of a unit hyperbola $x^2 - y^2 = 1$, with a hyperbolic triangle ABC.

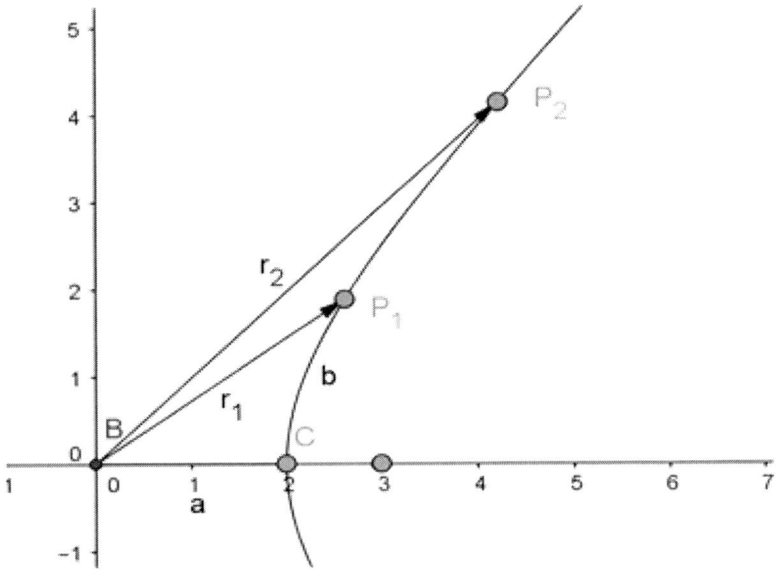

Figure 3. Two points P_1 and P_2 on the hyperbolic plane.

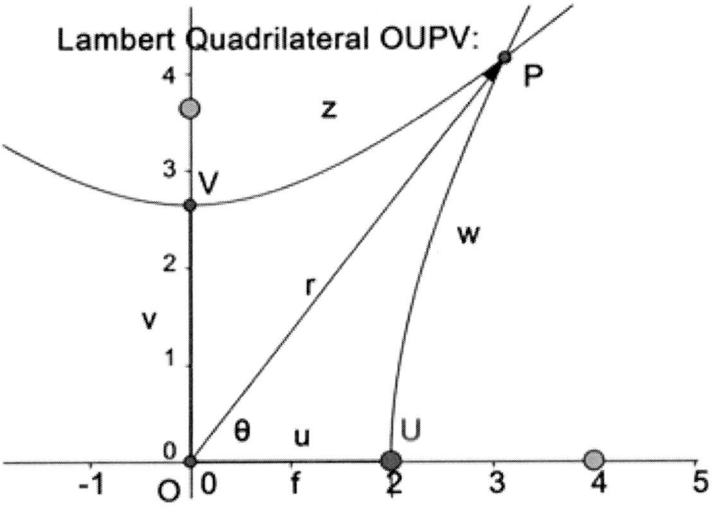

Figure 4. Lambert quadrilateral.

Using the definition of hyperbolic tangent, it can be verified that the mapping $u --> \tanh(u)$ is a one-to-one mapping from the real line into the interval (−1, 1).

Notice that a free online software Geogebra used to draw the following graphs [14] .

With a new advanced proposed postulate for a better unified approach in matter, energy, and momentum, we are challenged to connect the Complex Matter Theory to the two historical approaches of Relativity and Hyperbolic Geometry.

In this work, we conclude that in the new complex matter approach, not only it is a new theory consistent with others, but it describes all shadows and unclear relations by a very beautiful method.

- Matter is redefined with two components of mass and charge which are the intrinsic part of the matter.
- Complex Matter Space Theory: all three space-time regions are space-like, time-like, and light-like. A particle like moving faster than the speed of light can be described by an imaginary mass and a particle-like with imaginary mass can travel faster than the speed of light.
- The geometry of the Complex Matter Space is consistent with the Lorentz transformation, Minkowski, and hyperbolic geometry.
- Given two vector events $\boldsymbol{E}_1$ and $\boldsymbol{E}_2$, the angle between two events can be
- described by a known formula $\cosh(\beta) = \frac{\boldsymbol{E}_1 \cdot \boldsymbol{E}_2}{\|\boldsymbol{E}_1\|\|\boldsymbol{E}_2\|}$ in hyperbolic geometry.
- In every triangle OE_1E_2 when $\boldsymbol{E}_1$ and $\boldsymbol{E}_2$ are two event vectors with the angle that between two vectors is $\square E_1E_2 = \beta$ then

$$\|\boldsymbol{E}_1 - \boldsymbol{E}_2\|^2 = \|\boldsymbol{E}_1\|^2 + \|\boldsymbol{E}_2\|^2 - 2\|\boldsymbol{E}_1\|\|\boldsymbol{E}_2\|\cosh(\beta)$$

To compare these results with Euclidean space, one can see Greenberg (1980) [15] and Kay (2001) [16] .

REFERENCES

1. Hilbert, D. (1902) The foundations of Geometry. The Open Court Publishing Company, Chicago, IL.

2. Milnor, J. (1996) Hyperbolic Geometry: The First 150 Years. The Mathematics Heritage of Henry Poincare, Proceedings of Symposia in Pure Mathematics of the American Mathematical Society, 39, 25-40.
3. Taylor, J., Zafiratos, C. and Dubson, M.A. (2004) Modern Physics for Scientists and Engineers. 2nd Edition, Pearson Prentice Hall, New Jersey.
4. Richard, F. (1963) The Feynman Lectures on Physics. Addison Wesley, Reading, MA, Volume III, 1-6, 7-5.
5. NASA (2011) Gravity Probe-B. https://science.nasa.gov/science-news/science-at-nasa/2011/04may_epic
6. Lavenda, B.H. (2009) The Relativity of Hyperbolic Space. Universitadegli Studi, Camerino, Italy.
7. Ahangar, R.R. (2014) Foundation of Complex Matter Space and Special Theory of Relativity, a Unifying Approach. Journal of Nuclear and Particle Physics, 4, 147-153. http://article.sapub.org/10.5923.j.jnpp.20140405.03.html
8. Ahangar, R.R. (2014) Quantum Complex Matter Space. International Journal of Theoretical and Mathematical Physics, 4, 159-163. http://article.sapub.org/10.5923.j.ijtmp.20140404.04.html
9. Ahangar, R.R. (2014) Complex Matter Space and Relativistic Quantum Mechanics. Applied Mathematics, 5, No. 21. http://file.scirp.org/Html/11-7402454_52220.htm
10. Mayer, A.F. (2007) Wave Energy in Quantum Mechanics. Journal of Physics, Conference Series, 70, 012013. https://doi.org/10.1088/1742-6596/70/1/012013
11. Gauthier, R. (2009) The Dirac Equation and the Superluminal Electron Model. http://www.superluminalquantum.org
12. Naber, G.L. (1992) The Geometry of Minkowski Spacetime: An Introduction to the Mathematics of Special Theory of Relativity. Dover, New York.
13. Callahan, J. (2000) The Geometry of Spacetime: An Introduction to Special and General Relativity. Springer, New York.
14. Geogebra Software. For Hyperbolic Geometry See. http://www.geogebra.org http://www.geogebra.org/material/simple/id/7005
15. Greenberg, M.J. (1980) Euclidean and Non-Euclidean Geometries, Development and History. 2nd Edition, WH Freeman, New York.
16. Kay, D.C. (2001) College Geometry a Discovery Approach. 2nd Edition, Addison Wesley Longman, Boston.

CHAPTER 6

On q-Deformed Calculus in Quantum Geometry

Olaniyi S. Maliki[1], Emmanuel I. Ugwu[2]

[1]*Department of Industrial Mathematics and Applied Statistics, Ebonyi State University, Abakaliki, Nigeria*

[2]*Department of Industrial Physics, Ebonyi State University, Abakaliki, Nigeria*

ABSTRACT

The relation between noncommutative (or quantum) geometry and the mathematics of spaces is in many ways similar to the relation between quantum physics and classical physics. One moves from the commutative algebra of functions on a space (or a commutative algebra of classical observable in classical physics) to a noncommutative algebra representing a noncommutative space (or a noncommutative algebra of quantum observables in quantum physics). The object of this paper is to study the basic rules governing q-calculus as compared with the classical NewtonLeibnitz calculus.

Keywords: Quantum Geometry, q-Numbers, q-Factorials, q-Calculus

1. INTRODUCTION

There exists an intimate relationship between classical geometry and physics, and to appreciate this we will consider the Einstein field equations (EFE) of special relativity written as:

$$G_{\mu\nu} = 8\pi T_{\mu\nu} \text{ (EFE)}$$

where $G_{\mu\nu} = R_{\mu\nu} - \frac{1}{2}g_{\mu\nu}R$ is the Einstein tensor, $R = g^{\mu\nu}R_{\mu\nu}$ is the Ricci tensor. $T_{\mu\nu}$ is the energy-momentum tensor.

For the moment we are only interested in three basic properties of the above equation.

1) The equation (EFE) is a tensor equation. This is necessarily so, since the principle of invariance under coordinate transformations must hold, in other words the equations of physics must look the same in any frame of reference.

2) We can interpret equation (EFE) more simply as

Tensor representing geometry of space = Tensor representing energy content of space

i.e. it is the presence of matter in space that distorts the neighbouring geometry. Most equations of mathematical physics can be interpreted similarly.

3) The solution to equation (EFE) is a geometrical object, namely a line element given by

$$ds^2 = g_{\mu\nu}dx^\mu dx^\nu$$

where $g_{\mu\nu}$ is the metric tensor to be solved for in (EFE).

1.1. Quantum Geometry

Every geometry is associated with some kind of space. Quantum (or noncommutative) geometry [1] [2] deals with quantum spaces, including the classical concept of space as a very special case. In classical geometry spaces are always regarded as collections of points equipped with the appropriate additional structure (as for example a topological structure given by the collection of open sets, or a smooth structure given by the atlas). In contrast to classical geometry, quantum spaces are not interpretable in this way. In general, quantum spaces have no points at all! They exhibit non-trivial quantum fluctuations' of geometry at all scales.

In generalizing classical geometry to the non-commutative level, there are two important conceptual steps:

1) Translation of geometry into a commutative algebra format;

2) Non-commutative generalizations.

1.2. Reformulating Basic Geometrical Concepts

It turns out that the geometrical structure on any given topological space X is always completely expressible in the language of some associated *-algebra [3] [4] .

Points

Let X be a compact topological space, and let $A = C(X)$ be the *-algebra of continuous complex-valued functions on X. Every element $x \in X$ naturally gives rise to a linear functional $\chi = \chi_x\colon\ A \to \mathbb{C}$ defined by

$$\chi(f) = f(x)$$

This map is multiplicative in the sense that

$$\chi(fg) = \chi(f)\chi(g), \qquad \forall f, g \in A$$

Furthermore, χ is Hermitian in the sense that $\chi(f^*) = \chi(f)^*$, it is also non-zero, i.e. χ is a character on A. Conversely, consider an arbitrary character $\chi : A \to \mathbb{C}$, then it can be shown that there exists a unique point $x \in X$ such that $\chi = \chi_x$. In other words, we have a natural bijection between points of X and characters of A. It is important to note that this characterization of points also remains valid at the smooth level, in which X could be a compact smooth manifold and the associated *-algebra consists of smooth functions on X.

1.3. The Gelfand-Naimark Theorem

The algebra $A = C(X)$ of complex-valued functions on a compact topological space X, equipped with the maximum norm

$$|f| = \max_{x \in X} |f(x)|$$

is a commutative C*-algebra [2] . The classical theorem of Gelfand and Naimark characterizes the algebras of the form $A = C(X)$, as commutative unital C*-algebras. This means that for every commutative unital C*-algebra A there exists (up to homeomerphisms) a unique compact topological space X such that $A \cong C(X)$.

As we have earlier observed, the points of the space X are recovered as characters of the associated algebra A. In terms of this identification, the topology on X coincides with the weak*-topology, induced from the dual space A^*, consisting of continuous linear functionals on A. It turns out that homomorphisms between C*-algebras are automatically continuous, in particular characters are continuous linear functionals.

2. THE QUANTUM PLANE

A simple example of quantum geometry is the quantum plane (Figure 1). Usually, a plane is described by two coordinate functions x, y. Naturally, the functions xy and yx are the same since it does not matter whether you measure x first and then y or y first and then x. This is precisely what is lost in the quantum world.

In the quantum plane we replace the property $xy = yx$ by $xy = qyx$ where q is some parameter. We no longer have points, however we can continue to work algebraically with x and y.

Define $[x, y] = xy - yx$, the commutator bracket. Hence $xy = qyx$ can be rewritten as

$$[x, y] = (q-1) yx$$

The commutative case is obtained when $q = 1$.

2.1. q-Deformed Calculus

It is interesting to know that one can really do geometry in this setting, where coordinates do not commute. This is the remarkable discovery in recent times. For example, we can follow the approach of Newton-Leibnitz defining differentiation by

$$f'(y) = x^{-1}\left(f(x+y) - f(y)\right)\Big|_{x\to 0}$$

But when $xy \neq yx$ and in particular $xy = qyx$ we get instead

$$\left[f'(x)\right]_q = \frac{f(x) - f(qx)}{(1-q)x}$$

Thus for example the derivative of the function $f(x) = x^n$ in this non-commutative setting would be

$$\left[f'(x)\right]_q = \frac{x^n - q^n x^n}{(1-q)x} = \left(\frac{1-q^n}{1-q}\right)x^{n-1} = \left(1 + q + q^2 + \cdots + q^{n-1}\right)x^{n-1}$$

We observe here that when $q = 1$ the derivative of x^n for the commutative case is obtained, i.e. $f'(x) = nx^{n-1}$.

2.2. Basic Notions of q-Calculus

The mathematical study of noncommutative geometry is intimately related to the so-called q-calculus (qnumbers, q-factorials, q-differentials and integrals, basic q-hypergeometric functions, and q-orthogonal polynomials). Here we give a brief introduction to q-numbers and q-factorials which will be required in the subsequent sections.

2.2.1. q-Numbers and q-Factorials

For any nonzero complex number q, the q-number $[a]_q$, $a \in$ □, is defined by

$$[a]_q \equiv [a] := \frac{q^a - q^{-a}}{q - q^{-1}} = \frac{e^{ah} - e^{-ah}}{e^h - e^{-h}} = \frac{\sinh ah}{\sinh h}, \qquad q = \exp h \tag{2.1}$$

We observe that, $\lim_{q\to 1}[a]_q = a$. The following expression which is easily proved will prove useful.

$$[a]_q \equiv [a] := \frac{1-q^a}{1-q} = 1 + q + q^2 + \cdots + q^{a-1} = q^{(a-1)/2}\,[a]_{q^{1/2}} \tag{2.2}$$

Thus given $f(x) = x^n$, as shown previously $[f'(x)]_q = [n]x^{n-1}$.

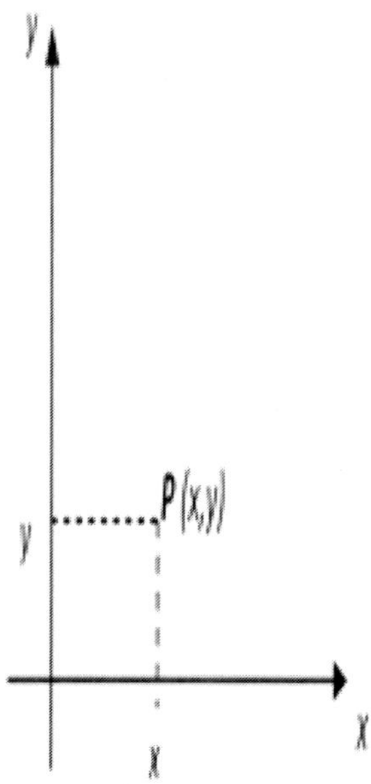

Figure 1. The quantum plane (xy ≠ yx).

2.2.2. Proposition

The q-numbers satisfy the following relations derived from the property of the exponential function

1) $[r] = q^{r-1} + q^{r-3} + \cdots + q^{-(r-1)}$

2) $[r+s] = q^s[r] + q^{-r}[s] = q^{-s}[r] + q^r[s]$

3) $[r-s] = q^s[r] - q^r[s] = q^{-s}[r] - q^{-r}[s]$

4) $[r][s-t] + [s][t-r] + [t][r-s] \equiv 0$

5) $[r]=[2][r-1]-[r-2]$

The proof of 1) is easy to see from the fact that;

$$[r]=\frac{q^{r}-q^{-r}}{q-q^{-1}}=\frac{q^{r}\left(1-q^{-2r}\right)}{q\left(1-q^{-2}\right)}=q^{r-1}\left(\frac{1-\left(q^{-2}\right)^{r}}{1-\left(q^{-2}\right)}\right)$$

Let

$$u=q^{-2},\ \therefore\ \frac{1-\left(q^{-2}\right)^{r}}{1-\left(q^{-2}\right)}=\frac{1-u^{r}}{1-u}=1+u+u^{2}+\cdots+u^{r-1}$$

Hence;

$$[r]=q^{r-1}\left(1+q^{-2}+q^{-4}+\cdots+q^{-2(r-1)}\right)=q^{r-1}+q^{r-3}+\cdots+q^{-(r-1)}\quad\square$$

The rest of the identities can be proved similarly. It is important to note that the relations 1)-5) remain valid when q is considered an indeterminate. Consequently any q-number $[n]_q, n\in \square$, belongs to the space $\square\left[q,q^{-1}\right]$ of Laurent polynomials [2] in q with integral coefficients.

Suppose $n\in \square$, then we define the q-factorial $[n]_q!\equiv[n]!$ by setting;

$$[n]!=[1][2]\ldots[n],\qquad [0]!:=1$$

The following expression is quite useful in the theory of hypergeometric functions as well as in combinatorics.

$$\{a:q\}_n:=\prod_{j=1}^{n}\left(1-aq^{j-1}\right)=(1-a)(1-aq)\left(1-aq^{2}\right)\cdots\left(1-aq^{n-1}\right),\quad \{a:q\}_0=1. \tag{2.3}$$

It is now possible for us to relate the above with the q-factorials. We observe that;

$$[n]! = \frac{q^{-n(n-1)/2}}{(1-q^2)^n}\{q^2 : q^2\}_n \tag{2.4}$$

From equation (2.3) we note that $\{q^{-m} : q\}_n = 0,\ \forall n = m+1, m+2, \cdots$
For $|q| < 1$, define

$$\{a : q\}_\infty := \lim_{n\to\infty}\{a : q\}_n = \prod_{j=1}^{\infty}\left(1 - aq^{j-1}\right) \tag{2.5}$$

which converges $\forall a \in \mathbb{C}$, and defines an analytic function on $\mathbb{C}$.

2.2.3. Proposition

$$\forall a \in \mathbb{C},\ |q| < 1,\ \{a : q\}_n = \frac{\{a : q\}_\infty}{\{aq^n : q\}_\infty}.$$

Proof

$$\frac{\{a:q\}_\infty}{\{aq^n:q\}_\infty} = \frac{\prod_{j=1}^{\infty}\left(1-aq^{j-1}\right)}{\prod_{j=1}^{\infty}\left(1-aq^{n+j-1}\right)} = \frac{(1-a)(1-aq)(1-aq^2)(1-aq^3)\cdots(1-aq^{n-1})(1-aq^n)(1-aq^{n+1})\cdots}{(1-aq^n)(1-aq^{n+1})(1-aq^{n+2})\cdots(1-aq^{2n-1})\cdots}$$

$$= (1-a)(1-aq)(1-aq^2)(1-aq^3)\cdots(1-aq^{n-1}) = \{a:q\}_n$$

Remark: We can also define and show that;

$$\{a : q\}_{-n} = \frac{\{a : q\}_\infty}{\{aq^{-n} : q\}_\infty} = \frac{1}{\{aq^{-n} : q\}_n} \tag{2.6}$$

2.2.4. q-Binomial Coefficients

The q-binomial coefficients are defined by the formula;

$$\begin{Bmatrix} n \\ m \end{Bmatrix}_q \equiv \begin{Bmatrix} n \\ m \end{Bmatrix} := \frac{\{q:q\}_n}{\{q:q\}_m \{q:q\}_{n-m}} = \frac{[n]_{q^{1/2}}!\, q^{(n-m)m/2}}{[m]_{q^{1/2}}!\,[n-m]_{q^{1/2}}!}, \qquad \forall n,m \in \mathbb{N} \cup \{0\},\ n \geq m. \tag{2.7}$$

Remark: There exist a close analogy between the classical binomial coefficients $\binom{n}{m} = \frac{n!}{(n-m)!m!}$ and their q-analogues. Many of the identities satisfied by the former have their counterparts for the q-binomial coefficients. For example the classical identity;

$$\binom{n+1}{m} = \binom{n}{m-1} + \binom{n}{m}$$

simply translates to;

$$\begin{Bmatrix} n+1 \\ m \end{Bmatrix}_q = \begin{Bmatrix} n \\ m-1 \end{Bmatrix}_q + q^m \begin{Bmatrix} n \\ m \end{Bmatrix}_q = q^{n-m+1} \begin{Bmatrix} n \\ m-1 \end{Bmatrix}_q + \begin{Bmatrix} n \\ m \end{Bmatrix}_q \qquad 2.8$$

2.2.5. Proposition

Let x and y be noncommuting variables satisfying the relation $xy = qyx$, then we have

$$(x+y)^n = \sum_{m=0}^{n} \begin{Bmatrix} n \\ m \end{Bmatrix}_q y^m x^{n-m} = \sum_{m=0}^{n} \begin{Bmatrix} n \\ m \end{Bmatrix}_{q^{-1}} x^m y^{n-m} \tag{2.9}$$

In case q is a primitive p^{th} root of unity, and p is odd, then

$$(x+y)^p = x^p + y^p \tag{2.10}$$

Proof

Equation (2.9) can be established by induction on n, and employing the first identity in (2.8). The second assertion follows directly from (2.9). Observe that:

$$(x+y)^p = \sum_{m=0}^{p} \begin{Bmatrix} p \\ m \end{Bmatrix}_q y^m x^{p-m} = \begin{Bmatrix} p \\ 0 \end{Bmatrix}_q x^p + \begin{Bmatrix} p \\ p \end{Bmatrix}_q y^p + \sum_{m=1}^{p-1} \begin{Bmatrix} p \\ m \end{Bmatrix}_q y^m x^{p-m}$$

From (2.7)

$$\begin{Bmatrix} p \\ m \end{Bmatrix}_q = \frac{\{q:q\}_p}{\{q:q\}_m \{q:q\}_{p-m}} = \frac{[p]_{q^{1/2}}!\, q^{(p-m)m/2}}{[m]_{q^{1/2}}!\,[p-m]_{q^{1/2}}!} = 0, \quad m = 1,2,\ldots,p-1.$$

and $\begin{Bmatrix} p \\ 0 \end{Bmatrix}_q = \begin{Bmatrix} p \\ p \end{Bmatrix}_q = 1$.

3. Q-DIFFERENTIAL AND Q-INTEGRAL OPERATORS

The following are important basic notions derived from their analogue in classical calculus, and will be employed subsequently.

3.1. The q-Differential Operator

For $1 \neq q \in \square$, we define the q-differential operator D_q by:

$$D_q f(x) = \frac{f(x) - f(qx)}{(1-q)x} \tag{3.1}$$

Note that $D_q \to d/dx$ as $q \to 1$.

Proposition:

$$D_q f(x) = \sum_{r=0}^{\infty} \frac{(q-1)^r}{(r+1)!} x^r \frac{d^{r+1}}{dx^{r+1}} f(x), \quad \forall 1 \neq q \in \square \tag{3.2}$$

Provided the expression on the right hand side exists.

Proof

Let $x+h=qx$, then by Taylor's series, $f(x+h)=f(x)+hf'(x)+\frac{h}{2!}f''(x)+\cdots$

$$\therefore \quad f(x+h)-f(x)=hf'(x)+\frac{h}{2!}f''(x)+\cdots=\sum_{r=0}^{\infty}\frac{h^{r+1}}{(r+1)!}\frac{d^{r+1}}{dx^{r+1}}f(x)$$

Setting $h=(q-1)x$, we have:

$$D_q f(x)=\frac{f(qx)-f(x)}{(q-1)x}=\frac{1}{(q-1)x}\sum_{r=0}^{\infty}\frac{\left[(q-1)x\right]^{r+1}}{(r+1)!}\frac{d^{r+1}}{dx^{r+1}}f(x)=\sum_{r=0}^{\infty}\frac{(q-1)^r x^r}{(r+1)!}\frac{d^{r+1}}{dx^{r+1}}f(x)$$

The formula for the product of two functions is given by

$$D_q\{u(x)v(x)\}=\frac{u(x)v(x)-u(qx)v(qx)}{x(1-q)}=\frac{v(x)\{u(x)-u(qx)\}+u(qx)\{v(x)-v(qx)\}}{x(1-q)}$$

We can now define the q-analogue of the Newton-Leibnitz rule:

$$D_q\{u(x)v(x)\}=v(x)d_q u(x)+u(qx)d_q v(x) \tag{3.3}$$

Thus,

$$\begin{aligned}D_q^2 f(x)=D_q\left(D_q f(x)\right)&=D_q\left(\sum_{r=0}^{\infty}\frac{(q-1)^r x^r}{(r+1)!}\frac{d^{r+1}}{dx^{r+1}}f(x)\right)\\&=\sum_{r=0}^{\infty}\frac{(q-1)^r}{(r+1)!}D_q\left\{x^r\frac{d^{r+1}}{dx^{r+1}}f(x)\right\}\\&=\sum_{r=0}^{\infty}\frac{(q-1)^r}{(r+1)!}\left\{\left(\frac{d^{r+1}}{dx^{r+1}}f(x)\right)D_q x^r+(qx)^r D_q\left(\frac{d^{r+1}}{dx^{r+1}}f(x)\right)\right\}\end{aligned}$$

By induction on n, it follows from (3.3) that:

$$D_q^n f(x)=\frac{x^{-n}q^{-n(n-1)/2}}{(q-1)^n}\sum_{j=0}^{n}\begin{Bmatrix}n\\j\end{Bmatrix}_q(-1)^j q^{j(j-1)/2}f\left(q^{n-j}x\right) \tag{3.4}$$

As a special case when $n=2$, $D_q^2 f(x)$ is evaluated to give:

$$D_q^2 f(x) = \frac{1}{q(q-1)^2 x^2}\left\{ f(q^2 x) - (1+q) f(qx) + q f(x) \right\} \tag{3.5}$$

3.2. The q-Integral Operator

The q-integral operator will be defined as the inverse of the q-differential operator.

Given $d_q F(x) = f(x)$, we have:

$$F(x) - F(qx) = (1-q) x f(x) \tag{3.6}$$

It then follows that; $F(q^j x) - F(q^{j+1} x) = (1-q) q^j x f(q^j x),\ j = 0, \ldots, (n-1)$.

Hence, summing these relations over j gives:

$$F(x) - F(q^n x) = (1-q) x \sum_{r=0}^{n-1} q^r f(q^r x) \tag{3.7}$$

Assuming $q \in (0,1)$ so that $F(q^n x) \to F(0)$ as $n \to \infty$, it follows that

$$F(x) - F(0) = (1-q) x \sum_{r=0}^{\infty} q^r f(q^r x)$$

We can now formally define the q-integral of a function $f(x)$ on a given interval $[0, \alpha]$ as

$$\int_0^{\alpha} f(x) \mathrm{d}x_q = \alpha(1-q) \sum_{r=0}^{\infty} q^r f(q^r \alpha) = \sum_{r=0}^{\infty} (x_r - x_{r+1}) f(x_r), \qquad x_r = \alpha q^r \tag{3.8}$$

On the semi-infinite interval $[\alpha, \infty)$, the q-integral of a function $f(x)$ is defined as:

$$\int_{\alpha}^{\infty} f(x)\mathrm{d}_q x = \alpha(1-q)\sum_{r=1}^{\infty} q^{-r} f\left(q^{-r}\alpha\right) \tag{3.9}$$

Over any closed interval $[\alpha, \beta]$, the q-integral of a function $f(x)$ is given by

$$\int_{\alpha}^{\beta} f(x)\mathrm{d}_q x = \int_{0}^{\beta} f(x)\mathrm{d}_q x - \int_{0}^{\alpha} f(x)\mathrm{d}_q x \tag{3.10}$$

We now define the integral over the interval $[0,\infty)$. This is achieved by setting $\alpha = 1$ in equations (3.8) and (3.9) and summing to get:

$$\int_{0}^{\infty} f(x)\mathrm{d}_q x = \int_{0}^{1} f(x)\mathrm{d}_q x + \int_{1}^{\infty} f(x)\mathrm{d}_q x = (1-q)\sum_{r=-\infty}^{\infty} q^{r} f\left(q^{r}\alpha\right) \tag{3.11}$$

The integration by parts formula of Newton-Leibnitz calculus is interpreted in the present noncommutative context as:

$$\int f_2(x) D_q f_1(x)\mathrm{d}_q x = f_1(x) f_2(x) - \int f_1(qx) D_q f_2(x)\mathrm{d}_q x$$

4. APPLICATION

There are a number of applications of the foregoing, we mention here just two, namely:

1) q-binomial formulae in two variables satisfying a quadratic relation, this has recently been published in [5] and [6] . These relations have applications in quantum group theory and non-commutative geometry.

2) A recent trend in modern physics is the study of the quantum anti-de Sitter space [7] [8] possibly in connection with q = root of unity [9] .

ACKNOWLEDGEMENTS

Olaniyi S. Maliki, Emmanuel I. Ugwu (AIMS) in Muizenberg South Africa, when the first author visited in 2010. I wish to thank the director Prof Fritz

Hahne for giving me a postdoctoral fellowship, and for providing a conducive environment which made this research possible.

REFERENCES

1. Connes, A. (1986) Non-Commutative Differential Geometry. Extrait des Publications Mathematiques-IHES, 62. (cited in: Qauntum Principal Bundles and Their Characteristic Classes (pdf), by MICO DURDEVIC, arXiv:q-alg/960505008vi (5 May 1996))
2. Connes A. (1994) Noncommutative Geometry. Academic Press New York.
3. Brateli O. and Robinson, D. (1979) Operator Algebras and Quantum Statistical Mechanics, Volumes 1/2. SpringerVerlag, Berlin.
4. Brown L.G., Douglas, R.G. and Filmore, P.G. (1977) Extensions of C*-Algebras and K-Homology. Annals of Mathematics, 105, 265-324.
5. Benaoum H.B. (1999) (q; h)-Analogue of Newton's Binomial Formula. Journal of Physics A: Mathematical and General, 32, 2037-2040.
6. Rosengren H. (1999) Multivariable Orthogonal Polynomials as Coupling Coefficients for Lie and Quantum Algebra Representations. Dissertation, Centre for Mathematical Sciences, Mathematics (Faculty of Science), Lund, 167.
7. Kowalski-Glikman J. (1998) Black Hole Solution of Quantum Gravity. Physics Letters A, 250, 62-66.
8. Chang Z. (1999) Quantum Anti-De Sitter Space. (reprint)
9. Steinacker H. (1998) Finite Dimensional Unitary Representations of Quantum Antide Sitter Groups at Roots of Unity. Communications in Mathematical Physics, 192, 687-706.

CHAPTER 7

Spatial and visual data fusion for capturing, retrieval, and modeling of as-built building geometry and features

Zhenhua Zhu* and Sara Donia

Department of Building, Civil and Environmental Engineering, Concordia University, Montreal H3G 1 M8, Canada

ABSTRACT

Background

As-built building information, including building geometry and features, is useful in multiple building assessment and management tasks. However, the current process for capturing, retrieving, and modeling such information is labor-intensive and time-consuming.

Methods

In order to address these issues, this paper investigates the potentials of fusing visual and spatial data for automatically capturing, retrieving, and modeling as-built building geometry and features. An overall fusion-based framework has been proposed. Under the framework, pairs of 3D point clouds are progressively registered through the RGB-D (Red, Green, Blue plus Depth) mapping. Meanwhile, building elements are recognized based on their visual patterns. The recognition results can be used to label the 3D points, which could facilitate the modeling of building elements.

Results

So far, two pilot studies have been performed. The results show that a high degree of automation could be achieved for the registration of building scenes captured from different scans and the recognition of building elements with the proposed framework.

Conclusions

The fusion of spatial and visual data could significantly facilitate the current process of retrieving, modeling, and visualizing as-built information.

Keywords: As-built information, Data fusion, RGB-D mapping, Visual recognition, Automation

1. INTRODUCTION

Three dimensional (3D) as-built building information records existing conditions of buildings, including their geometry and actual details of architectural, structural, and mechanical, electrical, and plumbing (MEP) elements (Institute for BIM in Canada IBC 2011). Therefore, such information is useful in multiple building assessment and management tasks. For example, as-built building information could be used to identify and quantify the deviations between design and construction, which could significantly reduce the amount of rework during the construction phase of a project (Liu et al. 2012). Also, the use of as-built building information could facilitate the coordination of the MEP designs, when renovating and retrofitting existing old buildings. This facilitation was expected to reduce almost 60% of the MEP field-to-finish workflow (ClearEdge3D 2012).

Although as-built building information is useful, the current process for capturing, retreiving, and modeling such information requires a lot of manual, time-consuming work. This labor-intensive and time-consuming nature increases the cost of using as-built building information in practice. As a result, it was not a value adding task for most general contractors and has not been widely used in the vast majority of construction and renovation/retrofit projects (Brilakis et al. 2011), unless the time and cost of the current process for capturing, retrieving, and modeling as-built building information could be significantly reduced, and the as-built building information could be constantly updated and closely reviewed (Pettee, 2005).

In order to automate the current manual process for capturing, retrieving, and modeling as-built building information, several research studies have been initiated. Most of them were built upon the remote sensing data captured by laser scanners or digital cameras. For example, (Okorn et al. 2010) presented an idea of creating as-built floor plans in a building by projecting the 3D point clouds captured by a laser scanner. (Fathi and Brilakis 2012) relied on the color images captured by digital cameras to model as-built conditions of metal roof panels. (Geiger et al. 2011) proposed an approach to build 3D as-built maps from a series of high-resolution stereo video sequences.

The sole reliance on one type of sensing data makes existing studies have inherent limitations. For example, when capturing, retrieving, and modeling 3D as-built building information in the building indoor environments, (Furukawa et al. 2009) noted that the prevalence of texture-poor walls, floors, and ceilings in the environments may fail those studies built upon the images or video frames captured by digital cameras. Also, (Adan et al. 2011) reported that a 3D laser scanner had to be set up at hundreds of locations in order to complete the scan of 40 rooms.

Compared with existing research studies, this paper investigates the potential of fusing two different types of sensing data (i.e. color images and depth images) to capture, retrieve, and model 3D as-built building information. The focus has been placed on building geometry and features. An overall fusion-based framework has been proposed in the paper. Under the framework, 3D point clouds are progressively registered through the RGB-D (Red, Green, Blue plus Depth) mapping, and building elements in the point clouds are recognized based on their visual patterns for 3D as-built modeling. So far, two pilot studies have been performed to show that a high degree of automation could be achieved with the proposed framework for capturing, retrieving, and modeling as-built building information.

2. BACKGROUND

The benefits of using 3D as-built building information have been well acknowledged by researchers and professionals in the architecture, engineering, and construction (AEC) industry. Meanwhile, the process for capturing, retrieving, and modeling such information has been identified as labor-intensive (Tang et al. 2010). Typically, the process starts with the collection of as-built building conditions using remote sensing devices, such as laser scanners or digital cameras. Then, the sensing data collected from multiple locations are registered (i.e. building scenes registration), and building elements in the sensing data are recognized (i.e. building elements recognition). This way, the semantic or high-level building geometric information in the sensing data could be used for domain related problem solving.

So far, a lot of manual work has been involved in the registration of building scenes and the recogniton of building elements. In order to address this issue, multiple research methods have been initiated. Based on the type of the sensing data they work on, the methods can be classified into three main categories. The methods in the first category were built upon the 3D point clouds directly captured by the devices, such as terrestrial laser scanners. The methods in the second category relied on the color images or videos taken by digital cameras or camcorders. Recently, the methods with the use of RGB-D data (i.e. pairs of color and depth images) have been proposed with the development of RGB-D cameras (e.g. Microsoft®. *Kinect*). Below are the details of these methods.

2.1 Point cloud based methods

In the point cloud based methods, terrestrial laser scanners are commonly used to capture the detailed as-built building conditions. One laser scan may collect millions of 3D points in minutes. Using this rich information, (Okorn et al. 2010) proposed an idea of creating as-built building floor plans by projecting the collected 3D points onto a vertical (z-) axis and a ground (x-y) plane. The projection results could indicate which points can be grouped. This way, the floor plans were created (Okorn et al. 2010). In addition to the generation of floor plans, (Xiong and Huber 2010) employed the conditional random fields to model building elements from 3D points, but their work was limited to those building elements with planar patches, such as walls and ceilings. The openings on the walls or ceilings could be further located by checking the distribution density of the points around the openings (Ripperda and Brenner, 2009) or using a support vector machine (SVM) classifier (Adan and Huber, 2011).

Although the detailed as-built building information can be captured with laser scanners, the scanners are heavy and not portable. Typically, at least two crews are needed to use one laser scanner to collect as-built conditions (Foltz, 2000). This non-portable nature makes laser scanners unconvenient to use, especially when capturing, retrieving, and modeling the as-built information in the building indoor environments. (Adan et al. 2011) mentioned that a laser scanner had to be set up at 225 different locations in order to scan 40 rooms, which was approximately 5.6 locations per room.

Also, the 3D points collected by laser scanners only record the spatial as-built building conditions, which limits the capability to recognize building elements. Existing recognition methods mainly relied on the points spatial features, which can be described globally or locally (Stiene et al. 2006). Global descriptors (Wang et al. 2007; Kazhdan et al. 2003) captured all the geometrical characteristics of a desired object. Therefore, they are discriminative but not robust to cluttered scenes (Patterson et al. 2008). Local descriptors (Shan et al. 2006; Huber et al. 2004) improved the recognition robustness, but they were computationally complex (Bosche and Haas 2008). Both global and local descriptors cannot recognize building elements with different materials. For example, it is difficult for the methods to differentiate between a concrete column and a wooden column just based on their 3D points, if both of them have the same shape and size.

2.2 Color based methods

The methods in the second category target on the visual data (i.e. color images or videos) from digital cameras or camcorders as an affordable and portable alternative for as-built modeling. So far, several critical techniques have been created. (Snavely et al. 2006) explored the possibility to create a 3D sparse point cloud from a collection of digital color images (i.e. structure-from-motion). (Furukawa and Ponce 2010) developed a multi-view stereo matching framework, which can generate an accurate, dense, and robust point clouds from stereo color images. In the work of (Durrant-Whyte and Bailey 2006), a robot was used to build a 3D map within an unknown environment while at the same time keeping track of its current locations (i.e. simultaneous localization and mapping). Based on these techniques, many image/videos based methods for capturing, retrieving, and modeling 3D as-built building information have been proposed, and the 3D reconstruction of the built environments has become possible. For example, the 3D models of built environments with high visual quality can be automatically derived even from single facade images of arbitrary resolutions (Müller et al. 2007). Also, (Pollefeys et al. 2008) used the videos from a moving vehicle to model the urban environments.

Compared with 3D point clouds, there are multiple recognition cues that can be extracted from images or videos, including color, texture, shape, and local invariant visual features. (Neto et al. 2002) used color values to recognize structural elements in digital images, while (Brilakis et al. 2006) presented the concept of "material signatures" to retrieve construction materials, such as concrete, steel, wood, etc. (Zhu and Brilakis 2010) combined two recogniton cues (shape and material) to locate concrete column surfaces. Although many recognition methods have been developed, the robust recognition of buildings elements has not been achieved yet. One main limitation lies in the fact that the recognition results from images or videos are limited to two-dimensional (2D) space. Therefore, it is difficult to directly use the recognition results for capturing, retrieving, and modeling 3D as-built building information.

In addition, digital cameras or camcorders are easy and convenient to use in building environments with multiple interior partitions. However, most existing image/videos based methods for capturing, retrieving, and modeling 3D as-built building information heavily relied on the extraction of visual features from images or video frames. Building environments are commonly prevalent with texture-poor walls, floors, and ceilings. This may fail the procedures of these methods, unless the scene-specific constraints for the environments are pre-created manually (Furukawa et al. 2009).

2.3 RGB-D based methods

Point cloud based methods and color based methods relied on one type of sensing data for capturing, retrieving, and modeling as-built building conditions. As a result, they have inherent limitations associated with the type of the sensing data they work on. Specifically, the point cloud based methods can accurately retrieve the detailed as-built building conditions, but the data capturing process is time-consuming. Most laser scanners are not portable and the setup of a laser scanner also takes time (Foltz, 2000). In addition, the recognition of building elements from the point clouds solely relied on their spatial features. Compared with the point cloud based methods, the recognition of building elements from color images or videos can be performed using multiple recognition cues, which could increase the recognition accuracy. The main limitation of the color based methods lies in their modeling robustness. Most color based methods may fail when building scenes have texture-poor elements. Table 1 summarizes the comparison results between the methods in these two categories.

The emergence of RGB-D cameras provides another idea to capture, retrieve, and model 3D as-built building information. The RGB-D cameras are novel sensing systems, which can capture pairs of mid-resolution color and depth images almost in real time. One example of the RGB-D cameras is Microsoft® *Kinect*. The resolution of the *Kinect* is up to 640 × 480, which is equivalent to the capture of 307,200 3D points per frame. Also, its sensing rate is up to 30 frames-per-second (FPS) sensing rate, which means almost 1 million points could be captured by the *Kinect* in one second. Considering these characteristics, (Rafibakhsh et al. 2012) once concluded that the RGB-D cameras, such as *Kinect*, have great potential for spatial sensing and modeling applications at construction sites. So far, several research studies (Weerasinghe et al. 2012; Escorcia et al. 2012) have been proposed. However, most of them focused on the recognition and tracking of construction workers, when the positions of the RGB-D cameras are fixed.

To the authors' knowledge, none of existing research studies has been focused on modeling as-built building information.

Table 1 Comparison of point cloud based methods and color based methods

Criteria	Point cloud based methods	Color based methods
Sensing Data Properties	Accurate and detailed	Not accurate and detailed
Scene Registration	Point clouds distance	Visual features matching
Building Elements Recognition	One recognition cue only (spatial feature)	Multiple recognition cues (color, texture, shape, etc.)

In addition to the research studies in the AEC domain, researchers in computer vision also investigated the use of RGB-D cameras. For example, (Henry et al. 2010) and (Fioraio and Konolige 2011) examined how a dense 3D map of building environments can be generated with RGB-D cameras (i.e. RGB-D mapping). More impressively, (Izadi et al. 2011) showed that the rapid and detailed reconstruction of building indoor scenes can be achieved; however, their work is currently limited to the reconstruction of a relatively small building scene (3 m cube) (Roth and Vona 2012).

3. OBJECTIVE AND SCOPE

Although the preliminary results from existing RGB-D based methods are promising, the full potentials of the fusion of the spatial and visual data have not been well investigated for capturing, retrieving, and modeling as-built building information. The objective of this paper is to fill this gap, and the focus is placed on how to automate the current process for capturing, retrieving, and modeling as-built building geometry and features. Specifically, an overall fusion-based framework for capturing, retrieving, and modeling as-built building geometry and features has been proposed. The framework mainly includes two parts: 1) 3D scenes registration and 2) 3D building elements recognition. First, the 3D point clouds captured from different scans are progressively registered through the RGB-D mapping. Then, the building elements in the 3D point clouds are recognized based on their visual patterns. The potential and effectiveness of the framework have been tested with two pilot studies. In the studies, the issues such as the accuracy and automation for 3D scene registration and elements recognition have been evaluated. Although the pilot studies use the RGB-D camera, Microsoft® *Kinect*, as the sensing device, it is worth noting that the fusion-based idea behind the proposed framework is also applicable for other stand-alone or combined sensing devices as long as the sensing devices could provide pairs of color and depth images with the one-on-one correspondence.

4. METHODS

As illustrated in Figure 1, the proposed fusion-based framework mainly includes two parts, 3D scene registration and 3D building elements recognition. The scene registration refers to the merge and alignment of the point clouds from multiple scans into one single point cloud under a pre-defined global coordinate system. The registration is always necessary in the process of capturing, retrieving, and modeling as-built building information, since one scan typically cannot capture all as-built conditions in a building environment. Here, the global coordinate system is defined as the coordinate system used by the point cloud in the first scan. The point clouds in other scans are aligned to the cloud in the first scan. The idea behind the RGB-D mapping (Henry et al. 2010) has been adopted to achieve the registration purpose. First, the visual feature detectors are used to determine the positions of the features in the color images. The features are distinctively described with feature descriptors. Then, the common features in consecutive color images. Based on the matching results, the pairs of 3D matching points can be determined by referring to the locations of the matched feature positions in the color images and their corresponding depth values. This way, the point clouds in the scans could be progressively registered.

A point cloud includes a set of 3D points with known coordinates. The coordinates indicate the positions of the points in the building environments without any semantic or high-level geometric as-built building information. In order to retrieve this information, building elements in the environments need to be recognized from the point cloud. This way, the geometries and dimensions of the elements can be estimated and modeled correspondingly. However, the direct recognition of the building elements from the point cloud has proven difficult, especially when the detailed prior information is not available (Bosche and Haas 2008).

Therefore, the visual features of building elements has been exploited here. Under the workflow of building elements recognition, the elements are first recognized in the color images based on their unique visual patterns. The patterns include the topological configurations of the elements' contour and texture features. For example, concrete columns in buildings are dominated by long near-vertical lines (contour features) at sides and concrete surfaces (texture feature) in the middle, no matter they are rectangular or circular. Therefore, they can be located by searching such cues in the color images. When building elements are recognized from the color images, the recognition results could be used to classify the 3D points, so that the points belonging to the same building elements could be grouped and modeled separately.

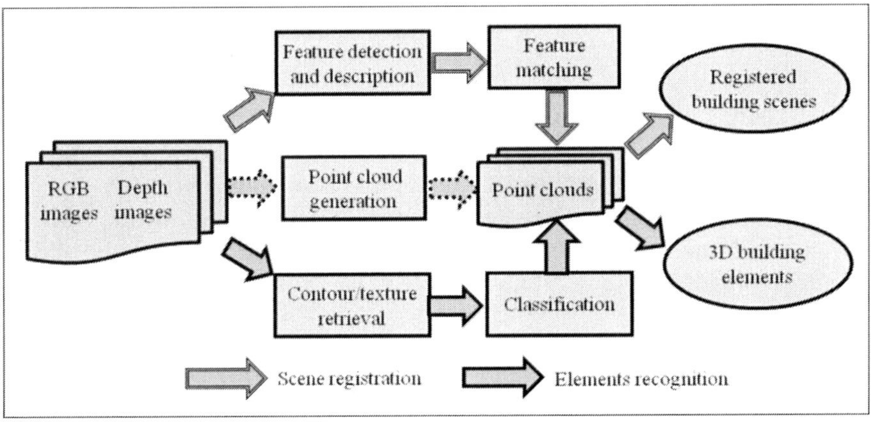

Figure 1. Fusion-based framework.

5. RESULTS

So far, two pilot studies have been performed to show the effectiveness of the proposed framework. In the studies, the sensing data collected by the Microsoft®. *Kinect* have been used to evaluate the framework. The *Kinect* is a novel hybrid sensing system, which can provide a stream of color and depth images with a resolution of 640 × 480 pixels and 2048 levels of depth sensitivity (Biswal 2011). The typical sensing range of the *Kinect* is limited from 1.2 meters to 3.5 meters with an angular field of view of 57° horizontally and 43° vertically (Biswal 2011). According to the report from (Rafibakhsh et al. 2012), the average sensing error of the *Kinect* was around 3.49 cm. The authors also compared the sensing data from the *Kinect* with the data from a Leica total station, and found that the average absolute and percentage errors were 3.60 cm and 2.898% (Zhu and Donia, 2013).

Although the *Kinect* has a limited sensing range and considerably lower accuracy than a high-end terrestrial laser scanner or total station, the main focus of the studies is not to evaluate the potential of the *Kinect* for construction applications. Instead, the emphasis is placed on evaluating the automation of the proposed framework for registering building scenes and recognizing building elements with the idea behind the fusion of visual and spatial data. Therefore, although the *Kinect* has been selected in the studies, other stand-alone or combined devices could be used, as long as they provide pairs of color and depth images. For example, the sensing data captured by the combination of a digital camera and a terrestrial laser scanner are also supported by the proposed framework, if the color images captured by the digital camera are well calibrated with the data captured by the laser scanner.

5.1 Scenes registration

The first pilot study is to test the 3D building scenes registration. Figure 2 shows the detection of the visual features in the color image with the proposed framework, where the ORB (Oriented FAST and Rotated BIEFF) detector and descriptor (Rublee et al. 2011) have been used.

(a) 2D positions of visual features

(b) 3D positions of visual features

Figure 2. 2D and 3D positions of visual features.

In the figure, the green points indicated the visual features detected in the color image, and their corresponding positions in the 3D point cloud could also be located, considering one-on-one correspondence between the pixels in the pair of the color and depth images. Figure 3 illustrates an example of matching common visual features in consecutive color images. The matching of visual features in consecutive color images was built upon the fast approximate nearest neighbor (FANN) searches. The detailed information about the FANN searches could be found in the work of (Muja and Lowe 2012) and (Muja and Lowe 2009), and the specific steps for feature detection, and description could be referred to the work of (Rublee et al. 2011). The matching results could indicate the common 3D points in the consecutive point clouds.

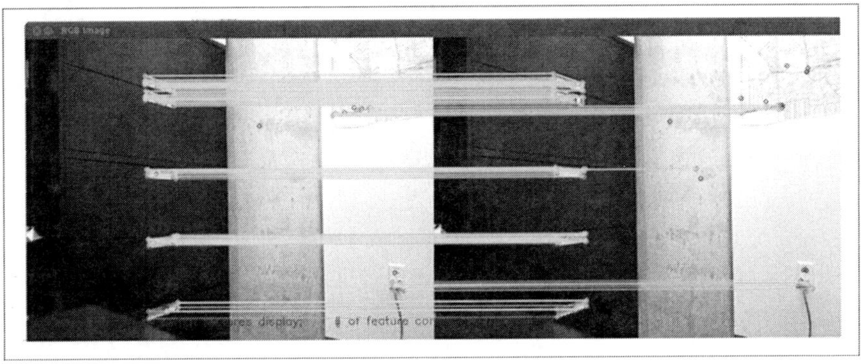

Figure 3. Matching of 2D visual features.

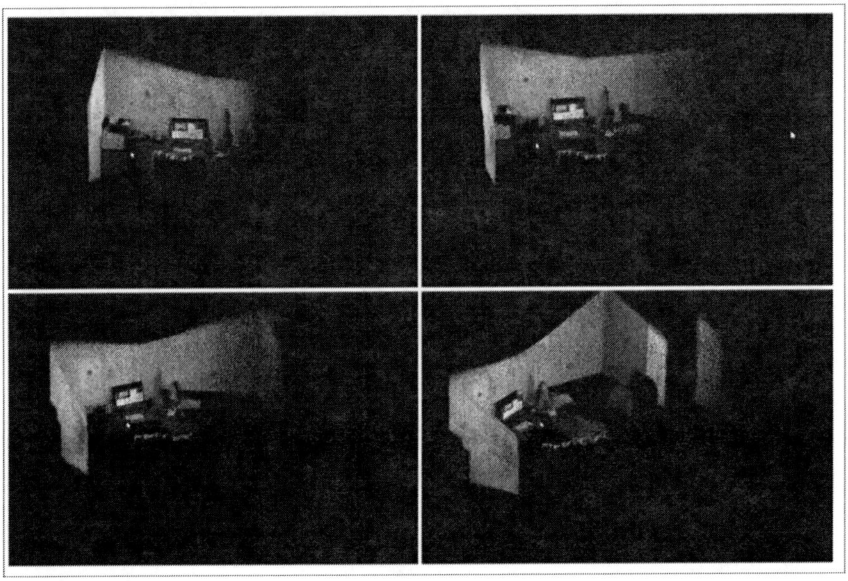

Figure 4. Progressive 3D scenes registration.

This way, the point clouds from different scans could be progressively and automatically registered (Figure 4). The final 3D point cloud after the registration is displayed in Figure 5.

Figure 5. Registered 3D point cloud of an office.

The accuracy of the scenes registration in the proposed framework has been tested with the common datasets prepared by (Sturm et al. 2012) in the Computer Vision Groups at the Technische Universität München. Here, the accuracy is measured by comparing the difference between the Kinect scan positions (i.e. estimated trajectory) with its real scan positions (i.e. ground truth). According to the test results, it was found that the average error of using the ORB detector and descriptor for the scenes registration was around 0.24 meter.

The error may come from the sensing error from the sensing device. For example, the average sensing error of the *Kinect* in one scan was around 3.49 cm (Rafibakhsh et al. 2012). In addition to the sensing error, the automatic pair-wise registration may also produce the registration error, since no explicit control reference points have been used during the registration process.

Such registration errors could not be completely eliminated but accumulated along the registration progress. A recent study indicated that the selection of appropriate visual feature detector and descriptor may significantly reduce the pair-wise registration error (Zhu, 2013). Therefore, more studies are needed to investigate the effectiveness of the color and texture information available in the building environments on the accuracy of the scenes registration.

5.2 Building elements recognition

The second pilot study is to test the 3D building elements recognition. Figures 6 and 7 show an example of 3D points segmentation and clustering with their visual cues. In this example, the color image is first converted from the red, green, and blue (RGB) space to the L*a*b* space, which includes a luminosity layer (L*) and two chromaticity layers (a* and b*). Then, the image is segmented and clustered with the k-means algorithm (Arthur and Vassilvitskii, 2007) based on the information in the two chromaticity layers. The luminosity layer is not considered here in order to remove the effect of the brightness on the segmentation and clustering results. This way, building elements, such as walls, doors, etc., in the color image could be identified (Figure 6). Moreover, the corresponding 3D points belonging to the same types of building elements could be extracted (Figure 7). Compared with the direct segmentation and clustering of 3D point cloud, it can be seen that the use of visual and spatial data could provide a convenient and fully automatic way to segment and cluster 3D point clouds.

(a) Original image (b) Segmentation result

Figure 6. Image segmentation.

Moreover, specific 3D building elements in the point cloud could be further recognized, when considering more visual cues. For example, concrete columns (rectangular or circular) in a color image are dominated by long near-vertical lines (contour cues) and concrete surfaces (color and texture cues). Therefore, they could be located by searching these cues in the color image. This idea of recognizing concrete columns has been implemented by the writer. More details can be found in the writer's pervious work (Zhu and Brilakis, 2010). When such cues are found in the image, the corresponding 2D column surface pixels could be mapped to the 3D points in the cloud. This way, the points belonging to the column surface can be classified and grouped to indicate the concrete column in 3D, as illustrated in Figure 8.

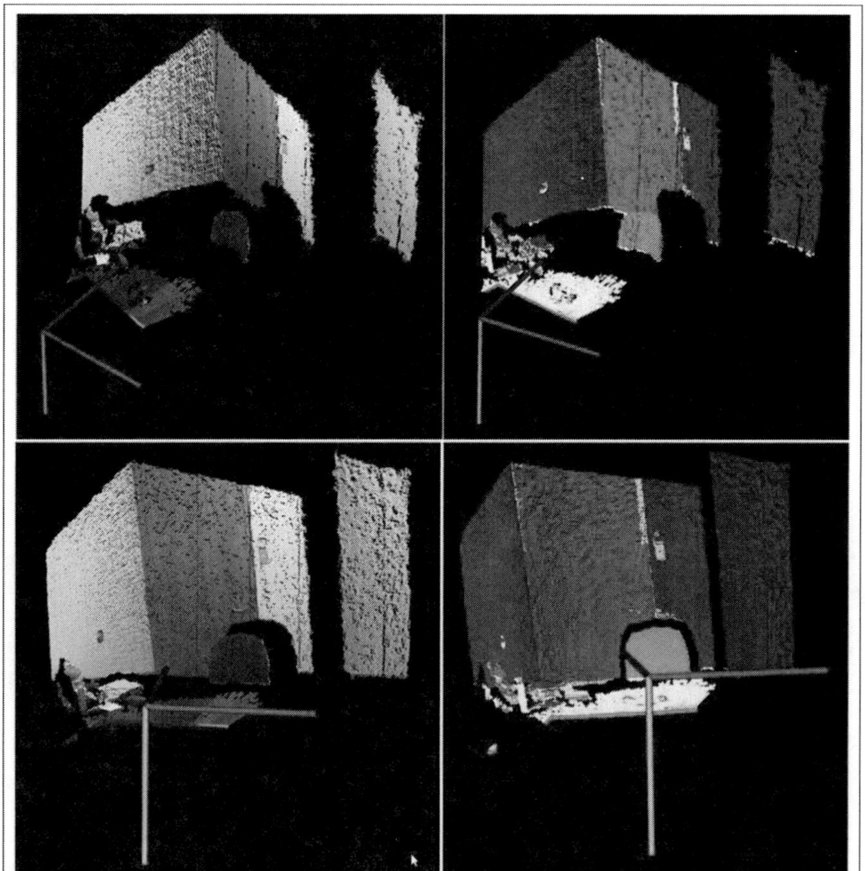

Figure 7. Elements segmentation in 3d point cloud.

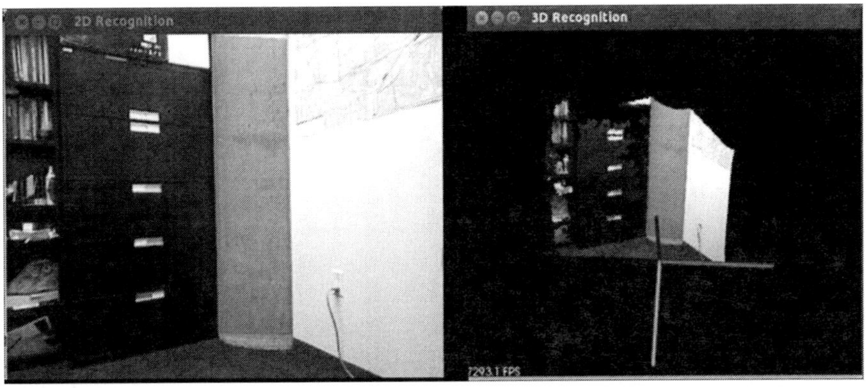

Figure 8. Example of concrete column recognition.

6. DISCUSSION

The main purpose of the two pilot studies is to show the degree of automation that the proposed framework could reach with the idea of fusing the spatial and visual data. The studies are not used to validate that more dense and/or accurate 3D points could be captured with the fusion of spatial and visual data. Although the sensing density and accuracy play an important role in the process of capturing, retrieving, and modeling as-built building information, they are not in the scope of this paper. Instead, the studies are used to indicate that a high degree of automation could be achieved in the current process of capturing, retrieving, and modeling as-built building information with the fusion of spatial and visual data. The automation has been highly valued to promote the use of 3D as-built building information, especially considering the current process of capturing, retrieving, and modeling as-built building information is labor-intensive and time-consuming.

The preliminary results from the two pilot studies have shown that building scenes could be progressively and automatically registered by detecting, describing, and matching visual features. Also, building elements in the environments could be automatically recognized based on their visual cues. In both studies, no specific user interventions are required for setting up the reference/control points in the building environments and editing the sensing data for building scenes registration and building elements recognition. A user just needs to hold the sensing device and keep scanning the building environments. The 3D information of the building environments could be automatically captured and retrieved. Meanwhile, the results could be fed back to the user almost simultaneously for his/her timely review. Compared with existing work in capturing and retrieving as-built building information, to the writers' knowledge, none of existing point cloud based methods or color based methods could reach such a high degree of automation. This is a significant advantage of fusing the spatial and visual data learned from the two pilot studies, and also the contribution of this paper to the existing body of knowledge in the areas of capturing, retrieving, and modeling as-built building information.

7. CONCLUSIONS

3D as-built building information, including building geometry and features, is useful in multiple building assessment and management tasks, but the current process for capturing, retrieving, and modeling as-built builidng information is labor-intensive and time-consuming. This labor-intensive and time-consuming nature limited the wide use of as-built building information in the AEC industry. In order to address this issue, several research studies have been proposed, but they were built upon one single type of sensing data. The sole reliance on one type of sensing data has several limitations.

This paper investigates the fusion of visual and spatial data to facilitate the process of capturing, retrieving, and modeling as-built building information. An overall fusion-based framework has been proposed. Two pilot studies have been performed. The first study tested the automation of building scenes registration with visual feature detection, description, and matching. It was found that the full automation of 3D building scenes registration could be achieved with the fusion of visual and spatial data, but the registration accuracy needs to be further improved. The second study tested the recognition of building elements in the 3D point cloud with their visual recognition cues. It was noted that the use of visual recogniton cues could facilitate the recognition of building elements in 3D.

So far, the proposed fusion-based framework rely on the visual features for 3D building scenes registration and building elements recognition. The precise extraction and description of visual features in the environments play an important role on the success of the proposed framework. That is why the selection of different visual feature extractors and descriptors may affect the scene registration accuracy as indicated in the pilot study 1. However, no matter what visual feature extractor and descriptor are selected, the visual features extraction and description could still be affected by the environmental conditions (e.g. lighting, lack of texture and color information, shadows, etc.), since the visual feature detectors and descriptors currently available are not ideal (Mikolajczyk and Schmid, 2005). Also, these environmental conditions might affect the segmentation and clustering of building scenes for building elements recognition. In the pilot study 2, although the RGB color space was converted to L*a*b* space and the luminosity layer (L*) was not considered in the segmentation and clustering process, the illumination effects on the segmentation and clustering results could still not be completely removed. Therefore, future work will focus on 1) how to increase the robustness of the proposed framework against common environmental conditions (e.g. lighting, lack of texture and color information, shadows, etc.), and 2) how to recognize other types of 3D building elements with different materials.

ACKNOWLEDGEMENT

This paper is based in part upon the work supported by the National Science and Engineering Research Council (NSERC) of Canada. Any opinions, findings, C and conclusions or recommendations expressed in this paper are those of the author(s) and do not necessarily reflect the views of the NSERC.

AUTHORS' CONTRIBUTION

ZZ carried out the spatial and visual data fusion study, participated the sequence alignment and drafted the manuscript, and directed the entire processes of this study. SD undertook the experiments to collect the spatial and visual data using a Kinect camera, and provided suggestions on the data collection and analysis. Both authors read and approved the final manuscript.

REFERENCES

1. Arthur D, Vassilvitskii S: **"K-MEANS++: THE ADVANTAGES OF CAREFUL SEEDING".** *Proceedings of the eighteenth annual ACM-SIAM symposium on discrete algorithms* 2007.
2. Adan A, Huber D: *"3D reconstruction of interior wall surfaces under occlusion and clutter".* Hangzhou, China: Proc. of 3DPVT 2011; 2011.
3. Adan A, Xiong X, Akinci B, Huber D: *"Automatic creation of semantically rich 3D building models from laser scanner data".* Seoul, Korea: Proc. of ISARC 2011; 2011.
4. Biswal S: *"Xbox 360 Kinect review, specification and price".* 2011. Accessed January 10, 2013 http://www.techibuzz.com/xbox-360-kinect-review/ Accessed January 10, 2013
5. Bosche F, Haas C: **AUTOMATED RETRIEVAL OF 3D CAD MODEL OBJECTS IN CONSTRUCTION RANGE IMAGES.** *Auto in Construction*2008,**17**(4):499–512. 10.1016/j.autcon.2007.09.001
6. Brilakis I, German S, Zhu Z: **VISUAL PATTERN RECOGNITION MODELS FOR REMOTE SENSING OF CIVIL INFRASTRUCTURE.** *Journal of Computing in Civil Engineering* 2011,**25**(5):388–393.

7. Brilakis I, Soibelman L, Shinagawa Y: **CONSTRUCTION SITE IMAGE RETRIEVAL BASED ON MATERIAL CLUSTER RECOGNITION.** *Journal of Advanced Engineering Informatics* 2006,**20**(4):443–452.
8. ClearEdge3D: *"3D imaging services reduces MEP workflow on Pharma project by 60% using edgewise plant 3.0"*. 2012. Accessed June 10, 2012 http://www.clearedge3d.com/ Accessed June 10, 2012
9. Durrant-Whyte H, Bailey T: **SIMULTANEOUS LOCALIZATION AND MAPPING.** *IEEE Robotics and Automation Magazine*2006,**13**(2):99–110.
10. Escorcia V, Golparvar-Fard M, Niebles J: *"Automated vision-based recognition of construction worker actions for building interior construction operations using RGBD cameras"*. West Lafayette, Indiana: ASCE Construction Research Congress; 2012:879–888.
11. Fathi H, Brilakis I: *"A videogrammetric as-built data collection framework for digital fabrication of sheet metal roof panels"*. Herrsching, Germany: The 19th EG-ICE Workshop on Intelligent Computing in Engineering; 2012. July 4–6, 2012 July 4–6, 2012
12. Fioraio N, Konolige K: **"REALTIME VISUAL AND POINT CLOUD SLAM".** *Proc. of the RGB-D workshop on advanced reasoning with depth cameras at robotics: Science and Systems Conf. (RSS)* 2011.
13. Foltz B: **APPLICATION: 3D LASER SCANNER PROVIDES BENEFITS FOR PENNDOT BRIDGE AND ROCKFACE SURVEYS.** *Professor Survey*2000,**20**(5):22–28.
14. Furukawa Y, Curless B, Seitz S, Szeliski R: *"Reconstructing building interiors from images"*. Kyoto, Japan: Proc. of ICCV; 2009.
15. Furukawa Y, Ponce J: **ACCURATE, DENSE, AND ROBUST MULTI-VIEW STEREOPSIS.** *IEEE Transactions on Pattern Analysis and Machine Intelligence* 2010,**32**(8):1362–1376.
16. Geiger A, Zeigler J, Stiller C: *Stereoscan: dense 3D reconstruction in real-time*. BadenBaden, Germany: Proc of IEEE Intelligent Vehicles Symposium; 2011.
17. Henry P, Krainin M, Herbst E, Ren X, Fox D: *RGB-D Mapping: using depth cameras for dense 3D modeling of indoor environments*. Delhi, India: Proc. of International Symposium on Experimental Robotics; 2010.
18. Huber D, Kapuria A, Donamukkala R, Hebert M: *"Parts-based 3D object recognition"*. Washington DC: Proc. of CVPR 2004; 2004.
19. Institute for BIM in Canada (IBC): *"Environmental scan of BIM tools and standards"*. Canadian Construction Association; 2011. Accessed Oct. 25 2012. http://www.ibc-bim.ca/documents/Environmental%20Scan%20of%20BIM%20Tools%20and%20Standards%20FINAL.pdf Accessed Oct. 25 2012.
20. Izadi S, *et al.*: *"KinectFusion: real-time 3D reconstruction and interaction using a moving depth camera"*. Santa Barbara, CA: ACM Symposium on User Interface Software & Technology (UIST); 2011.
21. Kazhdan M, Funkhouser T, Rusinkiewicz S: *Proceedings of the 2003 Eurographics/ACM SIGGRAPH symposium on Geometry processing (SGP '03)*. Aire-la-Ville, Switzerland: Eurographics Association; 2003:156–164.
22. Liu X, Eybpoosh M, Akinci B: *"Developing as-built building information model using construction process history captured by a laser scanner and a camera"*. West Lafayette: Construction Research Congress (CRC); 2012:1232–1241. IN., May, 21–23, 2012
23. Mikolajczyk K, Schmid C: **A PERFORMANCE EVALUATION OF LOCAL DESCRIPTORS.** *IEEE Transactions on Pattern Analysis and Machine Intelligence* 2005,**27**(10):1615–1630.
24. Muja M, Lowe D: **"** ***Fast matching of binary features*** **".** In *Proceedings of the 2012 Ninth Conference on Computer and Robot Vision (CRV '12)*. Washington, DC, USA: IEEE Computer Society; 2012:404–410.
25. Muja M, Lowe D **INTERNATIONAL CONFERENCE ON COMPUTER VISION THEORY AND APPLICATIONS.** In *Fast approximate nearest neighbors with automatic algorithm configuration*. Lisboa, Portugal; 2009:331–340.
26. Müller P, Zeng G, Wonka P, Van Gool L: **"IMAGE-BASED PROCEDURAL MODELING OF FACADES".** *ACM Transactions on Graphics*2007, **26:** 1–9. No. 3, Article 85
27. Neto JA, Arditi D, Evens MW: **USING COLORS TO DETECT STRUCTURAL COMPONENTS IN DIGITAL PICTURES.** *Computer Aided Civil Infrastructure English* 2002, **17:** 61–67.

28. Okorn B, Xiong X, Akinci B, Huber D: *"Toward automated modeling of floor plans"*. Paris, France: Proc. of 3DPVT 2010; 2010.
29. Patterson A IV, Mordohai P, Daniilidis K: *"Object detection from large-scale 3D datasets using bottom-up and top-down descriptors"*. Marseille, France: Proc. of ECCV 2008; 2008.
30. Pettee S **CM EJOURNAL, FIRST QUARTER: 1–19.** *"As-builts - problems and proposed solutions"* 2005. (June 10, 2012) http://cmaanet.org/files/as-built.pdf (June 10, 2012)
31. Pollefeys M, Nister D, Frahm JM, Akbarzadeh A, Mordohai P, Clipp B, Engels C, Gallup D, Kim SJ, Merrell P, Salmi C, Sinha S, Talton B, Wang L, Yang Q, Stewenius H, Yang R, Welch G, Towles H: **DETAILED REAL-TIME URBAN 3D RECONSTRUCTION FROM VIDEO.** *International Journal of Computer Vision* 2008,**78**(2):143–167.
32. Rafibakhsh N, Gong J, Siddiqui M, Gordon C, Lee H: *"Analysis of XBOX Kinect sensor data for use on construction sites: depth accuracy and sensor interference assessment"*. West Lafayette: Constitution Research Congress; 2012:848–857. IN. May 21–23
33. Ripperda N, Brenner C: *Application of a formal grammar to facade reconstruction in semiautomatic and automatic environments*. Hannover, Germany: Proc. of AGILE conference on geographic information science; 2009.
34. Roth H, Vona M: *"Moving volume KinectFusion"*. Surrey, UK: British Machine Vision Conf. (BMVC); 2012. September 2012
35. Rublee E, Rabaud V, Konolige K, Bradski G: *"ORB: an efficient alternative to SIFT or SURF"*. Barcelona, Spain: Proc. of ICCV 2011; 2011:2564–2571.
36. Shan Y, Sawhney H, Matei B, Kumar R: **SHAPEME HISTOGRAM PROJECTION AND MATCHING FOR PARTIAL OBJECT RECOGNITION.***IEEE TransTransactions on Pattern Anal Mach IntellAnalysis and Machine Intelligence* 2006,**28**(4):568–577.
37. Snavely N, Seitz S, Szeliski R: *"Photo tourism: exploring image collections in 3D"*. MA, Boston: Proc of SIGGRAPH 2006.
38. Stiene S, Lingemann K, Nuchter A, Hertzberg J: *"Contour-based object detection in range images"*. Chapel-Hill, NC: Proc. of 3DPVT 2006.
39. Sturm J, Engelhard N, Endres F, Burgard W, Cremers D: **"A BENCHMARK FOR THE EVALUATION OF RGB-D SLAM SYSTEMS".***Proc. of the International Conference on Intelligent Robot Systems (IROS)* 2012. Oct. 2012
40. Tang P, Huber D, Akinci B, Lipman R, Lytle A: **AUTOMATIC RECONSTRUCTION OF AS-BUILT BUILDING INFORMATION MODELS FROM LASER-SCANNED POINT CLOUDS: A REVIEW OF RELATED TECHNIQUES.** *Auto in Construction* 2010,**19**(7):829–843.
41. Wang D, Zhang J, Wong H, Li Y **LECTURE NOTES IN COMPUTER SCIENCE: ADVANCES IN VISUAL INFORMATION SYSTEMS.** *"3D model retrieval based on multi-shell extended gaussian image"* 2007, 426–437.
42. Weerasinghe I, Ruwanpura J, Boyd J, Habib A: *"Application of microsoft kinect sensor for tracking construction workers"*. West Lafayette, Indiana: ASCE Construction Research Congress (CRC); 2012:858–867.
43. Xiong X, Huber D: *"Using context to create semantic 3D models of indoor environments"*. Aberystwyth, UK: Proc. of BMVC 2010.
44. Zhu Z, Brilakis I: **"CONCRETE COLUMN RECOGNITION IN IMAGES AND VIDEOS".** *Journal of Computing in Civil Engineering, American Society of Civil Engineers* 2010,**24**(6):478–487. November/December 2010
45. Zhu Z, Donia S: *"Potentials of RGB-D cameras in as-built indoor environments modeling"*. Los Angeles, CA: 2013 ASCE International Workshop on Computing in Civil Engineering; 2013. June 23–25, 2013
46. Zhu Z: **"EFFECTIVENESS OF VISUAL FEATURES IN AS-BUILT INDOOR ENVIRONMENTS MODELING".** In *13th International Conference on Construction Applications of Virtual Reality*. London, UK; 2013. October, 30-31 2013 (accepted)

CHAPTER 8

Positive-Definite Sparse Precision Matrix Estimation

Lin Xia[1], Xudong Huang[1], Guanpeng Wang[1], Tao Wu[2]

[1]*School of Mathematics and Computer Science, Anhui Normal University, Wuhu, China*

[2]*School of Mathematics and Statistics, Huazhong University of Science and Technology, Wuhan, China*

ABSTRACT

The positive-definiteness and sparsity are the most important property of high-dimensional precision matrices. To better achieve those property, this paper uses a sparse lasso penalized D-trace loss under the positive-definiteness constraint to estimate high-dimensional precision matrices. This paper derives an efficient accelerated gradient method to solve the challenging optimization problem and establish its converges rate as $O\left(\frac{1}{k^2}\right)$. The numerical simulations illustrated our method have competitive advantage than other methods.

Keywords: Positive-Definiteness, Sparsity, D-Trace Loss, Accelerated Gradient Method

1. INTRODUCTION

In the past twenty years, the most popular direction of statistics is high-dimensional data. In functional magnetic resonance imaging (FMRI), bioinformatics, Web mining, climate research, risk management and social science, it not only has a wide range of applications, but also the main direction of scientific research at present. In theoretical and practical, high-dimensional precision matrix estimation always plays a very important role and has wide applications in many fields.

Thus, estimation of high-dimensional precision matrix is increasingly becoming a crucial question in many field. However, estimation of high-dimensional precision matrix has two difficulty: 1) sparsity of estimator; (ii) the positive-definiteness constraint. Huang et al. [1] considered using Cholesky decomposition to estimate the precision matrix. Although the regularized Cholesky decomposition approach can achieve a positive-semidefiniteness, it can not guarantee sparsity of estimator. Meinshausen et al. [2] use a neigh- bourhood selection scheme in which one can sequentially estimate the support of each row of precision matrix by fitting lasso penalized least squares regression model. Peng et al. [3] considered a joint neighbourhood estimator by using the lasso penalization. Yuan [4] considered the Dantzig selector to replace the lasso penalized least squares in the neighbourhood selection scheme. Cai et al. [5] considered a constrained ℓ_1 minimization estimator for estimating sparse precision matrices. However, this methods mentioned are not always achieve a positive-definiteness.

To overcome the difficulty (ii), one possible method is using the eigen-decomposition of $\hat{\Theta}$ and designing $\hat{\Theta}$ to satisfy condition $\{\Theta \geq 0\}$. Assume that $\hat{\Theta}$ has the eigen-decomposition $\hat{\Theta} = \sum_{i=1}^{p} \lambda_i v_i^T v_i$ and then a positive semi-definite estimator was gained by setting $\hat{\Theta} = \sum_{i=1}^{p} \max(\lambda_i, 0) v_i^T v_i$. However, this strategy destroys the sparsity pattern of $\hat{\Theta}$ for sparse precision matrix estimation. Yuan et al. [6] considered the lasso penalized likelihood criterion and used the maxd et al. gorithm to compute the estimator. Friedman et al. [7] considered the graphical lasso algorithm for solving the lasso penalized Gaussian likelihood estimator. Witten et al. [8] optimized the graphical lasso. By the lasso or ℓ_1 penalized Gaussian likelihood estimator, thoses methods simultaneously achieve positive-definiteness and sparsity.

Recently, Zhang et al. [9] consider a constrained convex optimization frame-work for high-dimensional precision matrix. They used lasso penalized D-trace loss replace traditional lasso function, and enforced the positive-definite constraint $\{\Theta \geq \varepsilon I\}$ for some arbitrarily small $\varepsilon > 0$. In their work, focusing on solving problem as follow:

$$\hat{\Theta}^{+} = \arg\min_{\Theta \geq \varepsilon I} \frac{1}{2}\left\langle \Theta^2, \hat{\Sigma}_n \right\rangle - \mathrm{tr}(\Theta) + \lambda \|\Theta\|_{1,\mathrm{off}} \tag{1}$$

It is important to note that ε is not a tuning parameter like λ. We simply include ε in the procedure to ensure that the smallest eigenvalue of the estimator is at least ε. They developed an efficient alternating direction method of multipliers (ADMM) to solve the challenging optimization problem (1) and establish its convergence properties.

To gain a better estimator for high-dimensional precision matrix and achieve the more optimal convergence rate, this paper mainly propose an effective algorithm, an accelerated gradient method ([10]), with fast global convergence rates to solve problem (1). This method mainly basis on the Nesterov's method for accelerating the gradient method ([11] [12]), showing that by exploiting the special structure of the trace norm, the classical gradient method for smooth problems can be adapted to solve the trace regularized nonsmooth problems. However, for our problem (1), we have not trace norm, instead of is ℓ_1 norm form, but this method have the similar efficiently result for our problem. Numerical results show that this method for our problem (1) not only has significant computational advantages, but also achieves the optimal convergence rate as $O\left(\frac{1}{k^2}\right)$.

The paper is organized as follows: Section 2 introduces our methodology, including model establishing in Section 2.1; step size estimation in Section 2.2; an accelerate gradient method algorithm in Section 2.3; the convergence analysis results of this algorithm in Section 2.4. Section 3 introduced numerical results for our method in comparing with other methods. And discussion are made in Section 4. All proofs are given in the Appendix.

2. AN ACCELERATED GRADIENT METHOD

2.1. Model Establishing

According to introduction, our optimization problem D-trace Loss function as follow:

$$\min_{\Theta \geq \varepsilon I} F(\Theta) := \frac{1}{2}\left\langle \Theta^2, \hat{\Sigma}_n \right\rangle - \mathrm{tr}(\Theta) + \lambda \|\Theta\|_{1,\mathrm{off}} \tag{2}$$

where λ is a nonnegative penalization parameter, $\hat{\Sigma}_n$ is the sample covariance matrix. $\hat{\Sigma}_n = \frac{1}{n}\sum_{i=1}^{n} X_i X_i^T$, and $|\Theta|_1 = \|\Theta\|_{1,off} = \sum_{i \neq j} \|\Theta_{ij}\|$ is the ℓ_1 off-diagonal penalty. Defining $f(\Theta) = \frac{1}{2}\langle \Theta^2, \hat{\Sigma}_n \rangle - \mathrm{tr}(\Theta)$, and $f(\Theta)$ is a continuously differentiable function. Considering the gradient step

$$\Theta_k = \Theta_{k-1} - \frac{1}{t_k} \nabla f(\Theta_{k-1}) \tag{3}$$

where $t_k \geq 0$ is a stepsize, $\nabla f(\Theta) = \frac{1}{2}(\Theta \hat{\Sigma}_n + \hat{\Sigma}_n \Theta) - I$. The smooth part (3) can be reformulated equivalent as a proximal regularization of the linearized function $f(\Theta)$ at Θ_{k-1}:

$$\Theta_k = \arg\min_{\Theta \geq \varepsilon I} \Phi_{t_k}(\Theta, \Theta_{k-1}) \tag{4}$$

where

$$\Phi_{t_k}(\Theta, \Theta_{k-1}) = f(\Theta_{k-1}) + \langle \Theta - \Theta_{k-1}, \nabla f(\Theta_{k-1}) \rangle + \frac{t_k}{2} \|\Theta - \Theta_{k-1}\|_F^2 \tag{5}$$

Based on this equivalence relationship, solving the optimization problem (2) by the following iterative step:

$$\begin{aligned}
\Theta_k &= \arg\min_{\Theta \geq \varepsilon I} \Psi_{t_k}(\Theta, \Theta_{k-1}) \square \arg\min_{\Theta \geq \varepsilon I} \Phi_{t_k}(\Theta, \Theta_{k-1}) + \lambda |\Theta|_1 \\
&= \arg\min_{\Theta \geq \varepsilon I} \frac{1}{2}\langle \Theta_{k-1}^2, \hat{\Sigma}_n \rangle - \mathrm{tr}(\Theta_{k-1}) + \Big\langle \Theta - \Theta_{k-1}, \frac{1}{2}(\Theta_{k-1}\hat{\Sigma}_n + \hat{\Sigma}_n \Theta_{k-1}) \\
&\quad - I \Big\rangle + \frac{t_k}{2} \|\Theta - \Theta_{k-1}\|_F^2 + \lambda |\Theta|_1 \\
&= \arg\min_{\Theta \geq \varepsilon I} \frac{t_k}{2} \left\| \Theta - \left[\Theta_{k-1} - \frac{1}{t_k}\left(\frac{1}{2}(\Theta_{k-1}\hat{\Sigma}_n + \hat{\Sigma}_n \Theta_{k-1}) - I \right) \right] \right\|_F^2 + \lambda |\Theta|_1
\end{aligned} \tag{6}$$

with equality in the last line by ignoring terms that do not depend on Θ.

Defining $(C)_+$ as the projection of a matrix C onto the convex cone $\{C \geq \varepsilon I\}$. Assuming that C has the eigen-decomposition $\sum_{j=1}^{p} \lambda_j v_j^T v_j$, and then $(C)_+$ can be obtained as $\sum_{j=1}^{p} \max(\lambda_j, \varepsilon) v_j^T v_j$. Defining an entry-wise soft-thresholding rule for all the off-diagonal elements of a matrix

$$S(Z,\tau) = \{s(z_{j\ell},\tau)\}_{1\leq j,\ell\leq p} \text{ with}$$

$$s(z_{j\ell},\tau) = \operatorname{sign}(z_{j\ell})\max(|z_{j\ell}| - \tau, 0) \mathrm{I}_{\{j\neq\ell\}} + z_{j\ell}\mathrm{I}_{\{j=\ell\}}$$

Thus, the above problem can be summarized in the following theorem:

Theorem 1: Let $\mathrm{B} \in \mathbb{R}^{n\times n}$, and Θ is symmetric covariance matrix, then:

$$S(\mathrm{B}) = \arg\min_{\Theta\geq\varepsilon I} \left\{ \frac{1}{2}\|\Theta - \mathrm{B}\|_F^2 + \lambda|\Theta|_1 \right\} \tag{7}$$

is given by $S(\mathrm{B}) = (S(\mathrm{B},\lambda))_+$, where $S(\mathrm{B},\lambda) = \{s(\mathrm{B}_{j\ell},\lambda)\}_{1\leq j,l\leq p}$ with $s(\mathrm{B}_{j\ell},\lambda) = \operatorname{sign}(\mathrm{B}_{j\ell})\max(|\mathrm{B}_{j\ell}| - \lambda, 0)\ \mathrm{I}_{\{j\neq\ell\}} + \mathrm{B}_{j\ell} I_{\{j=\ell\}}$.

The proof of this theorem is easy by applying the soft-thresholding method.

2.2. Step Size Estimation

To guarantee the convergence rate of the resulting iterative sequence, Firstly giving the relationship between our proximal function Ψ_{t_k} and the objection function F at the certain point.

Lemma 1: Let

$$T_\mu(\tilde{\Theta}) = \arg\min_{\Theta\geq\varepsilon I} \Psi_\mu(\Theta, \tilde{\Theta}) \tag{8}$$

where Ψ is defined in Equation (6). Assuming the following inequality holds:

$$F\left(T_\mu\left(\tilde{\Theta}\right)\right) \leq \Psi_u\left(T_\mu\left(\tilde{\Theta}\right), \tilde{\Theta}\right) \tag{9}$$

then for any $\Theta \in \mathbb{R}^{n \times n}$, then:

$$F(\Theta) - F\left(T_\mu\left(\tilde{\Theta}\right)\right) \geq \frac{\mu}{2}\left\|T_\mu\left(\tilde{\Theta}\right) - \tilde{\Theta}\right\|_F^2 + \mu\left\langle \tilde{\Theta} - \Theta, T_\mu\left(\tilde{\Theta}\right) - \tilde{\Theta}\right\rangle \tag{10}$$

This lemma is proved in the Appendix.

At each iterative step of the algorithm, an appropriate step size μ is needed to satisfy $\Theta_k = T_u\left(\Theta_{k-1}\right)$ and

$$F\left(\Theta_k\right) \leq \Psi_\mu\left(\Theta_k, \Theta_{k-1}\right) \tag{11}$$

Since the gradient of f(·) satisfies Lipschitz continuous, according to Nesterov et al. [11] work, having follow lemma.

Lemma 2: Supposing that $f(X)$ is a convex function, and the gradient of $f(X)$ denote $\nabla f(X)$ is Lipschitz continuous with constant L, then:

$$f(X) \leq f(Y) + \left\langle X - Y, \nabla f(Y)\right\rangle + \frac{L}{2}\|X - Y\|_F^2 \quad \forall X, Y \in \square^n \tag{12}$$

so, we have

$$f\left(T_L\left(\tilde{\Theta}\right)\right) \leq f\left(\tilde{\Theta}\right) + \left\langle T_L\left(\tilde{\Theta}\right) - \tilde{\Theta}, \nabla f\left(\tilde{\Theta}\right)\right\rangle + \frac{L}{2}\left\|T_L\left(\tilde{\Theta}\right) - \tilde{\Theta}\right\|_F^2 \tag{13}$$

Hence, when $\mu \geq L$, then:

$$F\left(T_\mu\left(\tilde{\Theta}\right)\right) \leq \Phi_\mu\left(T_\mu\left(\tilde{\Theta}\right), \tilde{\Theta}\right) + \lambda\left|T_\mu\left(\tilde{\Theta}\right)\right|_1 = \Psi_\mu\left(T_\mu\left(\tilde{\Theta}\right), \tilde{\Theta}\right) \tag{14}$$

The above results show that the condition in Equation (11) is always satisfied when the update rule

$$\Theta_k = T_L\left(\Theta_{k-1}\right) \tag{15}$$

2.3. An Accelerate Gradient Method Algorithm

In practice, L may be unknown or it is expensive to compute. To use the following step size estimation method, usually, giving an initial estimate of L as L_0 and increasing this estimate with a multiplicative factor $\gamma > 1$ repeatedly until the condition in Equation (11) is satisfied. It is well known ([11] [12]) that if the objection function is smooth, then the accelerate gradient method can achieve the optimal convergence rate of $O\left(\frac{1}{k^2}\right)$. Recently, there have other similar methods applying in problems consisted a smooth part and a non-smooth part ([10] [13] [14] [15]). Then giving the accelerate gradient algorithm to solve the optimization problem in Equation (2).

Algorithm1:An accelerate gradient method algorithm for high-dimensional precision matrix

1) Initialize: L_0, γ, $\tilde{\Theta}_1 = \Theta_0 \in \mathbb{R}^{n\times n}$, $\alpha_1 = 1$

2) Iterate:

3) set $\bar{L} = L_{k-1}$

4) While $F\left(T_{\bar{L}}\left(\tilde{\Theta}_{k-1}\right)\right) > \Psi_{\bar{L}}\left(T_{\bar{L}}\left(\tilde{\Theta}_{k-1}\right), \tilde{\Theta}_{k-1}\right)$, set $\bar{L} := \gamma \bar{L}$

5) Set $L_k = \bar{L}$ and update

$$\Theta_k = T_{L_k}\left(\tilde{\Theta}_k\right)$$

$$\alpha_{k+1} = \frac{1 + \sqrt{1 + 4\alpha_k^2}}{2}$$

$$\tilde{\Theta}_{k+1} = \Theta_k + \left(\frac{\alpha_k - 1}{\alpha_{k+1}}\right)\left(\Theta_k - \Theta_{k-1}\right)$$

2.4. Convergence Analysis

In our method, two sequences Θ_k and $\tilde{\Theta}_k$ are updated recursively. In particular, Θ_k is the approximate solution at the kth step and $\tilde{\Theta}_k$ is called the search point ([11] [12]), which is constructed as a linear combination of the latest two approximate solutions Θ_{k-1} and Θ_{k-2}.

In this section, the convergence rate of the method can be showed as $O\left(\frac{1}{k^2}\right)$. This result is sum-

marized in the following theorem.

Theorem 2: Let $\{\Theta_k\}$ and $\{\tilde{\Theta}_k\}$ be the covariance matrices sequence generated by our algorithm. Then for any $k \geq 1$, having

$$F(\Theta_k)-F(\Theta^*)\leq\frac{2\gamma L\|\Theta_0-\Theta^*\|_F^2}{(k+1)^2} \tag{16}$$

where $\Theta^* = \underset{\Theta \geq \varepsilon I}{\arg\min} F(\Theta)$.

3. SIMULATION

In this section, providing numerical results for our algorithm which will show our algorithmic advantages by three model. In the simulation study, data were generated from $N(0,\Sigma^0)$, where $\Theta^0 = (\Sigma^0)^{-1}$.

And the sample size was taken to be n = 400 in all models, and let p =500 in Models 1 and 2, and p = 484 in Model 3, which is similar to Zhang et al. [9] . The numerical results of three models as follow:

Model 1: $\Theta^0_{i,i} = 1, \Theta^0_{i,j} = 0.2$ for $1 \leq |i-j| \leq 2$ and $\Theta^0_{i,j} = 0$ otherwise.

Model 2: $\Theta^0_{i,i} = 1, \Theta^0_{i,j} = 0.2$ for $1 \leq |i-j| \leq 4$ and $\Theta^0_{i,j} = 0$ otherwise.

Model 3: $\Theta^0_{i,i} = 1, \Theta^0_{i,i+1} = 0.2$ for $\mathrm{mod}\left(i, p^{1/2}\right) \neq 0$, $\Theta^0_{i,i+p^{1/2}} = 0.2$ and $\Theta^0_{i,j} = 0$ otherwise; this is the grid model in Ravikumar et al. [16] and requires $p^{1/2}$ to be an integer.

Simulation results based on 100 independent replications are showed in Table 1. This paper mainly compare the three methods in terms of four quantities: the operator risk $\mathrm{E}\left(\left\|\hat{\Theta} - \Theta^0\right\|_2\right)$, the matrix $\ell_{1,\infty}$ risk $\mathrm{E}\left(\left\|\hat{\Theta} - \Theta^0\right\|_{1,\infty}\right)$, and the percentages of correctly estimated nonzeros and zeros (TP and TN), where $\ell_{1,\infty}$ norm $\max_i\left(\sum_j\left|X_{i,j}\right|\right)$ is written as $\left\|X\right\|_{1,\infty}$. In the first two columns smaller numbers are better; in the last two columns larger numbers are better. In general, Table 1 shows that our estimator performs better than Zhang et al.'s method estimator and the lasso penalized Gaussian likelihood estimator.

Table 1. Comparison of our method with Zhang et al.'s method and graphical lasso.

	Operator	$\ell_{1,\infty}$	TP	TN
		Model 1		
our method	0.77	0.80	94.91	99.20
Zhang *et al.*'s method	0.77	1.06	88.80	98.77
Graphical lasso	0.78	1.26	88.12	97.65
		Model 2		
our method	1.59	1.60	70.02	98.40
Zhang *et al.*'s method	1.59	1.92	63.47	98.66
Graphical lasso	1.61	2.11	64.88	97.40
		Model 3		
our method	0.56	0.81	1	99.21
Zhang *et al.*'s method	0.56	0.91	99.41	98.57
Graphical lasso	0.58	1.06	99.76	97.48

4. CONCLUSION

This paper mainly estimate positive-definite sparse precision matrix estimation via lasso penalized D-trace loss by an efficient accelerated gradient method. The positive-definiteness and sparsity are the most important property of large covariance matrices, our method not only efficiently achieves these property, but also shows an better convergence rate. Numerical results have show that our estimator also have a better performance, comparing to Zhang et al.'s method and the Graphical lasso method.

FUNDING

This project was supported by National Natural Science Foundation of China (71601003) and the National Statistical Scientific Research Projects (2015LZ54).

REFERENCES

1. Huang, J., Liu, N., Pourahmadi, M. and Liu, L. (2006) Covariance Matrix Selection and Estimation via Penalised Normal Likelihood. Biometrika, 93, 85-98.
2. Meinshausen, N. and Bühlmann, P. (2006) High-Dimensional Graphs and Variable Selection with the Lasso. Annals of Statist, 34, 1436-1462.
3. Peng, J., Wang, P., Zhou, N. and Zhu, J. (2009) Partial Correlation Estimation by Joint Sparse Regression Models. Journal of the American Statistical Association, 104, 735-746.
4. Yuan, M. (2010) High Dimensional Inverse Covariance Matrix Estimation via Linear Programming. Journal of Machine Learning Research, 11, 2261-2286.
5. Cai, T., Liu, W. and Luo, X. (2011) A Constrained Minimization Approach to Sparse Precision Matrix Estimation. Journal of the American Statistical Association, 106, 594-607.
6. Yuan, M. and Lin, Y. (2007) Model Selection and Estimation in the Gaussian Graphical Model. Biometrika, 94, 19-35.
7. Friedman, J.H., Hastie, T.J. and Tibshirani, R.J. (2008) Sparse Inverse Covariance Estimation with the Graphical Lasso. Biostatistics, 9, 432-441.
8. Witten, D., Frienman, J.H. and Simon, N. (2011) New Insights and Faster Computations for the Graphical Lasso. Journal of Computational and Graphical Statistics, 20, 892-900.
9. Zhang, T. and Zou, H. (2014) Sparse Precision Matrix Estimation via Lasso Penalized D-Trace Loss. Biometrika, 101, 103-120.
10. Ji, S. and Ye, J. (2009) An Accelerated Gradient Method for Trace Norm Minimization. International Conference on Machine Learning, 58, 457-464.
11. Nesterov, Y. (1983) A Method for Solving a Convex Programming Problem with Convergence Rate $O(1/K^2)$. Soviet Mathematics Doklady, 27, 372-367.
12. Nesterov, Y. (2003) Introductory Lectures on Convex Optimization: A Basic Course. Kluwer Academic.

13. Toh, K.C. and Yun, S. (2010) An Accelerated Proximal Gradient Algorithm for Nuclear Norm Regularized Linear Least Squares Problems. Pacific Journal of Optimization, 6, 615-640.
14. Tseng, P. (2008) On Accelerated Proximal Gradient Methods for Convex-Concave Optimization. Siam Journal on Optimization.
15. Beck, A. and Teboulle, M. (2009) A Fast Iterative Shrinkage-Thresholding Algorithm for Linear Inverse Problems. Siam Journal on Imaging Sciences, 2, 183-202.
16. Ravikumar, P., Wainwrighr, M., Rasutti, G. and Yu, B. (2011) High-Dimensional Covariance Estimation by Minimizing l_1-Penalized Log-Determinant Divergence. Electronic Journal of Statistics, 5, 935-980.

CHAPTER 9

Helicoidal Surfaces and Their Relationship to Bonnet Surfaces

Paul Bracken

Department of Mathematics, University of Texas, Edinburg, TX, USA

ABSTRACT

An important question that arises is which surfaces in three-space admit a mean curvature preserving isometry which is not an isometry of the whole space. This leads to a class of surface known as a Bonnet surface in which the number of noncongruent immersions is two or infinity. The intention here is to present a proof of a theorem using an approach which is based on differential forms and moving frames and states that helicoidal surfaces necessarily fall into the class of Bonnet surfaces. Some other results are developed in the same manner.

Keywords: Surface, Fundamental Forms, Structure Equations, Mean Curvature, Bonnet, Helicoidal

1. INTRODUCTION

Surfaces that admit isometries which preserve principal curvatures have been studied since the time of O. Bonnet [1] . It was shown by Bonnet that all surfaces with constant mean curvature not including planes and spheres, can be isometrically deformed while preserving mean curvature or equivalently both principal curvatures. Let Σ denote the classes of smooth oriented, connected surfaces carrying a Riemannian metric which will be studied here. Let $H:\Sigma\to\mathbb{R}$ be a given smooth function, then depending on the surface and the function H, it is possible to find isometric immersions of Σ into $\mathbb{R}^3$ such that each image has mean curvature function H. It may be asked how many geometrically distinct, or noncongruent, immersions exists. This question has been studied at various levels both locally and globally. It has been concluded that the number of noncongruent immersions which is denoted here by N, may be $0,1,2$ and ∞. The case in which $N=\infty$ means there exists a one-parameter family of pairwise noncongruent isometric immersions with the same mean curvature function. When $N=2$ or ∞, these surfaces are the so-called Bonnet surfaces. When $N=0$ or 1, there are many well-known surfaces, so all of these numbers can be realized.

Some of the global results that have been reported up to this point should be reviewed first. 1) If H is constant and there is an isometric immersion of Σ into $\mathbb{R}^3$ of mean curvature H which is not a plane or sphere, then there is an isometric deformation of Σ through noncongruent surfaces with the same constant mean curvature, H. 2) If Σ is compact and H is nonconstant, there exists at most two geometrically distinct immersions of Σ in $\mathbb{R}^3$ with mean curvature H, so $N=0,1$ or 2. If Σ is homeomorphic to S^2, the 2-sphere, there is at most one isometric immersion, and N is 0 or 1. 3) If Σ is a helicoidal surface in $\mathbb{R}^3$, then $N=2$ or ∞. 4) If the Gaussian curvature K of Σ is a nonzero constant, and the mean curvature H is nonconstant, then $N=0,1$ or 2.

The surfaces in Euclidean space that admit a mean curvature preserving isometry which is not an isometry of the whole space form a special class of surface which has been studied by many people such as noted already Bonnet as well as Cartan and Chern [2] [3] [4] . These surfaces may be broken up into three classes or types which can be described as follows: 1) There are surfaces of constant mean curvature other than the plane or sphere 2) There are certain surfaces of nonconstant mean curvature which admit a one-parameter family of geometrically distinct nontrivial isometries, and finally 3) There are surfaces of nonconstant mean curvature that admit a single nontrivial isometry which is unique up to an isometry of the entire space.

A surface that belongs to one of the above types is called a Bonnet surface, that is, an $N=2,\infty$ type mentioned above [5] [6] [7] . By a nontrivial isometry of the surface is meant an isometry of the entire space. A helicoidal surface in Euclidean three-space E^3 is the locus of an appropriately chosen curve under a helicoidal motion, with so-called pitch in the interval $(0,\infty)$ [8] [9] . Such a motion can be described by a one-parameter group of isometries in E^3. The actual orbits of the motion through the initial curve foliate the surface.

The intention here is to prove that the helicoidal surfaces are necessarily Bonnet surfaces, and moreover represent all three types of surface outlined above. Although not all the theorems presented here are new, the objective is to present new proofs based on the systematic use of differential forms and the moving frame concept [7] . This type of result is useful to have since it provides an answer, in the negative, to conjectures such as the following: Let S be a Riemannian surface and $H : S \to \mathbb{R}$ a smooth function. If a nontrivial family of isometric immersions with mean curvature function H does not exist, then there must be at most two noncongruent ones. Then it may be conjectured: In the absence of a nontrivial family, the immersion must be unique. On the other hand, it seems that not all Bonnet surfaces of the third type are helicoidal surfaces. A helicoidal surface is determined by one real-valued function of one variable, whereas a Bonnet surface of the third kind depends on four functions of one variable and therefore has a greater degree of generality [10] .

2. STRUCTURE EQUATIONS

Over Σ there exists a system of orthonormal frames $\{x, e_1, e_2, e_3\}$ which is well defined such that $x \in \Sigma$, e_3 is the unit normal at x and e_1, e_2 located along principal directions. The fundamental equations for a surface have the form [11] ,

$$dx = \omega_1 e_1 + \omega_2 e_2, \quad de_1 = \omega_{12} e_2 + \omega_{13} e_3, \quad de_2 = -\omega_{12} e_1 + \omega_{23} e_3, \quad de_3 = -\omega_{13} e_1 - \omega_{23} e_2. \tag{2.1}$$

These equations can be differentiated exteriorly in turn and results in a large system of equations for the exterior derivatives of the ω_i and ω_{ij}, as well as a final equation which relates some of the forms. This choice of frame and Cartan's lemma allows for the introduction of the two principal curvatures at x which are denoted by a and c by writing

$$\omega_{12} = h\omega_1 + k\omega_2, \qquad \omega_{13} = a\omega_1, \qquad \omega_{23} = c\omega_2. \tag{2.2}$$

It suffices to suppose that $a > c$ in the following and the mean curvature of Σ will be denoted by H and the Gaussian curvature is denoted by K. They are defined in terms of the functions a and c to be

$$H = \frac{1}{2}(a+c), \qquad K = a \cdot c. \tag{2.3}$$

The forms which appear in (2.1) satisfy the fundamental set of structure equations

$$\begin{aligned} d\omega_1 &= \omega_{12} \wedge \omega_2, \quad d\omega_2 = \omega_1 \wedge \omega_{12}, \\ d\omega_{13} &= \omega_{12} \wedge \omega_{23}, \quad d\omega_{23} = \omega_{13} \wedge \omega_{12}, \\ d\omega_{12} &= ac\,\omega_2 \wedge \omega_1 = -K\omega_1 \wedge \omega_2. \end{aligned} \tag{2.4}$$

The second pair of equations in (2.4) is referred to as the Codazzi equation and the last equation is called the Gauss equation.

Exterior differentiation of the Codazzi equations in (2.4) and using (2.2) yields

$$\left(da - (a-c)h\omega_2\right) \wedge \omega_1 = 0, \qquad \left(dc - (a-c)k\omega_1\right) \wedge \omega_2 = 0. \tag{2.5}$$

Now Cartan's lemma can be applied to (2.5). There exist two functions u and v such that

$$\frac{1}{a-c}da - h\omega_2 = (u-k)\omega_1, \qquad \frac{1}{a-c}dc - k\omega_1 = (v-h)\omega_2. \tag{2.6}$$

Subtracting the pair of equations in (2.6) gives an expression for $d\log(a-c)$,

$$d\log(a-c) = (u-2k)\omega_1 - (v-2h)\omega_2. \tag{2.7}$$

It is natural from (2.7) to define a new variable J in terms of a and c a as

$$J = \frac{1}{2}(a-c) > 0. \tag{2.8}$$

Equation (2.7) can then be put in the form,

$$\mathrm{d}\log J = (u-2k)\omega_1 - (v-2h)\omega_2. \tag{2.9}$$

The differential forms ω_i constitute a linearly independent system. Two related coframes ϑ_i and α_i can be defined in terms of the ω_i and the functions u and v as follows

$$\vartheta_1 = u\omega_1 + v\omega_2, \quad \vartheta_2 = -v\omega_1 + u\omega_2,$$
$$\alpha_1 = u\omega_1 - v\omega_2, \quad \alpha_2 = v\omega_1 + u\omega_2. \tag{2.10}$$

These relations imply that $\vartheta_1 = 0$ is tangent to the level curves which are singled out by setting H equal to a constant and $\alpha_1 = 0$ is its symmetry with respect to the original directions.

The relation $2H = a + c$ is squared and subtracting the definition of the Gaussian curvature, $4K = 4ac$ yields the result $4(H^2 - K) = (a-c)^2$. The Hodge operator, denoted here by $*$, will play an important role in the following. It produces the following result on the basis forms ω_i in (2.2),

$$*\omega_1 = \omega_2, \qquad *\omega_2 = -\omega_1, \qquad *^2 = -1.$$

From these properties, the dual relations can be determined as

$$*\vartheta_1 = u\omega_2 - v\omega_1 = \vartheta_2, \qquad *\vartheta_2 = -v\omega_2 - u\omega_1 = -\vartheta_1, \tag{2.11}$$

$$*\alpha_1 = u\omega_2 + v\omega_1 = \alpha_2, \qquad *\alpha_2 = v\omega_2 - u\omega_1 = -\alpha_1.$$

Moreover, adding the expressions for da and dc given by (2.6), we obtain

$$\frac{1}{a-c}(\mathrm{d}a+\mathrm{d}c)=(u-k)\omega_1+h\omega_2+(v-h)\omega_2+k\omega_1=u\omega_1+v\omega_2=\vartheta_1. \tag{2.12}$$

Finally, there is the relation,

$$\alpha_1+2*\omega_{12}=u\omega_1-v\omega_2+2*(h\omega_1+k\omega_2)=(u-2k)\omega_1-(v-2h)\omega_2=d\log J. \tag{2.13}$$

Therefore, the Codazzi Equations (2.12) and (2.13) can be summarized in terms of the two functions H and J as follows,

$$dH=J\vartheta_1, \qquad d\log J=\alpha_1+2*\omega_{12}. \tag{2.14}$$

3. BONNET SURFACES

Suppose that Σ^* is a surface which is isometric to Σ such that the principal curvatures are preserved under the transformation. Denote all quantities which pertain to Σ^* by the same symbols, but with an asterisk,

$$a^*=a, \qquad c^*=c. \tag{3.1}$$

The same convention will be applied to the variables and forms which pertain to Σ and Σ^*. When Σ and Σ^* are isometric, the forms ω_i on Σ are related to the forms ω_i^* on Σ^* by means of the transformation

$$\omega_1^*=\cos\tau\,\omega_1-\sin\tau\,\omega_2, \qquad \omega_2^*=\sin\tau\,\omega_1+\cos\tau\,\omega_2. \tag{3.2}$$

The following theorem from [7] will be required.

Theorem 3.1: Under the transformation of coframe given in (3.2), the associated connection forms are related by

$$\omega_{12}^*=\omega_{12}-\mathrm{d}\tau. \tag{3.3}$$

There is a very important result which can be developed at this point. In the case that $a=a^*$ and $c=c^*$, the Codazzi equations imply that

$$\alpha_1+2*\omega_{12}=\mathrm{d}\log(a-c)=\mathrm{d}\log\left(a^*-c^*\right)=\alpha_1^*+2*\omega_{12}^*.$$

Now apply the operator * to both sides of this equation to give,

$$\alpha_2-2\omega_{12}=\alpha_2^*-2\omega_{12}^*.$$

Substituting for ω_{12}^* from Theorem 3.1, this assumes the form,

$$2\mathrm{d}\tau=\alpha_2-\alpha_2^*. \tag{3.4}$$

Define H_1 and H_2 to be the derivatives of the function H in directions such that $\mathrm{d}H$ can be expanded as

$$\mathrm{d}H=H_1\omega_1+H_2\omega_2.$$

Since $dH=J\vartheta_1$ and using ϑ_1 given by (2.10), Equation (2.13) produces

$$u\omega_1+v\omega_2=\frac{H_1}{J}\omega_1+\frac{H_2}{J}\omega_2,$$

Comparing coefficients of ω_1 and ω_2 on both sides, we can identify

$$u=\frac{H_1}{J},\qquad v=\frac{H_2}{J}.$$

This result implies that

$$\alpha_1=\frac{H_1}{J}\omega_1-\frac{H_2}{J}\omega_2,\qquad \alpha_2=\frac{H_2}{J}\omega_1+\frac{H_1}{J}\omega_2. \tag{3.5}$$

Since $\alpha_2^* = \sin(2\tau)\alpha_1 + \cos(2\tau)\alpha_2$, it follows from (3.4) that

$$\alpha_2 - \mathrm{d}(2\tau) = \sin(2\tau)\alpha_1 + \cos(2\tau)\alpha_2.$$

Solving this for $\mathrm{d}(2\tau)$ and substituting for α_1 and α_2, it follows that

$$\begin{aligned}\mathrm{d}(2\tau) &= (1-\cos(2\tau))\alpha_2 - \sin(2\tau)\alpha_1 \\ &= -\sin(2\tau)\left(\frac{H_1}{J}\omega_1 - \frac{H_2}{J}\omega_2\right) - \cos(2\tau)\left(\frac{H_2}{J}\omega_1 + \frac{H_1}{J}\omega_2\right) + \alpha_2 \\ &= -\sin(2\tau)\left(\frac{H_1}{J}\omega_1 - \frac{H_2}{J}\omega_2\right) - \cos(2\tau)\left(\frac{H_2}{J}\omega_1 + \frac{H_1}{J}\omega_2\right) + {}^*\alpha_1.\end{aligned}$$

Therefore, using the second equation of (2.14) for α_1 implies that

$$\mathrm{d}(2\tau) = -\sin(2\tau)\left(\frac{H_1}{J}\omega_1 - \frac{H_2}{J}\omega_2\right) - \cos(2\tau)\left(\frac{H_2}{J}\omega_1 + \frac{H_1}{J}\omega_2\right) + {}^*\mathrm{d}\log J + 2\omega_{12}. \tag{3.6}$$

The differential in (3.6) will play a role in the study of helicoidal surfaces.

4. HELICOIDAL SURFACES

Every helicoidal surface can be parametrized in terms of two parameters (s,t), where t can be thought of as time along orbits from a fixed $t = t_0$, and s is an arc-length of curves orthogonal to orbits. Then the curves $t = c$ are carried along the orbits by the helicoidal motion for c constant. They remain orthogonal to the orbits and foliate the surface. An orthonormal frame e_1, e_2 is determined along these coordinate curves. The corresponding coframe may be written as

$$\omega_1 = \mathrm{d}s, \qquad \omega_2 = q(s)\mathrm{d}t, \tag{4.1}$$

where q depends only on s. Since $\mathrm{d}\omega_1 = 0$, the equation $\mathrm{d}\omega_1 = \omega_{12} \wedge \omega_2$ implies that ω_{12} is proportional to $\mathrm{d}t$, say $f\mathrm{d}t$ and $\mathrm{d}\omega_2 = \mathrm{d}s \wedge \omega_{12} = f\,\mathrm{d}s \wedge \mathrm{d}t$ implies that

$$q'(s)\mathrm{d}s \wedge \mathrm{d}t = f\,\mathrm{d}s \wedge \mathrm{d}t,$$

so,

$$\omega_{12} = q'(s)\mathrm{d}t = \frac{q'(s)}{q(s)}\omega_2. \tag{4.2}$$

Hence, the ω_1-curves are geodesics, and the ω_2-curves or orbits, have geodesic curvature equal to

$$\mu(s) = \frac{\mathrm{d}}{\mathrm{d}s}\log\left(\|q(s)\|\right). \tag{4.3}$$

Thus along each orbit, the quantities a, c, μ and τ are constant and depend only on s. In this case, the derivative $H_2 = 0$. Also for the same reason, the differential form $\mathrm{d}J = J_1\omega_1 + J_2\omega_2$ of J implies that $J_2 = 0$, hence

$$\mathrm{d}\log J = \left(\frac{J_1}{J}\right)\omega_1. \tag{4.4}$$

Hence, the equation for $\mathrm{d}(2\tau)$ in (3.6) takes the form

$$\mathrm{d}(2\tau) = -\sin(2\tau)\frac{H_1}{J}\omega_1 - \cos(2\tau)\frac{H_1}{J}\omega_2 + *\frac{J_1}{J}\omega_1 + 2\mu(s)\omega_2,$$

that is,

$$\mathrm{d}(2\tau) = -\sin(2\tau)\frac{H_1}{J}\omega_1 - \cos(2\tau)\frac{H_1}{J}\omega_2 + \frac{J_1}{J}\omega_2 + 2\mu(s)\omega_2. \tag{4.5}$$

Writing $\mathrm{d}(2\tau)$ as a differential form in terms of $\mathrm{d}s$ and $\mathrm{d}t$ and then equating coefficients of $\mathrm{d}s$ and $\mathrm{d}t$ on both sides of (4.5) yields the following pair of equations,

$$\frac{\mathrm{d}(2\tau)}{\mathrm{d}s}=-\sin(2\tau)J^{-1}\frac{\mathrm{d}H}{\mathrm{d}s}, \qquad 2\mu(s)=\cos(2\tau)J^{-1}H_1-J^{-1}J_1. \tag{4.6}$$

The results in (4.6) are used in the proof of Theorem 5.2 which follows.

5. MAIN THEOREMS AND PROOFS

Now by what has been established so far, both functions a and c depend on the variable s, so this mapping is an isometry which preserves H, since H is the average of a and c. In general, an isometry is trivial if and only if it preserves the mean curvature and the principal directions. In this case, the above mapping is trivial if and only if τ is a multiple of $\pi/2$. Then we obtain that the orbits are plane curves. But this is impossible for a helicoidal surface. This proves the following result.

Theorem 5.1. For a helicoidal surface, the mapping $(s,t)\rightarrow(s,-t)$ is a nontrivial isometry which preserves the mean curvature H.

To prove the second theorem, the following result due to Chern is required [4] [7].

Proposition 5.1. (Chern) A surface M admits a nontrivial isometric deformation that keeps the principal curvatures fixed if and only if

$$\mathrm{d}\alpha_1=0, \qquad \mathrm{d}\alpha_2=\alpha_1\wedge\alpha_2.$$

Theorem 5.2. A helicoidal surface is a Bonnet surface of the second type if and only if the following relation is satisfied,

$$\frac{\mathrm{d}}{\mathrm{d}s}\left(\frac{1}{J}\frac{\mathrm{d}H}{\mathrm{d}s}\right)-\left(\frac{1}{J}\frac{\mathrm{d}H}{\mathrm{d}s}\right)^2(\cos\varphi(s))+\frac{1}{J}\frac{\mathrm{d}H}{\mathrm{d}s}\frac{\mathrm{d}\log|q(s)|}{\mathrm{d}s}=0. \tag{5.1}$$

with $H=H(s)$ nonconstant.

Proof: Set $\varphi = 2\tau$ and consider the principal coframe

$$\omega_1 = \cos\varphi(s)\mathrm{d}s + \sin\varphi(s)q(s)\mathrm{d}t, \qquad \omega_2 = -\sin\varphi(s)\mathrm{d}s + \cos\varphi(s)q(s)\mathrm{d}t. \tag{5.2}$$

Define H_1 and H_2 as the coefficients in the differential $\mathrm{d}H$ by putting $\mathrm{d}H = H_1\omega_1 + H_2\omega_2$ and let α_1 and α_2 be given by (3.5). Next we substitute α_1 and α_2 into the equations which appear in Chern's result given in Proposition 5.1. Since $H_2 = 0$ in the $(\mathrm{d}s, \mathrm{d}t)$ basis of forms, it follows that

$$\begin{aligned}\mathrm{d}\alpha_1 &= \mathrm{d}\left(\frac{H_1}{J}\right)\wedge\omega_1 + \frac{H_1}{J}\mathrm{d}\omega_1 = \frac{\mathrm{d}}{\mathrm{d}s}\left(\frac{H_1}{J}\right)\mathrm{d}s\wedge\omega_1 + \frac{H_1}{J}\frac{\mathrm{d}}{\mathrm{d}s}\left(\sin\varphi(s)q(s)\right)\mathrm{d}s\wedge\mathrm{d}t \\ &= \frac{\mathrm{d}}{\mathrm{d}s}\left(\frac{H_1}{J}\right)\sin\varphi q(s)\mathrm{d}s\wedge\mathrm{d}t + \frac{H_1}{J}\left(\cos\varphi\frac{\mathrm{d}\varphi}{\mathrm{d}s}q + \sin\varphi\frac{\mathrm{d}q}{\mathrm{d}s}\right)\mathrm{d}s\wedge\mathrm{d}t.\end{aligned}$$

Since the first equation is $\mathrm{d}\alpha_1 = 0$, this implies that

$$\frac{H_1}{J}\left(\cos\varphi(s)\frac{\mathrm{d}\varphi}{\mathrm{d}s}\right) = -\frac{H_1}{J}\sin\varphi(s)\frac{\mathrm{d}}{\mathrm{d}s}\log|q(s)| - \frac{\mathrm{d}}{\mathrm{d}s}\left(\frac{H_1}{J}\right)\sin\varphi. \tag{5.3}$$

Similarly, using (3.5), we have

$$\begin{aligned}\mathrm{d}\alpha_2 &= \frac{\mathrm{d}}{\mathrm{d}s}\left(\frac{H_1}{J}\right)\mathrm{d}s\wedge\omega_2 + \frac{H_1}{J}\mathrm{d}\omega_2 \\ &= \frac{\mathrm{d}}{\mathrm{d}s}\left(\frac{H_1}{J}\right)\cos\varphi(s)q(s)\mathrm{d}s\wedge\mathrm{d}t + \frac{H_1}{J}\frac{\mathrm{d}}{\mathrm{d}s}\left(\cos\varphi(s)q(s)\right)\mathrm{d}s\wedge\mathrm{d}t,\end{aligned}$$

$$\alpha_1\wedge\alpha_2 = \left(\frac{H_1}{J}\right)^2 q(s)\mathrm{d}s\wedge\mathrm{d}t.$$

Equating these two results as in the second of Chern's two equations, we obtain

$$\frac{\mathrm{d}}{\mathrm{d}s}\left(\frac{H_1}{J}\right)\cos\varphi(s) - \frac{H_1}{J}\sin\varphi(s)\frac{\mathrm{d}\varphi}{\mathrm{d}s} + \frac{H_1}{J}\cos\varphi(s)\frac{\mathrm{d}}{\mathrm{d}s}\log|q(s)| = \left(\frac{H_1}{J}\right)^2. \tag{5.4}$$

Multiplying (5.3) by $\sin\varphi$ and (5.4) by $\cos\varphi$, it is found that the folowing hold:

$$\frac{H_1}{J}\left(\sin\varphi\cos\varphi\frac{d\varphi}{ds}\right)=-\frac{H_1}{J}\sin^2\varphi\frac{d}{ds}\log|q(s)|-\frac{d}{ds}\left(\frac{H_1}{J}\right)\sin^2\varphi,$$

$$\cos^2\frac{d}{ds}\left(\frac{H_1}{J}\right)-\frac{H_1}{J}\left(\cos\varphi\sin\varphi\frac{d\varphi}{ds}\right)+\frac{H_1}{J}\cos^2\varphi\frac{d}{ds}\log|q(s)|=\left(\frac{H_1}{J}\right)^2\cos\varphi.$$

Adding these two equations, the desired result is obtained,

$$\frac{d}{ds}\left(\frac{H_1}{J}\right)-\left(\frac{H_1}{J}\right)^2\cos\varphi(s)+\frac{H_1}{J}\frac{d}{ds}\log|q(s)|=0. \tag{5.5}$$

Replacing H_1 by dH/ds in (5.5), equation (5.1) follows. □

Multiply (5.5) by $q(s)\sin\varphi(s)$ to obtain,

$$J\frac{d}{ds}\left(J^{-1}\frac{dH}{ds}\right)\sin\varphi q(s)-\frac{dH}{ds}\cos\varphi\sin\varphi\frac{dH}{ds}J^{-1}q+\frac{dH}{ds}\sin\varphi\frac{dq}{ds}=0. \tag{5.6}$$

Substituting the derivative for 2τ into (5.6), it becomes,

$$J\frac{d}{ds}\left(J^{-1}\frac{dH}{ds}\right)\sin\varphi q(s)+\frac{dH}{ds}\cos\varphi\frac{d\varphi}{ds}q(s)+\frac{dH}{ds}\sin\varphi\frac{dq}{ds}=0.$$

By means of the product rule, this can be put in the form,

$$\frac{d}{ds}\left[\frac{1}{J}\frac{dH}{ds}\sin\varphi(s)q(s)\right]=0. \tag{5.7}$$

This is trivial to integrate, so if C is the integration constant, we obtain that

$$\frac{1}{J}\frac{dH}{ds}\sin\varphi(s)\cdot q(s)=C, \tag{5.8}$$

with $H = H(s)$ nonconstant.

Since this relation may be viewed as an ordinary differential equation for the real-valued function which determines the helicoidal surface under helicoidal motion, the existence of such a surface is guaranteed by the local existence uniqueness theorem for solutions of such an ordinary differential equation.

From the first equation of (4.6) and the fact that the space curvature of orbits is either $\left[\mu^2 + \left(a(\cos^2\sigma) + c(\sin^2\sigma)\right)^2\right]^{1/2}$ or this with $\sin - \cos$ interchanged, the last result follows.

Theorem 5.3. A helicoidal surface has constant mean curvature if and only if its principal directions make an angle constant with the orbits.

Combining all of these results, the main result of this work can be stated in the form of the following Theorem.

Theorem 5.4. The helicoidal surfaces are necessarily Bonnet surfaces and they represent all three types of surface.

A conclusion that follows from these results then is an interesting new geometric characterization of such surfaces. Thus, a helicoidal surface has constant mean curvature if and only if its principal directions make an angle which is constant with the orbits.

Finally, it will be proved that for any surface of revolution in $\mathbb{R}^3$ which has nonconstant mean curvature function H it holds that either $N = 1$ or $N = \infty$.

Let $x = r(z) > 0$ be a plane curve in the $x - z$ plane and form the surface of revolution

$$r(z, \vartheta) = \left(r(z)\cos\vartheta, r(z)\sin\vartheta, z\right), \qquad -\infty < \vartheta < \infty. \tag{5.9}$$

The principal curvatures are calculated to be

$$a = -\frac{r_{zz}}{\left[r_z^2 + 1\right]^{3/2}}, \qquad c = \frac{1}{r\left[r_z^2 + 1\right]^{3/2}}. \tag{5.10}$$

If $a=c$ at $p\in r(z,\vartheta)$, of course the entire parallel through p consists of umbilic points, so $N=1$. Here r_z denotes the derivative of $r(z)$ with respect to z.

Theorem 5.5: Surfaces of revolution with nonconstant mean curvature that admit a one-parameter family of geometrically distinct nontrivial isometries preserving principal curvatures a,c are exactly those for which the function $x=r(z)$ satisfies a specific fourth order differential equation in z.

Proof: For the surface of revolution of the form (5.9), the principal coframe is given by

$$\omega_1=\sqrt{r_z^2+1},\qquad \omega_2=r(z)\mathrm{d}\vartheta. \tag{5.11}$$

Since principal curvatures a and c in (5.10) depend only on z and not on ϑ, the first equation of (2.14) implies that

$$(a+c)_z\,\mathrm{d}z=(a-c)\left(u\sqrt{r_z^2+1}\mathrm{d}z-vr(z)\mathrm{d}\vartheta\right). \tag{5.12}$$

Equating the coefficients of the differentials $\mathrm{d}z$ and $\mathrm{d}\vartheta$ on both sides gives u and v,

$$u=\frac{(a+c)_z}{(a-c)\sqrt{r_z^2+1}},\qquad v=0. \tag{5.13}$$

Then the forms α_1 and α_2 can be calculated from (2.10),

$$\alpha_1=\frac{(a+c)_z}{(a-c)}\mathrm{d}z,\qquad \alpha_2=\frac{(a+c)_z r(z)}{(a-c)\sqrt{r_z^2+1}}\mathrm{d}\vartheta. \tag{5.14}$$

Substituting (5.14) into the differential expressions of Proposition 5.1, it is clear that $\mathrm{d}\alpha_1=0$ must always hold since the coefficient of α_1 depends only on z and $\mathrm{d}z \wedge \mathrm{d}z = 0$. To develope the second equation of the pair, we calculate

$$\mathrm{d}\alpha_2 = \left[\frac{(a+c)_z}{(a-c)\sqrt{r_z^2+1}} r(z) \right]_z \mathrm{d}z \wedge \mathrm{d}\vartheta, \qquad \alpha_1 \wedge \alpha_2 = \frac{[(a+c)_z]^2}{(a-c)^2} \frac{r(z)}{\sqrt{r_z^2+1}} \mathrm{d}z \wedge \mathrm{d}\vartheta. \tag{5.15}$$

Equating these two expressions, the following fourth-order differential equa- tion for $r(z)$ is obtained,

$$\left[\frac{(a+c)_z}{(a-c)\sqrt{r_z^2+1}} r(z) \right]_z = \frac{[(a+c)_z]^2}{(a-c)^2} \frac{r(z)}{\sqrt{r_z^2+1}}. \tag{5.16}$$

In (5.16), the principal curvatures a and c are given in (5.10), and since a contains second derivatives of r with respect to z, equation (5.16) will be a fourth order equation in z. This is the equation mentioned in the Theorem. □

REFERENCES

1. Bonnet, O. (1867) Mémoire sur la theorie des surfaces applicables sur une surface donnée. J. l'école Pol., Paris, XLII Cahier, 72-92.
2. Cartan, é. (1942) Sur les couples de surface applicables avec conservation des courbures principales. Bull. Sc. Math., 66, 55-85.
3. Chern, S.S. (1955) An Elementary Proof of the Existence of Isothermal Parameters on a Surface. Proceedings of the American Mathematical Society, 6, 771-782.
4. Chern, S.S. (1985) Deformation of Surfaces Preserving Principal Curvatures. In: Rauch, H.E., Ed., Differential Geometry and Complex Analysis, Springer, Berlin, 155-163.
5. Roussos, I.M. (1999) Global Results on Bonnet Surfaces. Journal of Geometry, 65, 151-168.
6. Bracken, P. (2008) Cartan's Theory of Moving Frames and an Application to a Theorem of Bonnet. Tensor, 70, 261-274.
7. Bracken, P. (2017) An Intrinsic Characterization of Bonnet Surfaces Based on a Closed Differential Ideal. Manifolds-Current Research Areas, ed. P. Bracken, Intech Open Science Press, 21-43.
8. Roussos, I.M. (1988) The Helicoidal Surfaces as Bonnet Surfaces. Tohoku Mathematical Journal, Second Series, 40, 485-490.
9. Roussos, I.M. (1987) Principal Curvature Preserving Isometries of Surfaces in Ordinary Space. Boletim da Sociedade Brasileira de Matemática, 18, 95-105.
10. Bobenko, A. and Eitner, U. (1998) Bonnet Surfaces and Painlevé Equations. Journal für die reine und angewandte Mathematik, 499, 47-79.
11. Chern, S.S., Chen, W.H. and Lam, K.S. (1999) Lectures in Differential Geometry, Vol. 1 of Series on University Mathematics. World Scientific, Singapore.

CHAPTER 10

Group Variable Selection via a Combination of L_q Norm and Correlation-Based Penalty

Ning Mao, Wanzhou Ye

Department of Mathematics, College of Science, Shanghai University, Shanghai, China

ABSTRACT

Considering the problem of feature selection in linear regression model, a new method called LqCP is proposed simultaneously to select variables and favor a grouping effect, where strongly correlated predictors tend to be in or out of the model together. LqCP is based on penalized least squares with a penalty function that combines the L_q $(0 < q < 1)$ norm and correlation-based penalty that is CP norm. It can shrink some coefficients to exactly zero and additionally the CP term links strength of penalization to the correlation among the predictors. The simulation studies show the advantages of LqCP with the increase of noise variables and the case of $p > n$. In addition, a simulation about grouped variable selection is performed. Finally, The model is applied to two real data: US Crime Data and Gasoline Data. In terms of prediction error and estimation error, empirical studies show the efficiency of LqCP.

Keywords: Linear Regression, Variable Selection, Elastic Net, Adaptive Elastic Net

1. INTRODUCTION

Here the usual linear regression mode is considered in the paper given by:

$$y = X\beta + \epsilon, \tag{1}$$

where $y_{n\times 1}$ are the observations, $\beta_{p\times 1}$ is a vector of unknown parameters to be estimated, $X_{n\times p}$ is an $n\times p$ matrix of p predictor vectors of n observations and ϵ is a random error vector with $E(\epsilon)=0$ which often is assumed that it is subject to normal distribution independently. Ordinary least squares (OLS) estimates are very common which can be obtained by minimizing the sum of square residuals. In general, the criteria for evaluating the quality of a model from the following two aspects. One is prediction accuracy on test data and the other is to tend to select a simple model. In other words, less variables would be selected in the condition of same prediction effects. Variable selection is necessary especially when the number of predictors is large. There are many applications using variable selection to solve problems like knowledge discovery with high-dimensional data source [1] and it could greatly enhance the pre- diction performance of the fitted model. Traditional model selection method is best-subset selection and its step-wise variants. However, best-subset selection is computationally prohibitive when the number of predictors is large. As analyzed by Breiman (1996) [2] , subset selection is unstable; thus, the resulting model has poor prediction accuracy. To overcome the drawbacks of subset selection, many variable selection methods are appeared and the most popular recently is regulation method.

In recent years, regularization method has attracted a great attention. It is used in applications such as machine learning, denoising, inpainting, deblurring, compressed sensing, source separation and more. Generally, the square loss function of penalized least squares estimates:

$$L_{(\beta;\lambda)} = \|y - X\beta\|_2^2 + \lambda\|\beta\|_q^q, \tag{2}$$

where $\lambda > 0$ is a penalty parameter. $q > 0$ is considered in this paper.

$\|\beta\|_q^q = \sum_{j=1}^{p}|\beta_j|^q$. When $q=1$, it becomes Lasso procedure. The procedure of minimizing the objective function is called ridge regression when $q=2$. As a continuous shrinkage method, ridge regression achieves its better prediction performance through a bias-variance trade off. However, ridge regression can not produce parsimonious model, which means all the predictors are kept in the model. Lasso is proposed by [3] in 1996. It imposes the L_1 norm regularization to the loss function and becomes a widely popular regularization method. Further, Knignt and Fu (2000) [4] studied the asymptotic properties of lasso. Lasso can shrink some coefficients to zero to achieve the effects of variable selection. Due to this reason, lasso has gained popularity in high-dimensional data.

Although lasso is a popular method for variable selection, it still has several drawbacks. The first is the lack of oracle property. The oracle property means the probability of selecting the right set of nonzero coefficients converges to one, and the estimators of the nonzero coefficients have asymptotically normal distribution with the same means and covariances. Fan and Li (2001) [5] first pointed the parameters estimators of lasso are biased and don't have the oracle property. Then adaptive lasso proposed by Zou (2006) [6] overcomes this limitation of lasso. In adaptive lasso, weights are used for regularizing different coefficients in the L_1 norm regularization, which means different coefficients have different shrinkage factors. Secondly, lasso have very poor performance when there are highly correlated variables in high-dimensional data. What's more, lasso only selects n variables at most when the number of variables p is more than the number of sample observation size n, which is a big limitation in application areas. At last, there are grouping variables among genes for the microarray gene data. These genes mostly have a high pairwise correlation with each other while lasso can not select a group of correlated genes. Based on these limitations, Zou and Hastie [7] proposed the elastic net regularization which is a combination of L_1 norm and L_2 norm. Similarly, elastic net (ENET) lacks the oracle property even though it outperforms Lasso. Further, Zou and Zhang [8] proposed adaptive elastic net to achieve the oracle property and good estimation accuracy. In the base of that, Ghosh [9] further studied grouping effects in the adaptive elastic net by using the ordinary least squares as the initial weight in low dimension data for simplicity. However, the estimators of ordinary least squares is very bad in the case of large numbers predictors or high correlated variables. Additionally, elastic net and adaptive elastic net don't take into account the information about the correlations of variables.

In the same spirits, there existed other penalty based on methods for handling grouping effects. Penalizing least squares via combining L_1 and L_∞ named OSCAR is presented by Bondell and Reich (2008) [10] . The Oscar forces some coefficients to be identically equal, encouraging correlated variables that have similar effects on the response to form clusters represented by the same coefficients. However, the computation of Oscar estimation is slow for large p which is based on a sequential quadratic programming. Then, considering the infor- mation of correlations of variables, Tutz and Ulbricht (2009) researched the property of correlation-based penalty (CP) and pointed the grouping effect of CP term. In the article of Tutz and Ulbricht, blockwise boosting procedure is applied in the simulations, which updates at each step the coefficient of more than one variable. However, in practical implementation, the step length factor and the stopping number of iterations have to be determined. Sometimes, this may be difficult and affects the sparsity of the solution as well as the speed of convergence of the algorithm. Therefore, El Anbari and Mkhadri (2014) [11] proposed an alternative regularization procedure called L1CP by combing L_1 norm with CP term. The method performs automatic variable selection and has the ability of grouping effects.

In this paper, motivated by the sparsity and grouping effect especially the case of the pairwise correlations are very high, a new regulation procedure called LqCP is proposed in linear regression setting. It combines the L_q $(0<q<1)$ norm and CP penalty. Similar to the L1CP method, LqCP also performs automatic variables selection and allows to select or to remove highly correlated variables together. Section 2 first introduces the property of elastic net and adaptive elastic net and then demonstrates LqCP and its algorithm as well as the method of choosing regularization parameters. In Section 3, simulation studies give the estimation mean squared errors for different circumstances to show the parameters estimation effects among LqCP, elastic net (ENET), adaptive elastic net (AENET) and L1CP. Section 4 is mainly about the application examples to show the prediction accuracy of models. The conclusion of this paper is given in Section 5.

2. METHODS

2.1. The Elastic Net and Adaptive Elastic Net

Here the form of elastic net described in the above firstly is showed in the following. The naive elastic net estimator $\hat{\beta}_{(\text{naive})}$ is the minimizer of equation:

$$\hat{\beta}_{(\text{naive})} = \arg\min_{\beta} \|y - X\beta\|_2^2 + \lambda_1 \|\beta\| + \lambda_2 \|\beta\|_2^2 . \tag{3}$$

This method is called the naive elastic net which overcomes the limitations of Lasso in the case of $p>n$. The penalty of combining L_1 norm and L_2 norm which has been proved that it has grouping effect in Zou and Hastie (2005)

[7] . The estimator of elastic net is a two-stage procedure: for each fixed λ_2, they first find the ridge coefficients, and then do the Lasso-type shrinkage along the lasso coefficient solution paths. At last, through the criterion of model selection, the best λ_1 and λ_2 are obtained. It appears to incur a double amount of shrinkage. Double shrinkage does not help to reduce variances much and introduces unnecessary extra bias, compared with pure lasso or ridge shrinkage. Hence, the elastic net estimators are rescaled by $\hat{\beta}_{(enet)} = (1+\lambda_2)\hat{\beta}_{(naive)}$. Such a scaling transformation preserves the variable selection property of the naive elastic and improves the prediction performance by correcting this double shrinkage.

In a similar way to Lasso, the elastic net does not enjoy the oracle property. Combining the property of the adaptive Lasso, Zou and Zhang [8] proposed the adaptive elastic net which combines the adaptive Lasso and L_2 norm. Ad- ditionally, they established the oracle property of the adaptive elastic net in the condition of weak regularity. The form of the adaptive elastic net is as follows

$$\hat{\beta}_{(\text{Adaenet})} = \arg\min_{\beta} \|y - X\beta\|_2^2 + \lambda_1 \sum_{j=1}^{p} \hat{\omega}_j |\beta_j| + \lambda_2 \|\beta\|_2^2, \tag{4}$$

where $\hat{\omega}_j = (|\hat{\beta}_j|)^{-\gamma}, j = 1, 2, \cdots, p$, and γ is a positive constant. Choosing the

initial weight is crucial in the adaptive elastic net. Zou and Zhang [8] proposed using the elastic net as an initial weight either in low-dimensional data or high-dimensional data. In the paper of Ghosh [9] , the weights is donated by $\hat{\beta}_{(ols)}$. Ghosh [9] researched the ability of variable selection and grouping effect problem, proved that the adaptive elastic net also can automatically select the group variables and also had a better performance for prediction accuracy than elastic net. But the estimator of ordinary least squares is very bad in the case of high dimension or high correlations of variables. Therefore, based on the paper of Zou [8] , the elastic net estimators is regarded as the initial weight in simulations in this paper. To avoid the invalid values when the estimators $\hat{\beta}_j = 0$, the initial weights are defined by $\hat{\omega}_j = (|\hat{\beta}_j| + 1/n)^{-\gamma}$, where n is the sample size and $\gamma > 0$. For adaptive methods, while as $\gamma \to 0$ it becomes a biased procedure like lasso. Similarly for $\gamma \to \infty$ adaptive methods are un- biased. A good choice of γ is a compromise between shrinkage and bias and can be obtained via cross-validation. As described in Ghosh, γ should not be chosen too big a quantity and typically first should be given a range of values. Then for each fixed γ, the cross-validation is applied to select other tuning parameters. But the elastic net and adaptive elastic net does not take into account the correlations structure of variables. Therefore, according to the CP term proposed by Tutz and Ulbricht (2009) [12] , proposed new method is as follows.

2.2. The Proposed Method

2.2.1. Introduction of Proposed Model

In the context of linear regression problems, the following penalty function based on residual sum squares is considered. The L_q penalty on f is defined as

$$\|f\|_q^q = \sum_{j=1}^{p} |\beta_j|^q . \tag{5}$$

When $q=0$, the corresponding penalty is discontinuous at the origin and consequently is not easy to compute. Thus in this paper $q>0$ is designed. The least squares subject to the L_q penalty with $0<q<1$ is first studied by Frank and Friedman (1993) [13] which is known as bridge regression. Fu (1998) [14] and Knight and Fu (2000) [4] studied asymptotic properties and the com- putation of bridge estimators. When $q=2$, the solution $\hat{\beta}$ never becomes zero unless $\hat{\beta}=0$ and it is biased. For $q \le 1$, the bridge estimator tends to shrink small absolute coefficients to exact zeros and hence selects important variables. As pointed out by Theorem 2 in Knight and Fu (2000) [14] , when $q>1$ the amount of shrinkage towards zero increases with the magnitude of the regression coefficients being estimated. It suggests that if $q<1$, estimate nonzero regression parameters at the usual rate without asymptotic bias while shrinking the estimates of zero regression parameters to 0 with positive pro- bability. In practice, in order to avoid unacceptable large bias for large para- meters, the value of q is often chosen not too large. When $0<q<1$, the L_q penalty may achieve better sparsity than L_1 penalty because larger penalty is imposed on small coefficients than L_1 penalty.

So according to better sparsity property of L_q $(0<q<1)$ and the case of high correlations of variables, the model is defined by

$$\hat{\beta} = \arg\min_{\beta} \|y - X\beta\|_2^2 + \lambda_1 \|\beta\|_q^q + \lambda_2 P_c(\beta), \tag{6}$$

where

$$P_c(\beta) = \sum_{j=1}^{p}\sum_{i>j} \frac{(\beta_i - \beta_j)^2}{1-\rho_{ij}} + \frac{(\beta_i + \beta_j)^2}{1+\rho_{ij}}. \tag{7}$$

$0 < q \leq 1$, λ_1, λ_2 are positive constants. ρ_{ij} denotes the (empirical) correlation between the ith and jth predictors. Here the penalty $P_c(\beta)$ is introduced by Tutz and Ulbricht (2009) [12] and is defined assuming that $\rho_{ij} \neq 1$ for $i \neq 1$. When $q = 1$, the model changes to the combination of L_1 and $P_c(\beta)$, which is the model called L1CP proposed by El Anbari M. and Mkhadri A. (2014) [11] .

As introduced in the above, L_q penalty in the loss of residual square sum have a better sparsity, while the correlation-based penalty $P_c(\beta)$ will encourage grouping effect for highly correlated variables. In fact, it's easy to see that for strong positive correlation ($\rho_{ij} \approx 1$) in (7), the first term becomes the dominant having the effect that estimates for β_i and β_j are similar ($\hat{\beta}_i \approx \hat{\beta}_j$). For strong negative correlation ($\rho_{ij} \approx -1$), the second term becomes dominant and β_i will be close to β_j. The effect is grouping, highly correlated effects show comparable values of estimates ($|\hat{\beta}_i| \approx |\hat{\beta}_j|$) with the sign being determined by positive or negative correlation.

Moreover, assuming that $\rho_{ij} \neq 1$ for $i \neq j$, the penalty (7) can be written in a sample quadratic form $P_c(\beta) = W^{\mathrm{T}} \beta W$, where $W = (w_{ij}), 1 \leq i, j \leq p$, is a positive definite matrix with general term

$$W_{ij} = \begin{cases} 2\sum_{s\neq i} \frac{1}{1-\rho_{is}^2} & \text{if } i = j \\ -2\frac{\rho_{ij}}{1-\rho_{ij}^2} & \text{if } i \neq j \end{cases} \tag{8}$$

Putting $\gamma = \lambda_2/(\lambda_1 + \lambda_2)$,the optimization problem (6) is equivalent to

$$\hat{\beta} = \arg\min_{\beta} \|y - X\beta\|_2^2, \text{ s.t } (1-\gamma)\|\beta\|_q^q + \gamma P_c(\beta) \le v \text{ for } v \ge 0. \tag{9}$$

Proof of (8): In fact, the penalty $Q(\beta) = \lambda_1 \|\beta\|_q^q + \lambda_2 P_c(\beta)$ can be written as follows:

$$\begin{aligned} Q(\beta) &= \lambda_1 \|\beta\|_q^q + \lambda_2 P_c(\beta) \\ &= (\lambda_1 + \lambda_2)\left\{\frac{\lambda_1}{\lambda_1 + \lambda_2}\|\beta\|_q^q + \frac{\lambda_2}{\lambda_1 + \lambda_2} P_c(\beta)\right\} \\ &= \lambda\left\{(1+\gamma)\|\beta\|_q^q + \gamma P_c(\beta)\right\} \end{aligned} \tag{10}$$

where $\lambda = \lambda_1 + \lambda_2$. So, the problem (6) is equivalent to finding

$$\hat{\beta}(\lambda, \gamma) = \arg\min_{\beta} \|y - X\beta\|_2^2 + \lambda\left\{(1-\gamma)\|\beta\|_q^q + \gamma P_c(\beta)\right\} \tag{11}$$

which is equivalent to the optimization problem (6).

2.2.2. The Algorithm Procedure

The estimators of β can be computed via the Cyclic Descent Algorithm for l_q sparsity penalized linear regression problem [15] . The main idea is to transform the LqCP problem into an equivalent problem on augmented data.

Indeed, the optimization problem (6) can be written as

$$\hat{\beta}(\lambda_1, \lambda_2) = \arg\min_{\beta} \|y - X\beta\|_2^2 + \lambda_1 \|\beta\|_q^q + \lambda_2 \beta^{\mathrm{T}} W \beta, \tag{12}$$

where W is defined by (8), is a real symmetric positive-define square matrix, assuming that $\rho_{ij}^2 \ne 1,$ with Choleski decomposition $W = LL^{\mathrm{T}}$ and $L = W^{\frac{1}{2}}$,

which always exists. Now, let

$$X^*_{(n+p)\times p} = \begin{pmatrix} X \\ \sqrt{\lambda_2} L_p^{\mathrm{T}} \end{pmatrix}, y^*_{(n+p)} = \begin{pmatrix} y \\ 0 \end{pmatrix}. \tag{13}$$

The LqCP estimator is defined as

$$\hat{\beta} = \arg\min_{\beta} \left\| y^* - X^*\beta \right\|_2^2 + \lambda_1 \left\| \beta^* \right\|_q^q. \tag{14}$$

Let

$$\begin{aligned} J(\beta) &= \|y - X\beta\|_2^2 + \lambda_1 \|\beta\|_q^q + \lambda_2 \|\beta\|_2^2 \\ &= \left[y^{\mathrm{T}} y + \beta^{\mathrm{T}} X^{\mathrm{T}} X \beta - 2 y^{\mathrm{T}} X \beta \right] + \lambda_2 \beta^{\mathrm{T}} W \beta + \lambda_1 \|\beta\|_q^q \\ &= \left[y^{\mathrm{T}} y + \beta^{\mathrm{T}} \left(X^{\mathrm{T}} X + \lambda_2 L L^{\mathrm{T}} \right) \beta - 2 y^{\mathrm{T}} X \beta \right] + \lambda_1 \|\beta\|_q^q \\ &= \left\| y^* - X^* \beta \right\|_2^2 + \lambda_1 \|\beta\|_q^q. \end{aligned} \tag{15}$$

Note that the sample size in the augmented problem is $n+p$ and X^* has rank p. As described in the paper of Marjanovic and Solo [15] in 2014, L_q $(0<q<1)$ sparsity penalizing linear regression can make a better sparsity result and a quick rate of convergence than L_1 norm. Whats more, they proposed a new algorithm called Cyclic Descent and have proved that minimizing the problem of sparsity penalized L_q in linear regression has global minimizers. In the base of that, the algorithm is used to solve the Equation (15) and the procedure is as follows:

$$\begin{aligned} J(\beta) &= \left\| y^* - X^*\beta \right\|_2^2 + \lambda_1 \|\beta\|_q^q \\ &= \left[\sum_{i=1}^{n} \left(y_i^* - \sum_{j\neq k} x_{ij}^* \beta_j - x_{ik}^* \beta_k \right)^2 \right] + \lambda_1 \sum_{j\neq k} |\beta_j|^q + \lambda_1 |\beta_k|^q \\ &= [\sum_{i=1}^{n} \left(y_i^* - \sum_{j\neq k} x_{ij}^* \beta_j \right)^2 + \sum_{i=1}^{n} x_{ik}^{*2} \beta_k^2 - 2 \sum_{i=1}^{n} \left(y_i^* - \sum_{j\neq k} x_{ij}^* \beta_j \right) x_{ik}^* \beta_k] + \lambda_1 \sum_{j\neq k} |\beta_j|^q + \lambda_1 |\beta_k|^q \\ &= \sum_{i=1}^{n} x_{ik}^{*2} \left[\beta_k^2 + \frac{\sum_{i=1}^{n} \left(y_i^* - \sum_{j\neq k} x_{ij}^* \beta_j \right)^2}{\sum_{i=1}^{n} x_{ik}^{*2}} - 2 \frac{\sum_{i=1}^{n} \left(y_i^* - \sum_{j\neq k} x_{ij}^* \beta_j \right) x_{ik}^* \beta_k}{\sum_{i=1}^{n} x_{ik}^{*2}} \right] + \lambda_1 \sum_{j\neq k} |\beta_j|^q + \lambda_1 |\beta_k|^q \\ &= \sum_{i=1}^{n} x_{ik}^{*2} \left[\beta_k^2 - 2 z_k \beta_k + z_k^2 \right] + \lambda_1 |\beta_k|^q + C \end{aligned} \tag{16}$$

Here, the related values are defined that

$$z_k = \frac{\sum_{i=1}^{n}\left(y_i^* - \sum_{j\neq k} x_{ij}^* \beta_j\right) x_{ik}^*}{\sum_{i=1}^{n} x_{ik}^{*2}},$$

$$C = \sum_{i=1}^{n}\left(y_i^* - \sum_{j\neq k} x_{ij}^* \beta_j\right) - \frac{\left[\sum_{i=1}^{n}\left(y_i^* - \sum_{j\neq k} x_{ij}^* \beta_j\right) x_{ik}^*\right]^2}{\sum_{i=1}^{n} x_{ik}^{*2}} + \lambda_1 \sum_{j\neq k} \left|\beta_j\right|^q,$$

$$\lambda^* = \frac{\lambda_1}{\sum_{i=1}^{n} x_{ik}^{*2}}.$$

Then,

$$\min_{\beta} J(\beta_k) = \sum_{i=1}^{n} x_{ik}^{*2}\left[(z_k - \beta_k)^2 + \lambda^* |\beta_k|^q\right]. \tag{17}$$

Through the whole procedure, calculating the model (12) has been transformed to (15). Then, the algorithm procedure is stated below:

Initialize with β^1 and calculate the initial residual $r^1 := y^* - X^* \beta^1$. Denote the iteration counter by m and the coefficient index of an iterate by k. Set $m = k = 1$, and:

(a) Calculate the adjusted gradient $z_k^m := z(\beta_{-k}^m)$ by:

$$\begin{aligned} z_k^m &= \frac{X_{(k)}^{*\mathrm{T}}\left(y^* - X^* \beta_{-k}^m\right)}{X_{(k)}^{*\mathrm{T}} X_{(k)}^*} \\ &= \frac{X_{(k)}^{*\mathrm{T}}\left(y^* - X^* \beta^m + X_{(k)}^{*\mathrm{T}} X_{(k)}^* \beta^n\right)}{X_{(k)}^{*\mathrm{T}} X_{(k)}^*} \\ &= \frac{X_{(k)}^{*\mathrm{T}} r^m}{X_{(k)}^{*\mathrm{T}} X_{(k)}^*} + \beta_k^m \end{aligned} \tag{18}$$

(b) Use (a) computing the map:

$$\tau\left(z_k^m, \beta_k^m\right) := \begin{cases} 0 & \text{if } \left|z_k^m\right| < h_{\lambda^*,q} \\ \text{sign}\left(z_k^m\right) b_{\lambda^*,q} I\left(\beta_k^m \neq 0\right) & \text{if } \left|z_k^m\right| = h_{\lambda^*,q} \\ \text{sign}\left(z_k^m\right) \bar{\beta} & \text{if } \left|z_k^m\right| > h_{\lambda^*,q} \end{cases} \tag{19}$$

where $b_{\lambda^*,q} = \left[2\lambda^*(1-q)\right]^{\frac{1}{2-q}}$, $h_{\lambda^*,q} = b_{\lambda^*,q} + \lambda^* q b_{\lambda^*,q}^{q-1}$ and $\bar{\beta} > 0$ satisfies $\bar{\beta} + \lambda^* q \bar{\beta}^{q-1} = \left|z_k^m\right|$. There are two solutions to this equation and $\bar{\beta} \in \left(b_{\lambda^*,q}, \left|z_k^m\right|\right)$ is the larger one. It can be computed from the iteration, $b_{\lambda^*,q} \geq \bar{\beta}^{(0)} \leq \left|z_k^m\right|$:

$$\bar{\beta}^{(t+1)} = \rho\left(\bar{\beta}\right) \text{ where } \rho\left(\bar{\beta}\right) = \left|z_k^m\right| - \lambda^* q \bar{\beta}^{q-1} \tag{20}$$

(c) Update β^m by:

$$\beta^{m+1} = \beta_{-k}^m + \tau\left(z_k^m, \beta_k^m\right) e_k \tag{21}$$

where e_k has a 1 in the k-th position and 0's in the rest.

(d) Update the residual r^m with:

$$\begin{aligned} r^{n+1} &= y^* - X^* \beta^{m+1} = y^* - X^*\left(\beta_{-k}^m + \beta_k^{m+1}\right) \\ &= \left(y^* - X^* \beta_{-k}^m - \beta_k^m x_{(k)}^*\right) + \beta_k^m x_{(k)}^* - \beta_k^{m+1} X^* e_k \\ &= r^m - \left(\beta_k^{m+1} - \beta_k^m\right) x_{(k)}^* \end{aligned} \tag{22}$$

(e) Update the iteration counter m by $m = m+1$.

(f) Update the coefficient index k by:

$$k = \begin{cases} p & \text{if } 0 \equiv m \bmod p \\ m \bmod p & \text{otherwise} \end{cases} \tag{23}$$

(g) Go to (a)

2.2.3. Selection of Tuning Parameters

In practice, it is important to select appropriate tuning parameters in order to obtain a good prediction precision or estimation precision. There are three parameters $(\lambda_1, \lambda_2, q)$ which need to be chosen. As mentioned in the previous section, how to choose a proper q is important, which depends on the nature of data. If sparsity of model is the point of focus, smaller q tends to be proper. The aim of this paper is not only to research the sparsity of variables but also to study grouping effect about strongly correlated variables. Therefore, the best q $(0<q<1)$ should be chosen by experiment data and cross-validation.

Firstly, q $(0<q<1)$ is given a grid of values to be compared. The choice of (λ_1, λ_2) will be different for different q. For each fixed q, cross-validation is used to choose the best (λ_1, λ_2). Typically, given a grid of values for λ_2, 5-fold cross-validation is applied to getting the best λ_1 in terms of minimized mean squared error of models or prediction error for each fixed λ_2. In simulation studies, the mean square error are defined by

$$\text{MSE} = \left(\hat{\beta} - \beta\right)^{\mathrm{T}} E\left(X^{\mathrm{T}} X\right)\left(\hat{\beta} - \beta\right), \tag{24}$$

where $E(X^{\mathrm{T}}X)$ is approximated by the sample covariance matrix of X of out-of-sample data. The formula of MSE is mentioned by Tibshirani [3] can measure the estimation error of model for simulation studies. In real data, because of unknown true parameters, test error defined by

$$\text{Test Error} = \frac{1}{n}\sum_{i=1}^{n}\left(y_i - \hat{y}_i\right)^2 \tag{25}$$

can measure prediction error of model. At last, the chosen value (q,λ_1,λ_2) is the best to compare with other models.

3. SIMULATION STUDIES

In this section, simulation studies are presented the finite sample performance of LqCP. The results analysis are considered from variable selection ability, the estimation errors, grouping effect. Data is generated from the true model:

$$y = X\beta + \sigma\epsilon, \epsilon \sim N(0,1). \quad (26)$$

y is the response variable and X is an $n \times p$ matrix with p predictor vectors and n observations. ϵ is a random error vector with $E(\epsilon)=0$. β is p dimension parameters and σ expresses the volatility of y. Three methods in the simulations study: the elastic net (ENET), the adaptive elastic net (AENET) and the L1CP are listed to be compared. Because these methods have the ability of grouped variable selection. Data is divided into two data sets: training data and testing data. Training data is used to do model fitting and cross-validation. Testing data is used to evaluate the error of models.

For each estimator $\hat{\beta}$, its estimation accuracy is measured by the mean squared error (MSE) in the testing data. The variables selection performance is gauged by C and IC, where C is the number of zero coefficients that are correctly estimated by zero and IC is the number of non-zero coefficients that are incorrectly estimated by zero. In addition, the algorithm's stopping criterion is $\left\|\beta^{n+1}-\beta^{n}\right\|_2 \le 10^{-4}$. In these simulation studies, q is given five different values which are $q=\left(0.1,\frac{1}{3},\frac{1}{2},\frac{2}{3},0.9\right)$. The results in the following tables are chosen by comparing MSE among different q. ENET and AENET are computed by "glmnet" package and L1CP is computed by "lars" package in R language .

Example 1. 100 data sets are generated with sample size 100 observations from the linear regression model $y = X\beta + \sigma\epsilon, \epsilon \sim N(0,1)$, The true regression coefficients vector β is $\beta=(3,-2,1.5,0,\cdots,0)$. Different dimension levels are considered as below. Smaller value $\sigma=2$ is more proper and can make all results stable through the simulation experiment. The correlation matrix is given by $\rho(x_i, x_j) = 0.5^{|i-j|}$. The observation sample size of training data is 70 and testing data is 30. Furthermore, considering the problem of sparsity, noise variables increase by increasing the number of zero in β to generate four different dimensional data. The dimensions of data are respectively 10 (30%), 30 (10%), 60 (5%), 100 (3%). The number of noise variables are respectively 7, 27, 57, 97. That is to say, this part concerns that how the results change with the increase of noise variables. The simulation results are in Table 1 and Table 2.

Example 2. This example is about the case of $p > n$. Similar to the example of Wu, Shen and Geyer (2009), corresponding to high dimensional simulation scenarios with correlated groups, large p and small n, the X_i are simulated from $N(0, \Sigma)$, where the jk-th element of Σ is $0.5^{|j-k|}$ and $\beta = (3,3,3,3,3,-2,-2,-2,-2,-2,1.5,1.5,1.5,1.5,1.5,1,1,1,1,1,0_{180})$.

So there are 20 grouped relevant predictors and 180 noise predictors and $n = 100$. Also, 100 data sets are generated and split data 70/30 into two parts for training data and testing data.

Example 3. About the grouping effect, 100 sample size described by 40 predictors is considered. The true parameters are $\beta = (2,2,2,2,2,3,3,3,3,3,4,4,4,4,4,0_{25})$

and $\sigma = 6$ which is selected to see the performance of models in bigger volatility. The predictors are generated as:

$$x_i = Z_1 + 0.2\epsilon_i, Z_1 \sim N(0,1) \text{ and } i = 1, \cdots, 5$$

$$x_i = Z_2 + 0.2\epsilon_i, Z_2 \sim N(0,1) \text{ and } i = 6, \cdots, 10$$

$$x_i = Z_3 + 0.2\epsilon_i, Z_3 \sim N(0,1) \text{ and } i = 11, \cdots, 15$$

x_i are independently identically distributed $N(0,1)$ for $i = 16, \cdots, 40$, where ϵ_i are independently distributed $N(0,1)$ for $i = 1, \cdots, 15$. In this data, three equally important groups have pairwise correlation $\rho \approx 0.96$, and there are 25 pure noise features. Also, the data is split as 70 observations for training data and 30 observations for testing data.

Table 1. Median mean squared errors of four methods based on 100 data replications with the standard errors estimated by using the bootstrap with B = 500 resamplings on the 100 mean squared errors for different numbers of noise variables.

Methods	30% level	10% level	5% level	3% level
ENET	0.521 (0.0314)	0.957 (0.0722)	1.501 (0.1342)	1.998 (0.1744)
AENET	0.166 (0.0280)	0.242 (0.0288)	0.535 (0.0680)	0.618 (0.1467)
L1CP	0.495 (0.0462)	0.824 (0.0543)	1.007 (0.0463)	1.336 (0.0724)
LqCP	0.200 (0.0309)	0.248 (0.0323)	0.372 (0.0573)	0.382 (0.1472)

Table 2. Median number of C and median number of IC based on 100 data replications for different numbers of noise variables. The number of true noise variables are 7, 27, 57, 97.

	30% level		10% level		5% level		3% level	
Methods	C	IC	C	IC	C	IC	C	IC
ENET	2	0	18	0	44	0	79.5	0
AENET	7	0	26	0	57	0	97	0
L1CP	3	0	17	0	46	0	82	0
LqCP	7	0	27	0	57	0	97	0

From Table 1, It is anticipated that MSE of all methods increase with the increase of noise variables. In 30% and 10% levels, AENET is better than other methods. But in in 5% and 3% levels, LqCP has the minimized MSE, which indicates LqCP performs better in low sparsity levels. In addition, LqCP is much better than ENET and L1CP for all levels. Table 2 illustrates that LqCP can estimate all true zero coefficients and the number of incorrect selection for true non-zero coefficients is always 0 with the increase of noise variables. Especially relative to ENET and L1CP, both of them can not select all true zero coefficients in four circumstances.

In the case of $p > n$ showed in Table 3, LqCP performs best and are more stable. As expressed in the example, the number of true noise variables is 180 and the result from LqCP is 179, which achieves a better selection of zero coefficients than other methods. From Table 4, LqCP performs with the minimized MSE and the minimized standard error which also has a better variable selection ability for true non-zero coefficients and zero coefficients in case of high correlations $\rho \approx 0.96$. It states better estimation effect of parameters of LqCP for grouping effect than other methods.

Table 3. Median mean squared errors of four methods based on 100 data replications with the standard errors estimated by using the bootstrap with B = 500 resamplings on the 100 mean squared errors. The dimension is 200 and the number of true noise variables is 180.

Methods	Median of MSE	C	IC
ENET	3.080 (0.2201)	151	0
AENET	1.795 (0.1956)	177	0
L1CP	2.172 (0.1124)	153	0
LqCP	1.456 (0.0767)	179	0

Table 4. Median mean squared errors of four methods based on 100 data replications with the standard errors estimated by using the bootstrap with B = 500 resamplings on the 100 mean squared errors. The number of true noise variables is 25.

Methods	Median of MSE	C	IC
ENET	8.351 (0.5367)	17	1
AENET	7.870 (0.4619)	19	2
L1CP	9.937 (0.5615)	20	4
LqCP	6.535 (0.4605)	23	0

4. REAL DATA SETS EXPERIMENT

This part is about the performances of LqCP for two real world data sets: the US Crime and Gasoline described by $p=15$ and $p=401$ explanatory predictors respectively. The dimension p of US Crime data set is smaller than the sample size ($n=47$), while the number of variables of Gaoline exceeds largely the sample size $n=69$. Because the true parameters in application is unknown and the concern is prediction accuracy of response variable. Test Error is mentioned in the above as the criterion comparing among models. The selection of q is also based on the minimized test error for $q=\left(0.1,\frac{1}{3},\frac{1}{2},\frac{2}{3},0.9\right)$.

4.1. US Crime Data

This data set is taken from R package "MASS" which contains 47 observations and 15 variables as well as one response variable. Criminologists are interested in the effect of punishment regimes on crime rates which has been studied using aggregate data on 47 states of the USA for 1960. This data set contained the following 16 columns: percentage of males aged 14 - 24 (M), indicator variable for a Southern state (So), mean years of schooling (Ed), police expenditure in 1960 (Pol), police expenditure in 1959 (Po2), labor force participation rate (LF), numbers of males per 1000 females (M.F), state population (Pop), number of non-whites per 1000 people (NW), unemployment rate of urban males 14 - 24 (U1), unemployment rate of urban males 35 - 39 (U2), gross domestic product per head (GDP), income inequality (Ineq), probability of imprisonment (Prob), average time served in state prisons (Time) and the outcome is the rate of crimes in a particular category per head of population (y). The data is split 100 times with a training set of 24 observations and a test set of 23 observations. The results are listed in Table 5.

Clearly, Table 5 shows that LqCP selects 7 variables and has the minimized test error, which performs better than other methods followed by AENET on Crime data. Although the standard error of Test Error is not the lowest, it is very close to the standard error of AENET. The selected variables are percentage of males aged 14 - 24 (M), mean years of schooling (Ed), police expenditure in 1960 (Pol), numbers of males per 1000 females (M.F), number of non-whites per 1000 people (NW), income inequality (Ineq), probability of imprisonment (Prob). These variables can also be selected by other methods. Seeing from prediction accuracy and sparsity effect of models, LqCP is the best.

Table 5. US Crime Data-Median test errors of four methods based on 100 random splits with the standard errors estimated by using the bootstrap with B = 500 resamplings on the 100 test errors. The median number of selected variables by each method is also reported.

Methods	Median of Test Error	Median no. of selected variables
ENET	0.530 (0.0163)	13
AENET	0.534 (0.0143)	10
L1CP	0.537 (0.0180)	9
LqCP	0.518 (0.0145)	7

4.2. Gasoline Data

This data set "Gasoline" comes from R package "pls". It is about infrared spectrum, which contains 69 observations. Recently, infrared spectrum is based on the function of diffuse reflecting degree measured by interval 2 nm from 900 nm to 1700 nm. Gasoline data have 401 prediction variables and the correlations of variables are very high that are almost 0.99. Similarly, the data set is split 100 times into a training set of 40 observations and a test set of 29. The prediction results are reported in Table 6.

Table 6. Gasoline Data-Median test errors of four methods based on 100 random splits with the standard errors estimated by using the bootstrap with B = 500 resamplings on the 100 test errors. The median number of selected variables by each method is also reported.

Methods	Median of Test Error	Median no. of selected variables
ENET	0.0213 (0.0006)	32
AENET	0.0217 (0.0009)	10
L1CP	0.0225 (0.0010)	11
LqCP	0.0191 (0.0007)	8

In the circumstance of high dimension ($p = 401$) and small sample observations size($n = 69$), LqCP is the winner in term of test error, which also gets the least number of variables. Significantly, in this application, the correlation of variables is very high and approaches 1. The result shows that AENET, L1CP and LqCP have similar variable selection effect but ENET is the worst. Therefore, this application proved the efficiency of LqCP from the aspect of $p > n$ and highly correlated variables.

5. CONCLUSION

In this paper, motivated by variable selection and grouped selection property in linear regression problems, a new method called LqCP is proposed, which is a regularization procedure based on the penalized least squares with a mix of L_q $(0 < q < 1)$ norm and correlation-based penalty. Firstly, this paper discusses the current models that have grouping effect including elastic net, adaptive elastic net and L1CP. Similar to them, LqCP can also encourage grouping effect, where strongly correlation among predictors tend to be in or out of the model together. Through the simulation studies in the above, LqCP has better per- formances in terms of variable selection ability with large numbers of noise variables, high dimension $p > n$ and grouped variable selection automatically for high correlations of variables. Additionally, two real data proved LqCP's efficiency from the aspects of $p < n$ and $p > n$ through comparing prediction error with other models. Moreover, the oracle property is important in statistics area. One of future works is to pay attention to prove the oracle property of LqCP and also LqCP can be expended to general regression like logistic regression, quantile regression to solve some regression problems or multi-calss classification problems especially for data with highly correlated variables.

REFERENCES

1. Fan, J. and Li, R. (2006) Statistical Challenges with High Dimensionality: Feature Selection in Knowledge Discovery. Marta Sanz Solé, 595-622.
2. Breiman, L. (1996) Heuristics of Instability and Stabilization in Model Selection. Annals of Statistics, 24, 2350-2383.
3. Tibshirani, R. (1996) Regression Shrinkage and Selection via the Lasso. Journal of the Royal Statistical Society: Series B, 58, 267-288.
4. Knight, K. and Fu, W.J. (2000) Asymptotics for Lasso-Type Estimators. Annals of Statistics, 28, 1356-1378.
5. Fan, J. and Li, R. (2001) Variable Selection via Nonconcave Penalized Likelihood and Its Oracle Properties. Journal of the American Statistical Association, 96, 1348-1360.
6. Zou, H. (2006) The Adaptive Lasso and Its Oracle Properties. Journal of the American Statistical Association, 101, 1418-1429.
7. Zou, H. and Hastie, T. (2005) Regularization and Variable Selection via the Elastic Net. Journal of the Royal Statistical Society: Series B, 67, 301-320.
8. Zou, H. and Zhang, H.H. (2009) On the Adaptive Elastic Net with a Diverging Number of Parameters. Annals of Statistics, 37, 1733-1751.
9. Ghosh, S. (2011) On the Grouped Selection and Model Complexity of the Adaptive Elastic Net. Statistics and Computing, 21, 451-462.
10. Bondell, H.D. and Reich, B.J. (2008) Simultaneous Regression Shrinkage, Variable Selection and Clustering of Predictors with OSCAR. Biometrics, 64, 115-123.
11. EL Anari, M. and Mkhadri, A. (2014) Penalized Regression Combining the L_1 Norm and a Correlation Based Penalty. Sankhya B, 76, 82-102.
12. Tutz, G. and Ulbricht, J. (2009) Penalized Regression with Correlation Based Penalty. Statistics and Computing, 19, 239-253.
13. Frank, I.E. and Friedman, J.H. (1993) An Statistical View of Some Chemometrics Regression Tools. Technometrics, 35, 109-135.
14. Fu. W.J. (1998) Penalized Regressions: The Bridge vs the Lasso. Journal of Computational and Graphical Statistics, 7, 397-416.
15. Marjanovic, G. and Solo, V. (2014) l_q Sparsity Penalized Linear Regression With Cyclic Descent. IEEE Transactions on Signal Processing, 62, 1464-475.

CHAPTER 11

From Fractal Geometry to Fractal Analysis*

Gabriele A. Losa[1#†], Dušan Ristanović[2†], Dejan Ristanović[3], Ivan Zaletel[4], Stefano Beltraminelli[1]

[1]*Institute of Scientific Interdisciplinary Studies, Locarno, Switzerland*

[2]*Department of Biophysics, Faculty of Medicine, Belgrade, Serbia*

[3]*PC Press, Belgrade, Serbia*

[4]*Department of Histology and Embriology, Faculty of Medicine, Belgrade, Serbia*

ABSTRACT

Euclidian geometry pertained only to the artificial realities of the first, second and third dimensions. Fractal geometry is a new branch of mathematics that proves useful in representing natural phenomena whose dimensions (fractal dimensions) are non-integer values. Fractal geometry was conceived in the 1970s, and mainly developed by Benoit Mandelbrot. In fractal geometry fractals are normally the results of an iterative or recursive construction using corresponding algorithm. Fractal analysis is a nontraditional mathematical and experimental method derived from Mandelbrot's Fractal Geometry of Nature, Euclidean geometry and calculus. The main aims of the present study are: 1) to address the dimensional imbalances in some texts on fractal geometry, proving that logarithm of a physical quantity (e.g. length of a segment) is senseless; 2) to define the modified capacity dimension, calculate its value for Koch fractal set and show that such definition satisfies basic demands of physics, before all the dimensional balance; and 3) to calculate theoretically the fractal dimension of a circle of unit radius. A quantitative determination of the similarity using the set of Koch fractals is carried out. An important result is the relationship between the modified capacity dimension and fractal dimension obtained using the log-log method. The text includes some important modifications and advances in fractal theory. It is important to notice that these modifications and quantifications do not affect already known facts in fractal geometry and fractal analysis.

Keywords: Capacity Dimension, Dimensional Balance, Fractal Dimension, Log-Log Method, Similarity

1. INTRODUCTION

Geometry is a branch of mathematics that concerns itself with questions of shape, size, relative position of figures and properties of space. The foundation of Euclidean geometry is a group of elementary notions and axioms. In Euclidean geometry, which has been posed in modern terms by David Hilbert [1] , the elementary notions are the point, straight line and plane. His 20 axioms logically connect the elementary notions. In fractal geometry the elementary notions and axioms are not introduced. Instead, some basic notions are posed, the most important being the fractal set. These basic notions can be defined by means of mathematics. Using the Hausdorff (or capacity) dimensionformula, for every fractal set a fractal dimension can be calculated. We show that this value for the fractal dimension can be obtained using log-log method on a specific manner. The same method can be used for calculating the fractal dimension of nonfractal sets and natural objects as a problem of fractal analysis. Euclidian geometry relates to the artificial realities of the first, second and third dimensions. Fractal geometry is a branch of mathematics concerned with irregular patterns made of parts that are in some way similar to the whole. Unlike conventional geometry which is concerned with regular shapes and whole-number dimensions, fractal geometry deals with shapes that have non-integer or fractal dimensions. This geometry has proven to be a useful tool in analyzing various phenomena [2] since its meaning received a great deal of attention in many natural sciences and engineering. Now it has become widespread and multidisciplinary, and its applications have evolved into many disciplines. Compared to Euclidean geometry, which has a long history for more than 2000 years, fractal geometry is very new. Mandelbrot's famous book "The Fractal Geometry of Nature" was published relatively recently, in 1982. Although conventional approaches based on Euclidean geometry can be well applied to describe regular geometries, it fails to describe the complex geometries. Fractal analysis consists of several methods (most of them being computer-based) useful to assign a fractal dimension and other fractal characteristics to a dataset in all areas of science. Although fractal geometry has been widely applied and used for quantitative morphometric studies of objects, mainly using fractal analysis and calculating the fractal dimensions of sets of objects [3] , there are still some unresolved issues that need to be addressed. Since some concepts of fractal geometry and fractal analysis are determined descriptively and/or qualitatively, this paper offers for some of them the more exact mathematical determination or explanation. Fractal geometry developed as a new geometry of nature [4] -[8] has been conceived with the aim to describe and evaluate the complexity and irregularity of shapes and processes in nature [3] . Up to now fractal geometry and fractal analysis are used in diverse research areas [9] -[12] . The goal of fractal geometry is to provide a platform for the demonstration of the importance of the fractal disciplines and to continuously explore the world of fractals not only in mathematics but also in sciences and technologies. One of the aims of the present paper is to show how fractal geometry gives rise to the fractal analysis confining our consideration to fractal sets and mathematics.

2. METHODS AND RESULTS

2.1. The Geometrical Fractal Set

In Fractal geometry the geometrical fractal set should be considered as an infinite ordered series of geometrical objects defined on a metric space. To determine a fractal set we need specify three things [2] [13] [14] : 1) the shape of a starting object; (the initiator), 2) the iterate algorithm enabling its iterative application on the initiator and then, repeatedly, on all obtained geometrical objects (the generators), and 3) the conditions which these generators should satisfy, before all the properties of geometrical similarity (see Subsection 2.3). In that case, such geometrical object (generator) is called a prefractal [2] [4] [6] . The final result of such infinite procedure is the limit fractal or simply fractal [5] [8] . The initiator, prefractals and fractal represent the geometrical fractal set [13] [15] .

2.2. Koch Fractal Set

Basic definitions and laws of fractal planimetry can be demonstrated on some classical fractal models [3] [13] such as Cantor dust, Sierpinski gasket, Koch triadic curves, Mandelbrot set, etc. For that purpose we chose the triadic (snowflake) Koch curve set. The sequential construction of the this set can begin with the initiator which is an equilateral triangle of the edges length r_0(Figure 1(a)). The iterate algorithm to generate the set of Koch curves (prefractals) consists of recursive reduction of the straight line segments (or the scales) by 1/3 exchanging repeatedly the middle third of each side of the initiator, or a preceding generator, with two sides of a smaller, equilateral triangle whose side is one-third the length of the previous side. The result after the first iteration (the stage of construction z = 1) is shown in Figure 1(b), and that after the second iteration (the stage of construction z = 2), in Figure 1(c). For the Koch prefractals, the length of a segment at the zth stage of construction (r_z) and the number of segments at the same stage (N_z) are respectively,

$$r_z = \frac{r_0}{3^z}; \quad N_z = 3 \cdot 4^z. \tag{1}$$

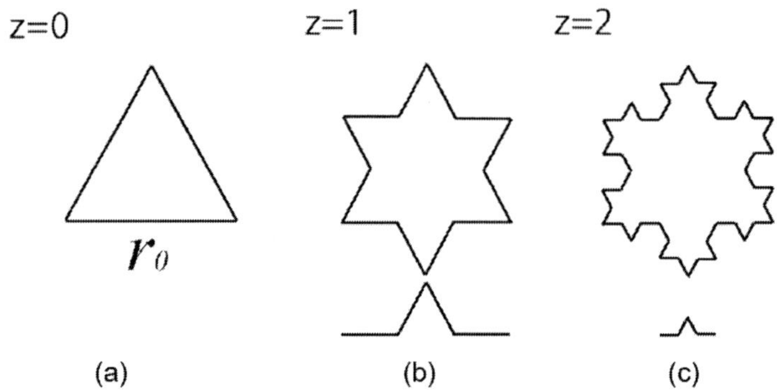

Figure 1. Koch fractal set. (a) Initiator-equilateral triangle, r_0 is its side length; (b) The first stage of construction (z = 1), detail beyond the drawing is the corresponding generating element; (c) The second stage of construction (z = 2), detail beyond the drawing is the corresponding generating element.

2.3. Geometrical Similarity

Similarity is a typical property of fractal sets. To define this concept we introduce a generating element of a generator. A generator is usually made up of straight-line segments (for example, see Figure 1). A logical concatenation of some segments of a generator can be thought of as the generating element [13] of a generator if the whole object can be completely built with such elements by their translations and/or rotations. For example, the drawing in Figure 1(b) can be subdivided into 3 generating elements, that in Figure 1(c) into 12 generating elements, and so on. In our example shown in Figure 1 the generators of the Koch prefractals, at the first and second stages of construction, have the generating elements made of four equal segments each, as shown as details below the drawings in Figure 1(b) and Figure 1(c). Two successive generating elements of two generators of a set of objects can be geometrically similar or not. According to the definition of similarity in Euclidean planimetry, two generating elements of the generators at stages z and z + 1 (say, those in Figure 1(b) and Figure 1(c)) are similar to each other [16] if (1) the ratio of the measure of a segment of the generating element at stage z + 1 and the measure of the corresponding segment of the element at the stage z is constant for all pairs of corresponding segments (e.g., for the four pairs of segments of the two mentioned details in Figure 1) and for all z, and (2) the angles between the pairs of corresponding segments of the two generating elements are congruent.

2.4. The Capacity Dimension

Fractal dimension is a quantitative measure of morphological complexity of an object. There are many different definitions of fractal dimension [2] . Mandelbrot [8] thinks that the "plethora" of distinct definitions of the fractal dimension should be reduced to two: the similarity dimension and Hausdorff dimension. The similarity dimension is basic dimension related to all similar (fractal) sets but this dimension is not the subject of our study. The most important definitions are the Hausdorff and capacity dimensions [2] . These two dimensions are quite similar [2] , but the Hausdorff dimension is rather sophisticated, being a subject of mathematical measure theory. The capacity dimension can be given by [2]

$$D = \lim_{z \to \infty} \left[\frac{\log(N_z)}{\log\left(\frac{1}{r_z}\right)} \right]. \tag{2}$$

Since the logarithm is defined only for dimensionless values (numbers), not for physical quantities like 1/r (where r is a length), we define the modified capacity dimension as:

$$D = \lim_{z \to \infty} \left[\frac{\log(N_z)}{\log\left(\frac{r_0}{r_z}\right)} \right] \tag{3}$$

where r_0 is a reference scale [17] [18] (for example, r_0 can be the length r_0 of the scale of an initiator of the Koch set, Figure 1). The quantity r_0 (Equation (1)) could be included into the definition of the capacity dimension (Equation (3)) because the ratio r_0/r_z in Equation (3) is a dimensionless quantity (a number).

If we submit Equation (1) into the last definition (Equation (3)) and put z to tend to infinity, one can see that the expression in Equation (3) gives $\log(4)/\log(3) = 1.262$:

$$\lim_{z\to\infty} \frac{\log(N_z)}{\log\left[\frac{r_0}{r_z}\right]} = \lim_{z\to\infty}\left(\frac{1}{z} + \frac{\log 4}{\log 3}\right) = D = 1.262. \tag{4}$$

Fractal dimension D is the main quantifier to measure complexity of a set of geometrical and natural objects. The larger the D, the higher the complexity of the set is. For example, since the fractal dimension of the Sierpinski set is $\log(3)/\log(2) = 1.585$, it means that the last set is more complex than the Koch set (Equation (4)).

2.5. Log-Log Method

Considering Equation (3) it would be interesting to analyze a relation between N_z and r_z of the Koch fractal set in a log-log coordinate system. If along the horizontal coordinate axis we put the values $r_z = 1/3^z$ for $z = 0,1,2,3,\cdots$ (the value of r_0 is set to 1for visual clarity) and along the vertical coordinate axis the values of $N_z = 3\cdot 4^z$ also for $z = 0,1,2,3,\cdots$ and fit the power function to these data using Microsoft Excel software, the decreasing straight line graph shown in Figure 2 is obtained. The fitting power function is:

$$N_z = 3\cdot r_z^{-1.262} \tag{5}$$

with the coefficient of determination $R^2 = 1$. The fitting parameter is −1.262 which is equal to the negative fractal dimension of the Koch fractal set already obtained using the formula for modified capacity dimension (Equation (3)). This procedure can be thought of as the log-log method. This method is also used in many other fractal techniques, particularly in computer-based ones.

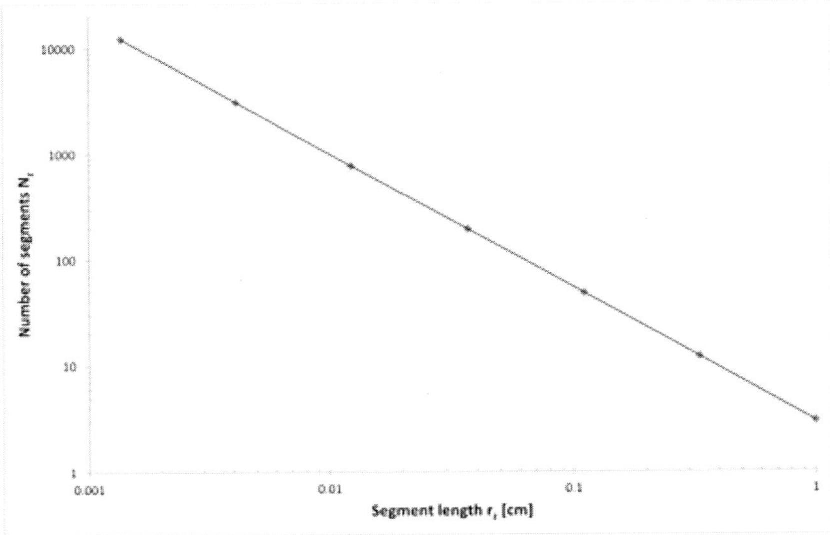

Figure 2. Log-log method applied to the Koch construction (the beginning of the construction is shown in Figure 1). The fitting graph of relationship between the number of segments and segment sizes (in cm) is a decreasing straight line with the high coefficient of determination R^2.

2.6. A Nonfractal Set

Following the fractal methodology, we inscribe an equilateral triangle (as the initiator) in a circle of unit radius (Figure 3(a)) and state the following iterate algorithm: from the middle of each triangle side the normal is erected to the sections with the circle. These three sections are connected to the nearest vertices of the triangle sides forming a hexagon inserted in the circle (Figure 3(b)). The result of the next iteration (Figure 3(c)) is a dodecagon. The next polygons can be constructed using the same algorithm. Inspecting the details below the drawings in Figure 3(b) and Figure 3(c), which are the generating elements for these polygons (generators), it is obvious that they are not mutually similar since the angles between the corresponding segments of the generating elements are different. The condition of side ratios constancy (geometrical similarity) also fails. Namely, it is known that for a circle of unit radius:

$$r_{2n} = \sqrt{2 - 2\sqrt{1 - 0.25 \cdot r_n^2}} \tag{6}$$

where r_n is the side length of a regular polygon inserted in the circle, n is the number of sides of the polygon and r_{2n} is the side length of the polygon with 2n sides. From this expression it is obvious that the ratio r_{2n}/r_n is not constant for every n (e.g., $r_6/r_3 = 0.577$, $r_{12}/r_6 = 0.518$ etc.), so that the class of such regular polygons inscribed in a circle cannot be considered as a set of prefractals.

If n tends to infinity, the polygons tend to the circle, which is not a limit fractal. Therefore this set of regular polygons inscribed in a circle cannot be a fractal set.

On the other hand, if we apply the log-log method to the data of the polygonal set obtained, a straight-line graph is again obtained (Figure 4). The equation of this graph is

$$N_z = 5.814 \cdot r_z^{-1.027} \tag{7}$$

which is analogous to Equation (6). From this equation the fitting parameter −1.027 corresponds to the negative value of the fractal dimension D of the nonfractal set considered. This is the fractal dimension of a circle of unit radius (which is the limit object). Using the mass method Smith et al. [3] have measured the fractal dimension of a circle and found D = 1.02. Using mathematics (with no measurements) we found for a circle of unit radius a more precise value D = 1.027 (Equation (7)).

2.7. Properties of Fractal and Nonfractal Sets

The main properties of geometrical fractal sets are geometrical self-similarity and scale invariance. The object's property known as self-similarity was first coined by Mandelbrot [5] . It can be geometrical or statistical. A fractal pattern is said to be geometrically self-similar if every small piece of an object resembles the whole object. This definition of the concept "geometrical self-similarity" should be quantified since small pieces that constitute geometrical or natural objects are rarely copies of the whole object [14] .

We have offered a more exact interpretation of this descriptive definition introducing a generating element of a generator as a "small piece". Therefore every fractal set (which must satisfy geometrical similarity) should satisfy the property of geometrical self-similarity since by definition every small generating element is always similar to any other large generating element.

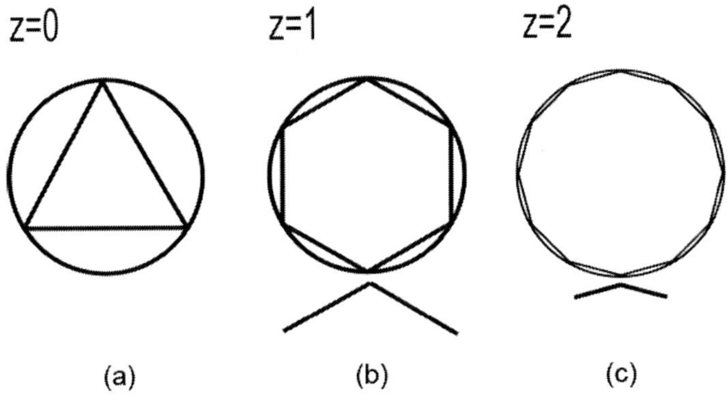

Figure 3. Nonfractal set of regular polygons. (a) Initiator- equilateral triangle inscribed in a circle of unit radius; (b) The first stage of construction (z = 1) is a hexagon; the detail beyond the drawing is its generating element; (c) The second stage of construction (z = 2) is a dodecagon with the corresponding generating element.

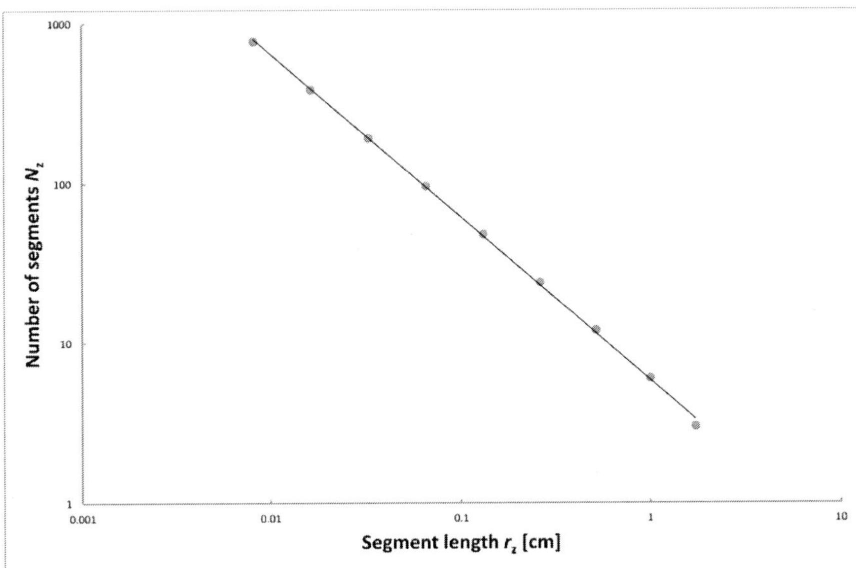

Figure 4. Log-log method applied to the set of polygons (Figure 3). The fitting graph is a decreasing straight line (see Figure 2 for comparison).

Scale invariance is a feature of objects or laws that they do not change if scales of length or other variables are multiplied by a common factor. For example, this common factor for the Koch fractal set is 1/3. If r_0 is 1 cm (Figure 1(a)), the length of the segment in Figure 1(b) is 1/3 cm, that in Figure 1(c) is 1/9 cm, etc. This means that if we multiply the length of each segment of the generating element in Figure 1(b) by 1/3, the same but reduced generating element of the prefractal shown in Figure 1(c) will be obtained. Nonfractal sets cannot satisfy geometrical similarity and properties of fractal sets.

2.8. A Fractal Set versus a Nonfractal Set

We showed that Koch set is a fractal set because it satisfies the geometrical similarity. It means that the ratio r_{z+1}/r_z is the same for every z. If z tends to infinity, we obtained the limit fractal which cannot be drawn or imagined. Nevertheless this fractal exists because we calculated the fractal dimensions of the limit fractal using both the formula for modified capacity dimension (Equation (3)) and log-log method. In both cases we found a real and finite value D = 1.262. We conducted the same procedure using the set of polygons inscribed in a circle. Although the polygonal set is not a fractal set we calculated the fractal dimension of this limit object using the log-log method. In fact, we calculated the fractal dimension of the circle which corresponds to the limit object in which the set of polygons is inscribed. Besides, the initiator in Fractal geometry is included into the fractal iterations as a part of prefractals while the starting object (the circle) in fractal analysis does not change its integrity during iterations.

The circle is a closed line, but the log-log method can be applied to any open line (Figure 5). We found that in Fractal geometry the values of segment lengths were strongly determined by the property of geometrical similarity, while in the fractal analysis values and shapes of open lines are not restricted (Figure 5 and Figure 6). Fractal analysis represents a formation of Fractal geometry consisting in using log-log and others methods [3] , and ignoring the property of geometrical similarity. Therefore we can use any set of regular polygons and always calculate the fractal dimension of the object in which these polygons are inscribed. For example, fractal analysis enables to calculate the fractal dimension of the border of any natural object (not of the set of objects) inscribing a set of straight lines of known and constant lengths along the object's border and count the number of segments. These data in a log-log coordinate system give the fractal dimension of the natural object (coastlines, borders between nations, dendrites of neurons etc.). This method has been the basis of the Richardson's method of fractal dimension measurement of natural objects [19] . It may be opportune to underline that an unlimited extension of log-log coordinate system is not valid when natural and biological objects are investigated. As previously reported, the fractal dimension could be reliable evaluated within a defined scale interval, called "fractal window" [17] [18] .

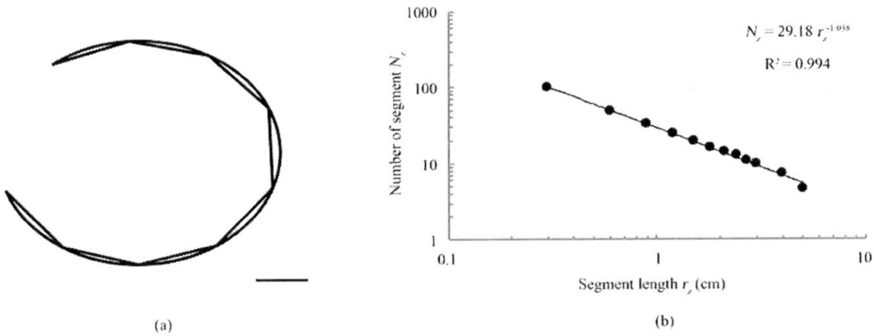

Figure 5. The fractal dimension of an open circle-like curved line. (a) The circle-like line considered, the length of a straight line segment is 1.5 cm, the bar is 1 cm; (b) Relation between N and r represented in a log-log coordinate system, the graph is a straight line whose exponent represents a negative value of the fractal dimension of the curve, R^3 is the coefficient of determination.

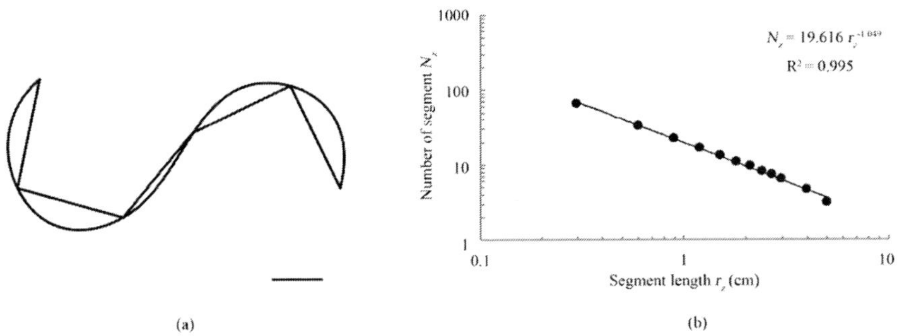

Figure 6. The fractal dimension of an open wavy line. (a) The curved line considered, the length of a straight line segment is 2.2 cm, the bar is 1 cm; (b) Relation between N and r, the graph is a result of application of the log-log method and re- presents a decreasing straight line whose exponent is a negative value of the fractal dimension of the curve in (a).

2.9. Richardson's Method

The polygonal method needs a group of about ten printed copies of a considered line and in each of them inscribes different regular polygon (Figure 5 and Figure 6). This method is rather tedious and time-consuming. A simpler but still time-consuming method is presented in Richardson's paper [20] . Richardson, as a physicist and Quaker, wondered if there are more wars between nations that share longer common border. Answering this question required measurement of the national borders lengths. To measure the length of the border drawn on a map of Atlas, he used as a ruler a divider having sharp points at the ends of both arms. For a given distance between the sharp points of the divider the total length of the border was given by the number of divider's steps along the border multiplied by the divider step size.

Since the length of a polygon's segment (Figure 5 and Figure 6) is the same as the corresponding divider's step size, all the measurements can be done on one printed line. We used this method when preparing Figure 2(b), Figure 4(b), Figure 5 and Figure 6.

The term fractal was introduced by Mandelbrot based on his paper on self-similarity [21] . In that paper, Mandelbrot cited previous work by Richardson [20] and noted that the diagrams presented in Richardson's paper lead to the conclusion that approximate length of the border between two nations can be given by the relationship [8]

$$L(r) \propto F \cdot r^{1-D} \tag{8}$$

where the value of the exponent D seems to depend upon the shape of the border chosen. Topology fails to discriminate between different borders [8] . Since Equation (9) can be rewritten as

$$L(r) = r \cdot Fr^{-D} \tag{9}$$

it is obvious that the factor Fr^{-D} corresponds to the number of divider's steps. For a given r the border of smaller length has smaller number of steps (smaller F) in relation to a longer border of the same shape (the same D) which has larger number of steps (larger F).

To Richardson, the D was a fitting parameter of no particular significance. Having in mind "unearthed" Richardson's work in which he claimed that his lines' slopes had no theoretical interpretation [8] , Mandelbrot claimed that the exponent D should be interpreted as a fractal dimension. We find that among many fractal analysis techniques, only Richardson's method enables correct calculation of the length of an object's border or irregular line.

3. DISCUSSION

When the physical or biological problem is stated in mathematical terms, dimensional balance should be a routine part of the solution of any problem. Exponents and logarithms must always be dimensionless [22] . When dimensioned quantities appear in exponents or in logarithms, they must be combined with the same dimensioned quantities so that their quotient or product is dimensionless [17] [18] . It seems that some of the authors do not take care of these facts. The problem is particularly distinct in defining the fractal dimensions in fractal geometry [2] [13] [23] [24] . One of the main aims of the present study is to define and discuss physically correct of modified capacity dimension. Bassingthwaighte et al. [2] noted that the capacity dimension tells us how much balls needed to cover the object when the size of the balls (r) decreases. In the analytical definition of capacity dimension there exists log (1/r). West and Deering [13] have defined the similarity dimension as the ratio of two logarithms log N/log (1/r). In defining the box-counting dimension Falconer [23] showed lower and upper box- counting dimension by the expression containing log δ where δ is the diameter of a corresponding set. Edger [24] also used log (1/r) in the definition of the upper box-counting dimension, where r is a diameter of the set. Richardson's method of measuring the fractal dimension is robust with very high correlation coefficients but it is, at the same time, tedious and time-consuming. Therefore, the need for more handsome methods for that measurement emerges. The conventional and most popular box-counting (computer-based) method is one suitable to measure the fractal dimensions of geometrical and natural objects. This method is commonly used among other fractal techniques [3] -[5] . Unfortunately, it seems that this method demonstrates some mistakes. Using this method we showed that the fractal dimension of a straight line segment is 0.927 not 1. The fractal dimension of a circle of the radius 1 cm obtained using purely mathematical method, is 1.027, while that measured using the box-counting method is larger than 1.15. The fractal dimension increases with the resolution of a figure and depends on its position on the screen. We showed that Richardson's method does not depend on these factors.

REFERENCES

1. Hilbert, D. (1899) Grundlagen der Geometrie. Teubner, Leipzig.
2. Bassingthwaighte, J.B., Liebovitch, L.S. and West, B.J. (1994) Fractal Physiology. Oxford University Press, New York.
3. Smith Jr., T.G., Lange, G.D. and Marks, W.B. (1996) Fractal Methods and Results in Cellular Morphology: Dimensions, Lacunarity and Multifractals. Journal of Neuroscience Methods, 69, 123-136.
4. Panico, J. and Sterling, P. (1995) Retinal Neurons and Vessels Are Not Fractal but Space-Filling. Journal of Comparative Neurology, 361, 479-490.
5. Fernández, E. and Jelinek, H.F. (1998) Use of Fractal Theory in Neuroscience: Methods, Advantages and Potential Problems. Methods, 24, 9-18.
6. Ristanovic, D., Nedeljkov, V., Stefanovic, B.D., Milosevic, N.T., Grgurevic, N. and Stulic, V. (2002) Fractal and Nonfractal Analysis of Cell Images: Comparison and Application to Neuronal Dendritic Arborization. Biological Cybernetics, 87, 278-288.
7. Losa, G.A., Di Ieva, A., Grizzi, F. and De Vico, G. (2011) On the Fractal Nature of Nervous Cell System. Frontiers in Neuroanatomy, 5, 51-62.
8. Mandelbrot, B.B. (1982) The Fractal Geometry of Nature. W Freeman & Co, New York. [Citation Time(s):6]
9. Bizzarri, M., Giuliani, M., Cucina, A., Anselmi, F.D., Soto, M.A. and Sonnenschein, C. (2011) Fractal Analysis in a System Biology Approach to Cancer. Seminars in Cancer Biology, 21, 175-182.
10. Thamrin, C., Stern, G. and Frey, U. (2010) Fractals for Physicians. Paediatric Respiratory Reviews, 11, 123-231.
11. Perrotti, G., Aprile, G., Degidi, M. and Piattelli, A. (2011) Fractal Analysis: A Novel Method to Assess Roughness Organization of Implant Surface Topography. The International Journal of Periodontics and Restorative Dentistry, 3, 633-639.
12. Talu, S. (2011) Fractal Analysis of Normal Retinal Vascular Network. Oftalmologia, 5, 11-16.
13. West, B.J. and Deering, W. (1994) Fractal Physiology for Physicists Levi Statistics. Physics Reports, 246, 1-100.
14. Grizzi, G., Grimaldi, G. and Dioguardi, N. (2001) Fractal Geometry: A Useful Tool for Quantifying Irregular Lesions in Human Liver Biopsy Specimens. Italian Journal of Anatomy and Embryology, 106, 337-346.
15. Feder, J. (1998) Fractals. Plenum Press, New York.
16. Milosevic, N.T. and Ristanovic, D. (2007) Fractal and Nonfractal Properties of Triadic Koch Curve. Chaos, Solitons and Fractals, 34, 1050-1059.
17. Losa, G.A. and Nonnenmacher, T.E. (1996) Self-Similarity and Fractal Irregularity in Pathologic Tissues. Modern Pathology, 9, 174-182.
18. Losa, G.A. (2002) Fractal Morphology of Cell Complexity. Biological Forum, 95, 239-258.
19. Ristanovic, D. and Milosevic, N.T. (2012) Fractal Analysis Methodologies for Biomedical Researchers. Theoretical Biology Forum, 105, 99-108.
20. Richardson, L.F. (1961) The Problem of Contiguity: An Appendix to Statistics of Deadly Quarrels. General System Yearbook, 6, 139-187.
21. Mandelbrot, B.B. (1967) How Long Is the Coast of Britain? Statistical Self-Similarity and Fractional Dimension. Science, 165, 636-638. http://dx.doi.org/10.1126/science.156.3775.636
22. Simon, W. (1972) Mathematical Techniques for Physiology and Medicine. Academic Press, New York, London.

23. Falconer, K. (2003) Fractal Geometry, Mathematical Foundations and Applications. 2nd Edition, John Wiley and Sons, New York and London. http://dx.doi.org/10.1002/0470013850
24. Edger, G. (2008) Measure, Topology and Fractal Geometry. 2nd Edition, Springer, New York. http://dx.doi.org/10.1007/978-0-387-74749-1

NOTES

*The present paper is dedicated to the imperishable memory of Professor Benoît Mandelbrot, who passed away October 10, 2010.

CHAPTER 12

General Topology of the Universe

Aalok Pandya[1,2]

[1]*Department of Physics, University of Rajasthan, Jaipur, India*

[2]*Department of Physics, JECRC University, Jaipur, India*

ABSTRACT

General topology of the universe is described. It is concluded that topology of the present universe is greater or stronger than the topology of the universe in the past and topology of the future universe will be stronger or greater than the present topology of the universe. Consequently, the universe remains unbounded. The general topological approach comprises of powerful techniques that could prove to be useful to prescribe mathematical constraints on the global character of the universe as well as on the manifold of space-time.

Keywords: General Topology, Manifold, Boundary, Causality, Euclidean Space, Discreteness, Limit Points

1. INTRODUCTION

General topology of the universe and its evolutionary aspects are discussed in this paper. In the scenario of universe evolving with time, we conclude that topology of the present universe is greater or stronger than the topology of the universe in the past and topology of the future universe will be stronger or greater than the present topology of the universe. Consequently, the universe remains unbounded. The conditions of whole space topology can be treated as mathematical constraints on the global topology of the universe. The present exercise draws motivation from the recent explorations [1] [2] of relativistic aspects of the causal evolution of world lines and the earlier work on causal set topology by Joshi [3] . The Null-Cone representation by means of causal set topology [1] [2] revives interests in topological analysis of the universe. In earlier attempts, topological aspects of the large-scale structures in the universe have been discussed by Hartle and Hawking [4] , Ellis and Hawking and others [4] - [6] . Topological aspects of the birth of the universe discussed by Miao Li [7] also lead to a substantial development of mathematical ideas on physical topology of the Universe. The geometry and topology of the universe in the physical realm have been discussed by Lachieze-Rey and Luminet [6] and Mellot [8] . All the care is taken to ensure that fundamental concepts of general topology are not violated. Moreover, the present discussion has been developed with a focus on the earlier studies such as Pankaj Joshi [3] , Penrose [9] and Vyas [10] wherein causal structure and general topology of the space-time and the global aspects in gravitation and cosmology were discussed rigorously. The computer simulation and animation in $3+1\ D$ space in the reference [2] motivates us to further investigate the unexplored aspects of the general topology of space-time.

It is felt that apart from the manifold of the universe, other general topological properties of the universe and the topological changes following them are equally important. The general topology of the universe can be uniquely described if its physical features are specified. A topological space is an abstract conception and thus it is always not possible to attribute physical properties to it. But, the other way round is certainly possible. The topology of a physical object with evident metricity that is inbuilt distance features can certainly be described. Moreover, it is pertinent to mention that the physical realities of the universe [3] - [8] are in perfect agreement with the mathematical conditions discussed in the present paper. Thus, it is good ground to assume that topological considerations could prove to be crucial mathematical constraints on the overall geometry of the universe.

We have several models of the universe at the cross-roads of scientific development, but we consider only one scenario wherein the universe is the single largest entity that encloses every physical object that exists in nature including space and time. The discussion in this paper is developed on two logical grounds such that firstly it justifies the observed facts and secondly it satisfies the mathematical conditions of the whole space topology. The universe without boundary has already been described in the framework of General Theory of Relativity (GTR) too. But, we feel that GTR and Einstein equation describe the universe in a specific physical scenario of a theoretical framework, whereas, general topology of the whole space is absolute truth, and thus it is not limited to any specific physical scenario. This discussion addresses not only the boundary of the universe, but also a detail spatial analysis of the universe with reference to open space, closed space, boundary, compactness, connectedness, limit points, frontier points, interior points, topological transitions and causality and topology. It is concluded in this paper that the general topology of the whole space could be a very good philosophical basis for describing geometry of the universe as it prescribes mathematical constraints on the global geometry of the universe.

2. OPENNESS, COMPACTNESS, AND THE BOUNDARY

The metric of the universe can be considered as the super metric or the universal metric defined on the whole space U. Then such a universal topology is the set of all possible sets and subsets including the null set. With the consideration that everything that exists in nature is contained in the universe, the complement of the universe is obviously nonexistent.

By definition universal topology being collection of all possible sets and subsets, is always discrete in nature. Thus, the topology of the universe is a discrete topology. Also, the whole space U is always open as well as closed. We explain this geometrical duality using both geometrical physical explanations. Since, $U \cap \{\phi\} = \{\phi\}$; U is closed if its complement $\sim U$ is open and U is open if its complement $\sim U$ is closed, as the null space ϕ is both open as well as closed.

Mathematically, if the closure of the space U is given by $\bar{U}$, the condition of the whole space implies that: $U = \bar{U}$.

Hence, if the closure of the universe is equal to the universe itself which is precisely true for the physical universe, the set U of the universe is said to be closed.

An important mathematical corollary [11] asserts that boundary of a space can be given by: $\text{Bd}\ U = \bar{U} - U$.

But, since $U = \bar{U}$, the boundary of the space vanishes as $\text{Bd}\ U = \phi$.

Thus, $\text{Bd}\ U = \phi \Leftrightarrow$ is both open and closed.

This agrees with the conclusion which we have already drawn but by another reason. It is interesting to cite here a mathematical theorem, which states that boundary of a boundary is always empty [11] :

$$\partial\partial M = \phi.$$

Furthermore, if d is a metric defined on space U, a subset A of U is said to be bounded if there is some number N such that $d(a_1, a_2) \leq N$; for every pair of points a_1, a_2 of A.

If A is bounded, the diameter of A is defined to be the number

$$\textbf{diam}(A) = \textbf{least upper bound } \{d(a_1, a_2) : a_1, a_2 \in A\}.$$

But, we know that for continuously expanding U, particularly in inflationary scenario, this condition cannot be fulfilled as for every pair a_1, a_2; the distance $d(a_1, a_2)$ cannot remain less than or equal to a fixed number N forever. The aforesaid conclusion is in agreement with the statement that the universe has no boundary, because a boundary represents in a sense the "edge" of the universe not detected by any astronomical observations [4] [6] . However, the metric of the universe is closed but it cannot be bounded, and therefore it is not at all compact.

3. CAUSALITY AND SPACE-TIME TOPOLOGY

An event p chronologically precedes another event q, denoted by $p \ll q$, if there is a smooth future directed time-like curve from p to q [1] - [3] [9] [10] .

If such a curve is non-space-like that is, either time-like or null, we say that p causally precedes q, or $p < q$. The chronological future $I^+(p)$ of p is the set of all points q such that $p \ll q$. The chronological past of p is defined dually [3] . Thus we have

$$I^+(p) = \{q \in M \,|\, p \ll q\}, \tag{1}$$

and

$$I^-(p) = \{q \in M \,|\, q \ll p\}. \tag{2}$$

The causal future (past) for p can be defined similarly

$$J^{+}(p)=\{q\in M\,|\,p<q\},\tag{3}$$

$$J^{-}(p)=\{q\in M\,|\,q<p\}.\tag{4}$$

We now observe yet another duality, which can be realized for all $p\ll q\ll r$, such that

$$I^{-}(r)\supset I^{-}(p),\tag{5}$$

but,

$$I^{+}(r)\subset I^{+}(p).\tag{6}$$

And similarly, causal future (past) could be defined dually as:

$$J^{-}(r)\supset J^{-}(p),\tag{7}$$

and

$$J^{+}(r)\subset J^{+}(p).\tag{8}$$

Proposition 3.1: If there is a smooth future directed time-like curve with all $p\ll q\ll r$ on it, then

$$I^{-}(r)\supset I^{-}(p) \text{ and } I^{+}(r)\subset I^{+}(p).$$

Proof: If we consider $p\ll q\ll r$ in the given case, then obviously

$$q\notin I^{+}(r), \text{ and } q\in I^{+}(p)-I^{+}(r).$$

Thus, the fact that $I^{+}(p)-I^{+}(r)\neq\phi$ implies $I^{+}(r)\subset I^{+}(p)$.

Similarly, one can easily assert that: $q \notin I^{-}(p)$ and $q \in I^{-}(r) - I^{-}(p)$.

And, the fact that $I^{-}(r) - I^{-}(p) \neq \phi$ implies $I^{-}(p) \subset I^{-}(r)$.

Hence is the proposition.

The computer simulation containing animation in $3+1\ D$ space in the reference [2] is reassuring that this discussion is on firm footing.

Also, this conveys the feature of open-ended future of the universe at any point of time. If the universe goes on expanding then the topology of the universe will tend to be greater or stronger topology. And the topology of the universe at any point of time in future will be stronger or greater than the present topology, and it will further tend to be greater and stronger. This is due to the fact that it will have more and more number of open sets, following this enhancement.

Proposition 3.2: The topology of the present universe is greater or stronger than the topology of the universe in the past and topology of the future universe will be stronger or greater than the present topology of the universe.

Proof: If we consider the topology of the universe defined on the whole space $(U, \Im)$, at some point of time in the past then we can construct the topologies of the universe at present and in the future, relative to this. The topology of the present universe can be given as:

$$\Im_{\text{present}} = \{U, \phi, \bigcup\{x_\mu\}\};$$

where, $\bigcup\{x_\mu\} \subset U$ is the collection of all space-time points accumulated in U, after having defined the topology $\Im_{\text{past}}$.

Similarly, we can define the topology of the universe of future as:

$\Im_{\text{future}} = \{U, \phi, \bigcup\{x_\nu\}\}$, where $\bigcup\{x_\nu\} \subset U$, and $\bigcup\{x_\nu\}$ is the collection of all space-time points accumulated in U, after having defined the topology $\Im_{\text{past}}$; such that $\bigcup\{x_\nu\} \supset \bigcup\{x_\mu\}$.

Hence, $\Im_{\text{past}} \subset \Im_{\text{present}} \subset \Im_{\text{future}}$.

In other words the topology of $U(\text{past})$ is smaller or weaker than the topology of $U(\text{present})$, and topology of $U(\text{future})$ will be greater or stronger than the topology of $U(\text{present})$.

Hence is the proposition.

This further guarantees that the universe will remain unbounded. We can alternatively explain the topological transition of the universe from a weaker or smaller to a stronger or greater topology in terms of matter-energy distribution. Considering the total energy of the universe to be conserved, the continuous expansion of the universe will lead to increase in the intergalactic distances and decline in the matter density over the space. This implies that the topology of the universe in past $U(\text{past})$ was coarser than the $U(\text{present})$, and topology of $U(\text{future})$ will be finer than the $U(\text{present})$.

4. LIMIT POINTS

Now, we look for the limit points of the universe. However, this part of the discussion may not be applicable to the consideration of multi-universe Cosmos. If the universe is not static and is continuously expanding, its farthest possible extents can be easily approximated. The radius of the expanding universe can be given by the well-known relation $R = \frac{c}{H}$, or alternatively $R \approx c\tau$; where H^{-1} is Hubble age, c is the velocity of light, and τ is characteristic time scale or the age of the universe. Alternatively, for the sake of generality, this distance L_{Horizon} that is (comoving) radius of the observable universe (with or without inflation) is given with a good precision as [6] :

$$L_{\text{Horizon}} \approx \frac{u}{(1-\gamma)} L_{\text{Hubble}}.$$

This follows from the logic that the size of the universe can be determined by means of the distance traversed by the fastest moving particle ever since the genesis of the universe. Obviously, this radius is only instantaneous radius and is increasing continuously. Moreover, the universe is considered to be homogeneous and isotropic. With the expansion of the universe, the continuous inclusion of more number of space (or space-time) points restricts the universe from being bounded and compact.

5. DISCUSSION

One can define a manifold over a topology if that topological space is locally homeomorphic to R^n, which means it is Hausdorff and possesses metric properties, and connectedness is well described in that. Moreover, a smooth manifold can be described only if continuity as well as differentiability is well defined on it. The manifold of the universe can only be described, if it satisfies these requisite conditions. And we know that it does satisfy them.

Also, if the origin of the universe is considered to have followed the single event say Big-Bang, then all the physical objects and events in the universe are causally connected. Thus, one can say that the metric of the space-time is connected too. Physically interesting known space-time examples such as the Schwarzschild geometry and Robertson-Walker space-times are topologically Hausdorff. Also, for the sake of the causal regularity of space-time, the continuum of the space-time structure is necessary which would ensure a globally well- behaved space-time [3] . In nut-shell, at least for the Minkowski space-time, the manifold is globally Euclidean with the topology of R^4 over the Lorentzian metric. In case of general theory of relativity, the principles of GTR effectively imply that it is the space-time metric, and the quantities derived from it, that must appear in the equations for physical quantities and that these equations must reduce to the flat space-time case when the metric is Minkowskian. This is the basic content of the GTR where the space-time manifold is allowed to have topologies other than R^4 too. However, locally and globally, there could be important differences in the causal structure due to a different local space-time topology, strong gravitational fields and so on [3] .

There is a reason to extend the understanding of topology of space-time to the universe at large. Once we understand the nitty-gritty of causal structure and topological aspects of space-time, the understanding of causal structure enables us to deliberate space-like, light-like and time-like events. And the universe at large is nothing but an assembly of space-like, light-like and time-like events. However, the manifold of the universe can have different possible topological features depending upon the physical conditions considered in a specific scenario the global aspects of the general topology of the universe as the whole space remain unchanged. Thus, a prominent suggestion emerges that the mathematical conditionality of the whole space topology can be used as a constraint to describe the overall topology of the universe. In other words, the general topological approach comprises of powerful techniques that could prove to be useful to prescribe mathematical constraints on the global character of the Universe as well as on the manifold of the space-time. In future we can explore frontier points, interior points and other topological notions too.

ACKNOWLEDGEMENTS

The author is grateful to Professor T. Pandmanabhan of Inter University Centre for Astronomy and Astrophysics (IUCAA) for his encouragement and patient reading of the manuscript.

REFERENCES

1. Aalok, P. (2012) Null Cone Representation of Interactions and Causal Structure. Physics Essays, 25, 228-232.
2. Aalok, P. (2013) Computational Investigations of the Null Cone Representation of Interactions. Physics Essays, 26, 24-26.
3. Joshi, P.S. (1993) Global Aspects in Gravitation and Cosmology. Oxford Science Publications, Oxford.
4. Hartle, J.B. and Hawking, S.W. (1983) Wave Function of the Universe. Physical Review D, 28, 2960-2975; Hawking, S.W. (1984) Quantum State of the Universe. Nuclear Physics B, 39, 257-276.
5. Hawking, S.W. and Ellis, G.F.R. (1973) The Large Scale Structure of Space-Time. Cambridge University Press, Cam-bridge.
6. Lachieze-Rey, M. and Luminet, J.P. (1995) Cosmic Topology. Physics Reports, 254, 135-214.
7. Li, M. (1986) On Topological Aspects of the Birth of the Universe. Physics Letters B, 173, 229-232.
8. Melott, A.L. (1990) The Topology of the Large Scale Structure in the Universe. Physics Reports, 193, 1-39.
9. Penrose, R. (1972) Techniques of Differential Topology in Relativity. A. M. S. Colloquium Publications, SIAM Philadelphia.
10. Vyas, U.D. and Joshi, P.S. (1989) Topological Techniques in General Relativity in Geometry and Topology. World Scientific, Singapore, 315-331.
11. Munkres, J.R. (1975) Topology: A First Course. Prentice-Hall Inc., Englewood Cliffs.

CHAPTER 13

Efficient Combination of Topology and Parameter Optimization

Yusheng Lin, Zheng Sun, Alexandru Dadalau, Alexander Verl

Institute for Control Engineering of Machine Tools and Manufacturing Units (ISW), University of Stuttgart, Stuttgart, Germany

ABSTRACT

This paper presents a combination method of Particle Swarm Optimization (PSO) and topology optimization. With this method a better result can be achieved compared with the sequential application of these two optimization methods. It inherits the ability in finding global optimum from PSO and also suits for discretized design domain. Some special schemes are used in order to provide higher computation efficiency. This method has only been tested with a convex optimization problem. The application in case of a concave problem will be a future study.

Keywords: Topology Optimization, Parameter Optimization, Combination Optimization, PSO, Structure Design

1. INTRODUCTION

Parameter optimization and topology optimization are two powerful tools in mechanical design, which affect remarkably the shape and the performance of a structure. Traditionally, the usage of these two procedures is separated and sequential. The structure is often parameter-optimized firstly, and then topology-optimized. However, the sequential method has potential insufficiencies. The general size of a structure, which is usually regarded as a design domain in topology optimization, is always changed due to the parameter optimization. The change in geometric size is sometimes useless or even conflicting as a preparation to topology optimization. In this paper, an efficient method of combined optimization is presented. It executes the parameter optimization with continuous or discretized variables and the topology optimization simultaneously, and thus achieves a better result than the sequential optimization. The paper is organized as follows: The theoretical basis of topology and parameter optimizations is introduced in Section 2. For a better presentation and explanation, an optimized case with variable bearing hole position is shown in Section 3. The implementation of the combined optimization and the methods for increasing the efficiency are proposed in Section 4. Finally, conclusion and future work are provided.

2. THEORETICAL BASIS

2.1. Finite Element Method

The finite element method (FEM) is a numerical method that divides a domain into small connected sub-do- mains (elements), such that differential equations can be solved by solving a set of polynomial equations. With the development of computer science, more and more complicated mechanic problems in different fields are now solvable with FEM. One feature of FEM we need here is the definition of parameters for each element individually. This is important for the implementation of topology optimization, since the design variables should be set for every single element locally.

2.2. Topology Optimization

Topology optimization is a mathematical approach that optimizes the material layout within a given design space, for a given set of load and boundary conditions. It is widely used in the field of shape or structure design, especially for the lightweight design without losing much stiffness. Different kinds of topology optimization algorithms have already been carried out. Xie and Steven introduced the Evolutionary Structural Optimization (ESO) to eliminate the less effecting elements of a structure [1] [2] . Querin et al. proposed Bi-directional Evolutionary Structural

Optimization (BESO) as an improvement of ESO that can also add elements in a design domain to compensate the structure defect caused by a single directional elimination of elements [3] . Bendsøe presented the Solid Isotropic Material with Penalization (SIMP) method to update the element-density-depen- dent material properties in order to reach the final optimum [4] . The penalization factor is applied to reduce the amount of intermediate densities. However, the introduction of penalization usually has a negative effect on both computational time and final results. In order to reduce such influences, Dadalau [5] introduced SIMP with an adaptive penalization scheme. In this paper the SIMP-model is used. As shown in Equation (1), it regards every element density ρ_e as a single optimization variable. All the variables are updated with a specific scheme which is derived by Lagrange multipliers [6] .

$$\rho_{k+1} = \begin{cases} \max\{(1-\zeta)\rho_k \quad \rho_{\min}\} & \text{if} \quad \rho_k B_k^{\eta} \le \max\{(1-\zeta)\rho_k \quad \rho_{\min}\} \\ \min\{(1+\zeta)\rho_k \quad 1\} & \text{if} \quad \rho_k B_k^{\eta} \ge \min\{(1+\zeta)\rho_k \quad 1\} \\ \rho_k B_k^{\eta} & \text{if} \quad \text{otherwise} \end{cases} \tag{1}$$

Here ρ_k denotes the density at k-th step. Variables η and ζ control the possible change in one iteration with the typical values of 0.5 and 0.2. This scheme adds material to the structure if $B_k > 1$ and removes it if $B_k < 1$ without violating the bound of ρ. B_k is given by

$$B_k = \Lambda_k^{-1} p\rho(x)^{p-1} E^0_{ijkl}\varepsilon_{ij}(u_K)\varepsilon_{kl}(u_K) \tag{2}$$

where B_k is the displacement at k-th step, which is calculated from the equilibrium function of the structure, Λ is a Lagrange multiplier and p is the penalization factor.

2.3. Parameter Optimization

Classical methods of parameter optimization, which are based on sensitivity calculation and the gradient descent method like the Newton method, are already well developed with high accuracy. However, the drawbacks are also obvious. It is expensive to calculate the needed sensitivities, if the variables are not explicit in the objective function. Moreover, in many discretized situations, the loss of continuity is usually problematic. Furthermore, the global optimum is difficult to reach, since it can be easily trapped by a local optimum. The Particle Swarm Optimization (PSO) method, known as a purpose optimization method, was originally proposed by Kennedy and Eberhart [7] . It optimizes a problem by looking for the best candidate solution in a search domain with a set of particles. The general equations for moving particles can be written as follows,

$$v_{i,k+1} = \omega v_{i,k} + \phi_p r_p \left(p_i - x_{i,k}\right) + \phi_g r_g \left(g_i - x_{i,k}\right)$$
$$x_{i,k+1} = x_{i,k} + v_{i,k+1} \tag{3}$$

where $v_{i,k}$ is the search velocity of i-th particle for k-th step, $x_{i,k}$ is the position of i-th particle for k-th step, ω is the inertia of particle, p_i is the best point i-th particle ever met, g_i is the global best point among personal bests, ϕ_p and ϕ_g are weight parameters. Particles are randomly generated at first and they find the next step with the influence of personal and global best points. After several iterations, all the particles could gradually converge to the optimal design point. Various researches about the behavior due to parameter selection in PSO have been presented, such as Shi and Eberhart [8] , Van Den Bergh [9] and Trelea [10] . As the choice of position for the next iteration is relatively random, this method can be easily applied in both the continuous and the discretized problem.

3. PROBLEM SETTING

Figure 1 shows an example case to be optimized. The length of this structure is 130, the height is 100. It has a bearing hole with a radius of 15 and the wall thickness of the hole is 10. The hole can be moved in a 51 × 51 rectangle. The structure is meshed with 2D bilinear quadrilateral elements (Q4) with an element size of 1. Thus the design domain is also discretized into 51 × 51 nodes. Along the lower half of the hole, a distributed force of 3 is added in −Y direction. At the right bottom corner, two unit forces in +X and +Y direction are applied. The left boundary is fixed in both X and Y direction.

The task is to reduce the volume by 50% and to optimize the hole position, such that the compliance is minimized. The volume of the hole is also included as a part of the deleted volume. Mathematically the optimization problem can be expressed as follows:

$$
\begin{aligned}
& \min c\left(\rho_e \quad x_c \quad y_c\right) \\
& s.t.: \sum_{e=1}^{N} v_e \rho_e \leq V, \quad 0<\rho_{\min} \leq \rho_e \leq 1, \\
& \qquad x_c \in [55 \quad 105], \\
& \qquad y_c \in [25 \quad 75]
\end{aligned} \tag{4}
$$

where c is the compliance as the objective function, x_c and y_c are the coordinates of the hole center, V is the maximum accepted volume after optimization, $\rho_{\min}$ is a value close to zero to avoid numerical problems due to the zero density and N is the element number. The density for each element ρ_e is related to the Young's modulus of each element with penalization factor.

In the sequential method the best hole position is found by parameter optimization before executing the topology optimization. The final result is calculated by applying topology optimization on this hole position. The question is, whether the best position found by parameter optimization is still the best one after the topology optimization. Two enumeration tests were made as verification, which cover all the possible results in this optimi- zation case. In the first test, the compliance of the structure is calculated with the hole position on each node of the domain without topology optimization. The value distribution is shown in Figure 2(a). The second test shown in Figure 2(b) is with topology optimization with penalization factor $p = 3$.

As can be seen from Figure 2(a), the best hole position is located on (86, 73) before the topology optimization, which means this position would be found by the sequential method and used as the hole position for further topology optimization. And from Figure 2(b) we can easily see that the final compliance on (86, 73) is 71.51. But the optimized position from Figure 2(b) is (94, 41) with the value of 63.63, which is significantly lower than the former one. This means, at least in some cases, the results from the sequential usage of parameter and topology optimization are far from the real optimum. For this reason, we proposed a new method to find the real optimum.

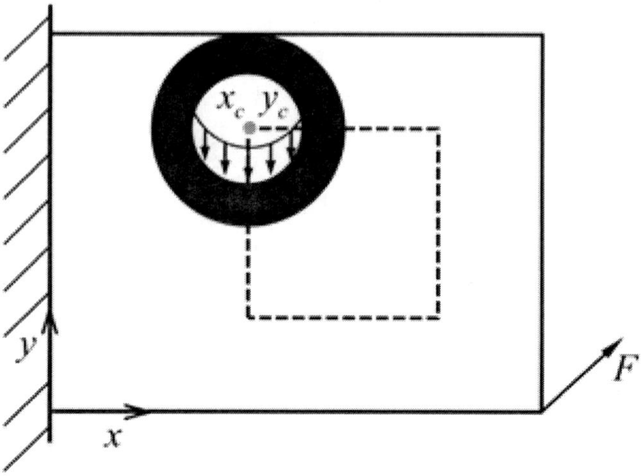

Figure 1.A simple structure with bearing hole to be optimized.

4. IMPLEMENTATION OF COMBINATION METHOD

The general idea of the combination method is based on parameter optimization with the PSO method. The topology optimization is used as a way for calculating compliance, which is also the objective function of PSO. The algorithm is shown in the flowchart in Figure 3. The method is implemented with a Matlab program based on the finite element example code in [11] and the topology optimization code in [12] . However, due to the low efficiency of Matlab with its default large-scale matrix operations, the matrix storage scheme and solving method is modified according to [13] to accelerate the whole process.

In the first step, the bearing hole is set randomly on several positions. On those positions, the fast topology optimization is executed in Step 2. "Fast" means less iterations are needed to converge compared with normal topology optimization by some special schemes, which are mentioned in the following sections.

The new positions are generated in Step 4 through Equation (3), until all the particles (hole positions) converge to the same position (Step 3). Finally, a slower topology optimization is applied on the global best position to achieve the optimum structure (Step 5).

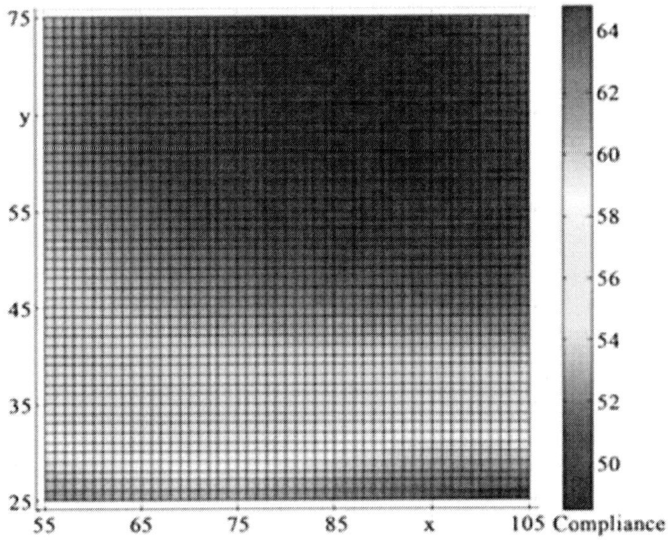

(a)

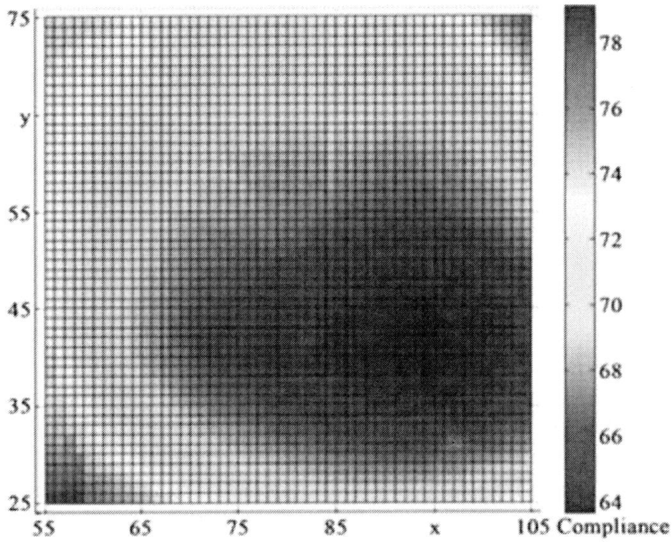

(b)

Figure 2. Results of the objective function raster. This is a mapping of compliance value within the design domain (zone within dash line in Figure 1). (a) stands for the compliance mapping before topology optimization while (b) stands for the one after topology optimization.

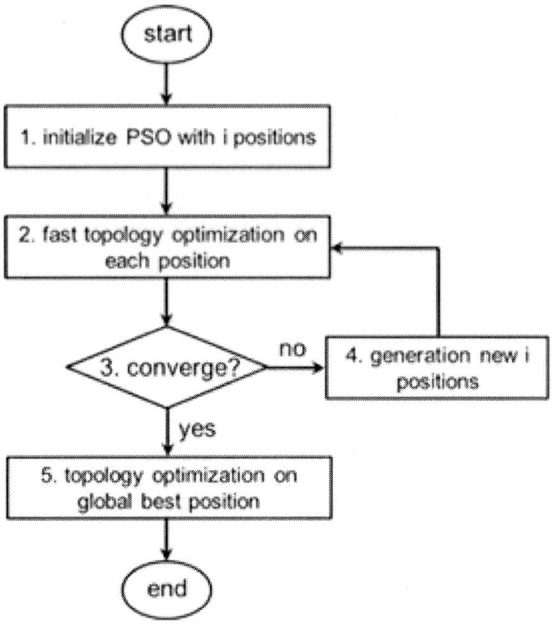

Figure 3. Flowchart of combination method.

4.1. Passive Element Method

The task of moving the hole position during the optimization process must be simple to handle. Normally, the empty area is not included in mesh. When the hole position is changed, meshing and assembling of the stiffness matrix need to be repeated, which decreases the efficiency greatly. This problem can be solved by the passive element method, with which the hole in the design domain can be defined as "zero" density element (ρ_{min} in the program) instead of a description of the geometric information. The mechanism of the passive element is to set the density in the hole area from the last iteration to one and set those in a new hole area to zero, when the position is changed during the searching procedure.

The nodes in the lower part of the hole can also be easily selected to add the load. With the passive element method, only the material density parameter is applied to evaluate the global stiffness matrix, the remeshing process is no longer needed.

4.2. Unified Penalization Factor

The penalization factor p plays an important role in topology optimization. By $p=1$, which means the density and the Young's modulus of an element are exactly linearly related, the optimality of the topology optimization is guaranteed. However, the optimized structure will be filled with large numbers of intermediate densities, which are not realistic to be manufactured. With a large penalization factor, such as $p=3$, the structure will be more realistic and closer to a 0-1 structure. Nevertheless, high penalization factors also have some disadvantages. Table 1 shows that the compliance by $p=3$ is worse, and it converges more slowly than that by $p=1$.

Due to the desired efficiency, the topology optimization during the PSO should be fast, thus the penalization factor within the fast topology optimization (Step 2 in Figure 3) is fixed to one.

4.3. Simplified Convergence Condition

The convergence condition also has a great effect on the convergence speed. Normally, the topology optimization is terminated when the largest change of the element density is smaller than a preset value [6] [12] . Under this condition, a very fine result can be achieved. Nevertheless, the change of density can fluctuate on some positions of the domain, which makes the process difficult or sometimes impossible to converge, although the change of compliance is already very small. The simplified convergence condition terminates the optimization loop, if the change of compliance is smaller than a preset value. According to Table 1, it decreases the convergence loops greatly without losing much accuracy. Through the simplified convergence condition and unity penalization, the fast topology optimization will converge in around 10 loops instead of almost 100 loops each time. Along with the flexibility of the program, the strict condition with a high penalization value can still be used to get the accurate final structure (Step 5 in Figure 3).

4.4. Test Results

The optimized hole position found by the combination method is converged on (93, 40) with the compliance value of 63.64, which is just one element width beside the best position of the reference (Figure 2(b)). Actually, due to the property of the PSO method, the converged result is not always the best but very close to it.

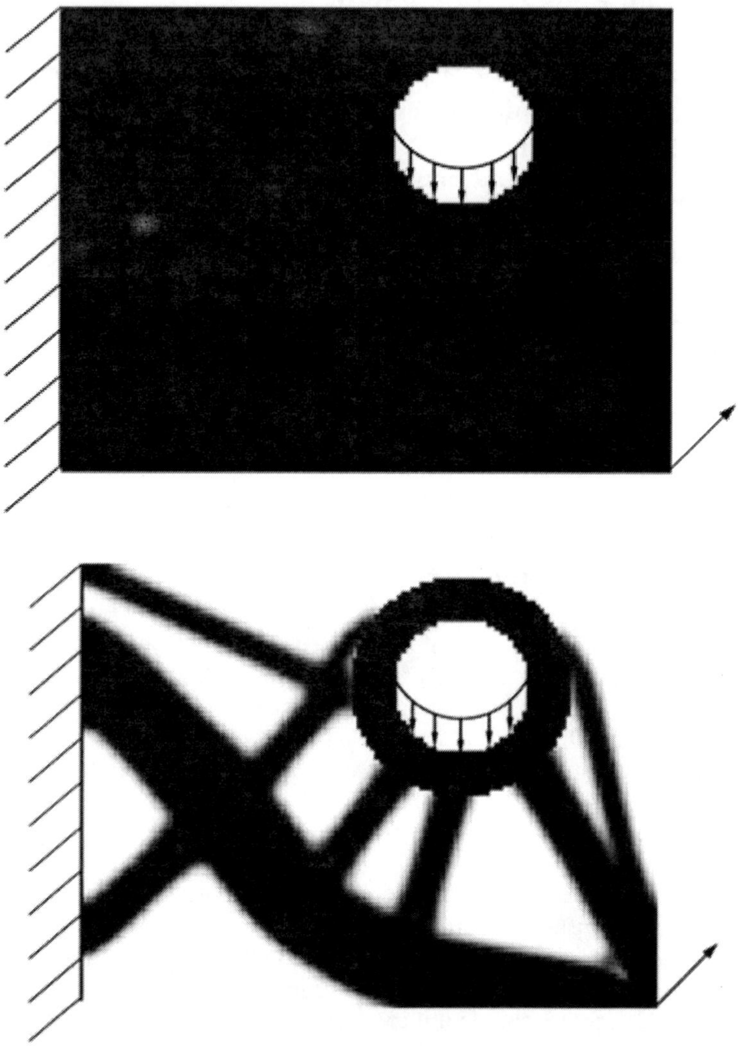

Figure 4. Structures before and after topology optimization in sequential method. The left figure shows the actual structure after getting the best hole position by parameter optimization. The right figure is the final structure acquired by sequential method.

Figure 4 describes the two structures after two separate steps in a sequential method. The final structure resulting from the combined method is shown in Figure 5.

Table 1. Compliance and convergence speed test.

Penalization factor	1	2	3
Compliance	61.17	63.31	63.63
Compliance with simplified condition	61.22	63.35	63.64
Convergence loops	35	Inf.	68
Convergence loops with simplified condition	12	20	25

Figure 5. Final structure from the combination optimization.

5. CONCLUSION

In this paper, an algorithm is implemented to combine the topology optimization and PSO-based parameter optimization. Different schemes are applied to increase the computation efficiency. This algorithm was tested by an example case with a movable bearing hole. It finds the acceptable optimum result, whose compliance after the topology optimization is significantly lower than the sequential optimization. As the calculation of compliance for each position in one PSO iteration is absolutely independent, this algorithm can be easily extended to parallel computation, with which the computation speed can be further increased. It is suitable for the structure design with a convex design domain. How to extend the algorithm to the case with a concave boundary will be studied in the future.

ACKNOWLEDGEMENTS

Yusheng Lin, Zheng Sun, Alexandru Dadalau, Alexander Verl (ISW) and by the excellence cluster SimTech Stuttgart. This support is highly appreciated

REFERENCES

1. Xie, Y.-M. and Steven, G.-P. (1993) A Simple Evolutionary Procedure for Structural Optimization. Computers & Structures, 49, 885-896.
2. Chu, D.-N., Xie, Y.-M., Hira, A. and Steven, G.-P. (1996) Evolutionary Structural Optimization for Problems with Stiffness Constraints. Finite Elements in Analysis and Design, 21, 239-202.
3. Querin, O.-M., Steven, G.-P. and Xie, Y.-M. (1998) Evolutionary Structural Optimization (ESO) Using a Bidirectional Algorithm. Engineering Computations, 15, 1031-1048.
4. Bendsøe, M.-P. (1989) Optimal Shape Design as a Material Distribution Problem. Structural Optimization, 1, 192-202.
5. Dadalau, A., Hafla, A. and Verl, A. (2009) A New Adaptive Penalization Scheme for Topology Optimization. Production Engineering. Research and Development, 3, 427-434.
6. Bendsøe, M.-P. and Sigmund, O. (2004) Topology Optimization: Theory, Methods and Applications. Springer Press, Berlin.
7. Kennedy, J. and Eberhart, R.-C. (1995) Particle Swarm Optimization. Proceedings of the IEEE International Conference on Neural Networks, 4, 1942-1948.
8. Shi, Y.-H. and Eberhart, R.-C. (1998) A Modified Particle Swarm Optimizer. Proceedings of the IEEE World Congress on Computational Intelligence, 69-73.
9. van den Bergh, F. and Engelbrecht, A.-P. (2002) A New Locally Convergent Particle Swarm Optimiser. Proceedings of the IEEE International Conference on System Man and Cybernetics, 3.
10. Trelea, I.-C. (2003) The Particle Swarm Optimization Algorithm: Convergence Analysis and Parameter Selection. Information Processing Letters, 85, 317-325.
11. Ferreria, A.-J.-M. (2008) Matlab Codes for Finite Element Analysis. Soild Mechanics and Its Applications Vol. 157, Springer Press, Berlin.
12. Andreassen, E., Clausen, A., Schevenels, M., Lazarov, B.-S. and Sigmund, O. (2011) Efficient Topology Optimization in Matlab Using 88 Lines of Code. Structural and Multidiscriplinary Optimization, 43, 1-16.
13. Davis, T. (2000) Creating Sparse Fimite-Element Matrices in Matlab. http://blogs.mathworks.com/loren/2007/03/01/creating-sparse-finite-element-matrices-in-matlab/

CHAPTER 14

On the Geometry of Curves in Minkowski 3-Space and Its Foldings

A. E. El-Ahmady[1,2], E. Al-Hesiny[1]

[1]*Mathematics Department, Faculty of Science, Taibah University, Madinah, KSA*

[2]*Mathematics Department, Faculty of Science, Tanta University, Tanta, Egypt*

ABSTRACT

We will introduce a new connection between some transformations and some aspects of differential geometry of some curves in Minkowski space. The concept of folding, retractions and contraction on some curves in Minkowski space will be characterized by using some aspects of differential geometry. Types of the deformation retracts of some curves in Minkowski 3-space are obtained. The relations between the foldings and the deformation retracts of some curves are deduced. The connections between some transformations and time like, space like, light like of some curves in Minkowski 3-space are also presented.

Keywords: Retractions; Deformation Retracts; Folding; Contraction; Minkowski 3-Space

1. INTRODUCTION AND DEFINITIONS

As is well known, the theory of deformation retract is always one of the interesting topics in Euclidian and Non-Euclidian space and it has been investigated from the various viewpoints by many branches of topology and differential geometry El-Ahmady [1-3].

Minkowski space is originally from the relativity in physics. In fact, a time like curve corresponds to the path of an observer moving at less than the speed of light, a light like curve corresponds to moving at the speed of light and a space like curve moving faster than light El-Ahmady [4,5].

The Minkowski 3-space E_1^3 is the Euclidean 3-space E_1^3 provided with the standard flat metric given by $g=-dx_1^2+dx_2^2+dx_3^2$, where (x_1, x_2, x_3) is a rectangular coordinate system of E_1^3. Since g is an indefinite metric, recall that a vector $v\in E_1^3$ can have one of three Lorentzian causal characters, it can be space like if $g(v,v)>$ 0 or $v=0$, time like if $g(v,v)<0$ and light like if $g(v,v)=0$ and $v\neq 0$. Similarly, an arbitrary curve $\alpha=\alpha(s)$ in E_1^3 can locally be space like, time like or light like, if all of its velocity vectors $\alpha'(s)$ are respectively, space like, time like or light like respectively. A curve in Lorentzian space L^n is a smooth map $\alpha : I\rightarrow L^n$, where I is the open interval in the real line $\mathbb{R}$. The interval has a coordinate system consisting of the identity map u of I. The velocity of α at $t\in I$ is $\alpha'=\left.\frac{d\alpha(u)}{du}\right|_t$.

A curve α is said to be regular if $\alpha'(s)$ does not vanish for all $t\in I$. $\alpha\in L^n$ is space like if its velocity vectors $\alpha'(t)$ are space like for all $t\in I$, similarly for timelike and null. If α is a null curve, we can reparametrize it such that $\langle\alpha'(t),\alpha'(t)\rangle=0$ and $\alpha'(t)\neq 0$ El-Ahmady [5].

Most folding problems are attractive from a pure mathematical standpoint, for the beauty of the problems themselves. The folding problems have close connections to important industrial applications. Linkage folding has applications in robotics and hydraulic tube bending. Paper folding has application in sheet-metal bending, packaging, and air-bag folding El-Ahmady [6]. Following the great Soviet geometer El-Ahmady [5], also, used folding to solve difficult problems related to shell structures in civil engineering and aero space design, namely buckling instability El-Ahmady [7]. Isometric folding between two Riemannian manifold may be characterized as maps that send piecewise geodesic segments to a piecewise geodesic segments of the same length [8,9]. For a topological folding the maps do not preserves lengths El-Ahmady [8-10] i.e. A map $\Im : M \to N$, where M and N are C^{∞}-Riemannian manifolds of dimension m, n respectively is said to be an isometric folding of M into N, iff for any piecewise geodesic path $\gamma : J \to M$, the induced path $\Im \circ \gamma : J \to N$ is a piecewise geodesic and of the same length as γ. If $\Im$ does not preserve length, then $\Im$ is a topological folding El-Ahmady [10-12].

A subset A of a topological space X is called a retract of X if there exists a continuous map $r : X \to A$ such that $r(a) = a$, $\forall a \in A$, where A is closed and X is open. El-Ahmady [7-12] and Gregory [13]. This can be restated as follows. If $i : A \to X$ is the inclusion map, then $r : X \to A$ is a map such that $ri = id_A$ Miles [14] and Martin [15]. If, in addition $ri \simeq id_x$, we call r a deformation retract and A a deformation retract of X Jeffrey [16] and John [17].

2. MAIN RESULT

Let $\alpha(s)$ be a curve in the space-time E_1^3, parameterized by arc length function s Lopez [18] and Formiga [19]. Then for the unit speed curve $\alpha(s)$ with non-null frame vectors the following Frenet equations are given in

$$\begin{bmatrix} T' \\ N' \\ B' \end{bmatrix} = \begin{bmatrix} 0 & k & 0 \\ \varepsilon_1 k & 0 & \varepsilon_2 \tau \\ 0 & \varepsilon_3 \tau & 0 \end{bmatrix} \begin{bmatrix} T \\ N \\ B \end{bmatrix} \tag{1}$$

We write following subcases.

1) If $\alpha(s)$ is space-like curve in E_1^3, then T is a space-like vector. Thus, we distinguish according to N.

Case 1: If N is space-like vector, then B is time-like vector, then $\varepsilon_i (1 \le i \le 3)$ read $\varepsilon_1 = -1$, $\varepsilon_2 = \varepsilon_3 = 1$. And T, N and B are mutually orthogonal vectors satisfying equations, $g(T,T) = g(N,N) = 1$, $g(B,B) = -1$.

Case 2: N is time-like vector, then $\varepsilon_i (1 \le i \le 3)$ read $\varepsilon_1 = \varepsilon_2 = \varepsilon_3 = 1$.

And T, N and B are mutually orthogonal vectors satisfying equations

$$g(T,T)=g(B,B)=1, g(N,N)=-1.$$

2) If $\alpha(s)$ is time-like curve in E_1^3, then T is timelike vector. Then $\varepsilon_i (1 \le i \le 3)$ read $\varepsilon_1 = \varepsilon_2 = 1,\ \varepsilon_3 = -1$. And T, N and B are mutually orthogonal vectors satisfying equations,

$$g(N,N)=g(B,B)=1, g(T,T)=-1.$$

3) If $\alpha(s)$ is light-like curve in E_1^3, then the following Frenet equations are given in

$$\begin{bmatrix} T' \\ N' \\ B' \end{bmatrix} = \begin{bmatrix} 0 & 1 & 0 \\ \tau & 0 & -1 \\ 0 & -\tau & 0 \end{bmatrix} \begin{bmatrix} T \\ N \\ B \end{bmatrix} \tag{2}$$

Also, if $\alpha(s)$ be a curve in the space-time E_1^4, parameterized by arc length function s. Then for the unit speed curve $\alpha(s)$ with non-null frame vectors the following Frenet equations are given in

$$\begin{bmatrix} T' \\ N' \\ B_1' \\ B_2' \end{bmatrix} = \begin{bmatrix} 0 & k & 0 & 0 \\ \mu_1 k & 0 & \mu_2 \tau & 0 \\ 0 & \mu_3 \tau & 0 & \mu_4 \sigma \\ 0 & 0 & \mu_5 \sigma & 0 \end{bmatrix} \begin{bmatrix} T \\ N \\ B_1 \\ B_2 \end{bmatrix} \tag{3}$$

Due to character of $\alpha(s)$, we write following subcases.

1) If $\alpha(s)$ is space-like curve in E_1^4, then T is a space-like vector. Thus, we distinguish according to N.

Case 1: If N is space-like vector, then B_1 can have two causal characters.

Case 1.1: B_1 is space-like vector, then $\mu_i (1 \le i \le 5)$ read, $\mu_1 = \mu_3 = -1,\ \mu_2 = \mu_4 = \mu_5 = 1$. And T, N, B_1 and B_2 are mutually orthogonal vectors satisfying equations

$$g(T,T)=g(N,N)=g(B_1,B_1)=1, g(B_2,B_2)=-1$$

Case 1.2: B_1 is time-like vector, then $\mu_i(1 \le i \le 5)$ read $\mu_1 = -1,\ \mu_2 = \mu_3 = \mu_4 = \mu_5 = 1$. And T, N, B_1 and B_2 are mutually orthogonal vectors satisfying equations

$$g(T,T) = g(N,N) = g(B_2,B_2) = 1, g(B_1,B_1) = -1$$

Case 2: N is time-like vector, then $\mu_i(1 \le i \le 5)$ read $\mu_1 = \mu_2 = \mu_3 = \mu_4 = 1,\ \mu_5 = -1$. And T, N, B_1 and B_2 are mutually orthogonal vectors satisfying equations

$$g(T,T) = g(B_1,B_1) = g(B_2,B_2) = 1, g(N,N) = -1.$$

2) If $\alpha(s)$ is time-like curve in E_1^4, then T is timelike vector. Then $\mu_i(1 \le i \le 5)$ read $\mu_1 = \mu_2 = \mu_4 = 1$, $\mu_3 = \mu_5 = -1$. And T, N, B_1 and B_2 are mutually orthogonal vectors satisfying equations

$$g(N,N) = g(B_1,B_1) = g(B_2,B_2) = 1, g(T,T) = -1$$

Hence, we can formulate the following theorems.

Theorem 1. Under the retraction, a spacelike curve and a timelike curve $r(\alpha(s))$ in the space-time E_1^3 parameterized by arc length function s, where $\dim r(\alpha(s)) > 0$ and with non-vanishing curvature, lies in ahyperplane if and only if the torsion vanishes identically.

Proof. Suppose the curve $r(\alpha(s))$ lies in a hyperplane. Let us assume that we can bring $r(\alpha(s))$ to lie in the $(x^0, x^1, 0)$-hyperplane. Then the parametric equations of $r(\alpha(s))$ are of the form

$x^a(s) = (x^0(s), x^1(s), 0)$. Let (e_0, e_1, e_2) denote the vectors of the canonical coordinate basis. Thus, in these coordinates,

$$T = \frac{dx^a(s)}{ds} = \frac{dx^0}{ds}e_0 + \frac{dx^1}{ds}e_1$$

and

$$\frac{dT}{ds} = \frac{d^2x^{\alpha}(s)}{ds^2} = \frac{d^2x^0}{ds^2}e_0 + \frac{d^2x^1}{ds^2}e_1.$$

From (1) we have $\frac{dT}{ds} = KN$. Since $k \neq 0$ then N has no components in the x^2 -direction, i.e.

$$N = \left(y^0(s), y^1(s), 0\right).$$

Thus, $\frac{dN}{ds} = \frac{dy^0}{ds}e_0 + \frac{dy^1}{ds}e_1$ hence from the equation $\frac{dN}{ds} = \varepsilon_1 kT + \varepsilon_2 \tau B$ we conclude that $\tau z^2(s) = 0$, where $B = \left(z^0(s), z^1(s), z^2(s)\right)$. However, $z^2(s)$ cannot be zero. Otherwise the set of vectors $\{T, N, B\}$ would not be linearly independent. Then τ must vanish.

Suppose that $\tau = 0$. Since $\frac{dB}{ds} = \varepsilon_2 \tau N$, then B is a constant vector. Let us conveniently choose our coordinate system in such a way that $e_2 = B$. Now, since T is orthogonal to B we must have $B \cdot T = 0$, which means that $r(\alpha(s))$ lies in the hyperplane $x^2(s) = \text{const}$.

Corollary 1. Under the folding, $f : \alpha(s) \to \alpha(s)$, a spacelike curve and a timelike curve $f(\alpha(s))$ in the space-time E_1^3 parametrized by arc length function s, with non-vanishing curvature, lies in a hyperplane if and only if the torsion vanishes identically.

Theorem 2. Under the retraction, if the curve is a lightlike curve $r(\alpha(s))$ in the space-time E_1^3 parameterized by arc length function s, where $\dim r(\alpha(s)) > 0$ then the curve is not lies in a hyperplane.

Proof. Suppose the curve $r(\alpha(s))$ lies in a hyperplane. Let us assume that we can bring $r(\alpha(s))$ to lie in the $(x^0, x^1, 0)$-hyperplane. Then the parametric equations of $r(\alpha(s))$ are of the form $x^{\alpha}(s) = \left(x^0(s), x^1(s), 0\right)$. Let (e_0, e_1, e_2)

denote the vectors of the canonical coordinate basis. Thus, in these coordinates,

$$T = \frac{\mathrm{d}x^{\alpha}(s)}{\mathrm{d}s} = \frac{\mathrm{d}x^{0}}{\mathrm{d}s} e_0 + \frac{\mathrm{d}x^{1}}{\mathrm{d}s} e_1$$

and

$$\frac{\mathrm{d}T}{\mathrm{d}s} = \frac{d^2 x^{\alpha}(s)}{\mathrm{d}s^2} = \frac{\mathrm{d}^2 x^{0}}{\mathrm{d}s^2} e_0 + \frac{\mathrm{d}^2 x^{1}}{\mathrm{d}s^2} e_1 .$$

From (2) we have $\frac{\mathrm{d}T}{\mathrm{d}s} = N$. Then N has no components in the x^2-direction, i.e. $N = \left(y^0(s), y^1(s), 0\right)$. Thus, $\frac{\mathrm{d}N}{\mathrm{d}s} = \frac{\mathrm{d}y^0}{\mathrm{d}s} e_0 + \frac{\mathrm{d}y^1}{\mathrm{d}s} e_1$, hence from the equation $\frac{\mathrm{d}N}{\mathrm{d}s} = \tau T - B$, we conclude $z^2(s) = 0$, where $B = \left(z^0(s), z^1(s), z^2(s)\right)$. But $z^2(s)$ cannot be zero because the set of vectors $\{T, N, B\}$ would not be linearly independent, then the curve is not lies in a hyperplane.

Corollary 2. Under the folding, $f : \alpha(s) \to \alpha(s)$ if the curve is a lightlike curve $f(\alpha(s))$ in the spacetime E_1^3 parameterized by arc length function s, then the curve is not lies in a hyperplane.

Theorem 3. Under the retraction, a spacelike curve and a timelike curve $r(\alpha(s))$ in the space-time E_1^4 parameterized by arc length function s, where $\dim r(\alpha(s)) > 0$ and with non-vanishing curvature, lies in ahyperplane if and only if the second torsion vanishes identically.

Proof. Let us start with the necessary condition. Suppose the curve $r(\alpha(s))$ lies in a hyperplane. Let us assume that we can bring $r(\alpha(s))$ to lie in the $\left(x^0, x^1, x^2\right)$-hyperplane. Then the parametric equations of $r(\alpha(s))$ are of the form $x^{\alpha}(s) = \left(x^0(s), x^1(s), x^2(s), 0\right)$. Let (e_0, e_1, e_2, e_3) denote the vectors of the canonical coordinate basis. Thus, in these coordinates,

$$T = \frac{\mathrm{d}x^{\alpha}(s)}{\mathrm{d}s} = \frac{\mathrm{d}x^{0}}{\mathrm{d}s} e_0 + \frac{\mathrm{d}x^{1}}{\mathrm{d}s} e_1 + \frac{\mathrm{d}x^{2}}{\mathrm{d}s} e_2$$

and

$$\frac{dT}{ds}=\frac{d^2x^{\alpha}(s)}{ds^2}=\frac{d^2x^0}{ds^2}e_0+\frac{d^2x^1}{ds^2}e_1+\frac{d^2x^2}{ds^2}e_2.$$

From (3) we have $\frac{dT}{ds}=N$. Since $k\neq 0$ then N has no components in the x^3-direction, i.e.

$$N=\left(y^0(s),y^1(s),y^2(s),0\right).$$

Thus, $\frac{dN}{ds}=\frac{dy^0}{ds}e_0+\frac{dy^1}{ds}e_1+\frac{dy^2}{ds}e_2$, hence from the equation $\frac{dN}{ds}=\mu_1 kT+\mu_2\tau B_1$ and we conclude that $\tau z^3(s)=0$, where $B_1=\left(z^0(s),z^1(s),z^2(s),z^3(s)\right)$. If $\tau=0$, then σ also must vanish, for in this case B_1 is chosen to be constant. If $\tau\neq 0$, then $z^3(s)=0$, hence $B_1=\left(z^0(s),z^1(s),z^2(s),0\right)$. Also, $\frac{dB_1}{ds}=\frac{dz^0}{ds}e_0+\frac{dz^1}{ds}e_1+\frac{dz^2}{ds}e_2$ and the third Serret-Frenet equation $\frac{dB_1}{ds}=\mu_3\tau N-\mu_4\sigma B_2$ we are led to conclude that $\sigma\varpi^3(s)=0$, where

$$B_2=\left(\varpi^0(s),\varpi^1(s),\varpi^2(s),\varpi^3(s)\right).$$

However, $\varpi^3(s)$ cannot be zero. Otherwise the set of vectors $\{T,N,B_1,B_2\}$ would not be linearly independent. Then σ must vanish.

Suppose that $\sigma=0$. Since $\frac{dB_2}{ds}=\mu_5\sigma B_1$, then B_2 is a constant vector. Let us conveniently choose our coordinate system in such a way that $e_3=B_2$. Now, since T is orthogonal to B_2 we must have $B_2\cdot T=0$, which means that $r(\alpha(s))$ lies in the hyperplane $x^3(s)=\text{const}$.

Corollary 3. Under the contraction, a spacelike curve and a timelike curve $C(\alpha(s))$ in the space-time E_1^4 parameterized by arc length function s, where $\dim C(\alpha(s))>0$ and with non-vanishing curvature, is lies in a hyperplane if the first and second torsions vanishes identically.

Corollary 4. Under the folding, a spacelike curve and a timelike curve $f(\alpha(s))$ in the space-time E_1^4 parameterized by arc length function s, with non-vanishing curvature, lies in a hyperplane if and only if the second torsion vanishes identically.

Corollary 5. Under the folding, $f:\alpha(s)\to\alpha(s)$, a spacelike curve and a timelike curve $f(\alpha(s))$ in the space-time E_1^4 parameterized by arclength function s, with non-vanishing curvature, is lies in a hyperplane if the first and second torsions vanishes identically.

Theorem 4. Given differentiable functions $K(s)>0$ and $\tau(s), s\in I$ such that s is the arc length, there exists a regular parameterized spacelike curve under the folding with the spacelike vector N, $f(\alpha(s))$, in the space-time E_1^3. Also, $K(s)$ is the curvature and $\tau(s)$ is the torsion of $f(\alpha(s))$. Moreover, any other spacelike curve $\bar{\alpha}(s)$ with the spacelike vector $\bar{N}$ satisfying the same conditions and $f(\alpha(s_0))=\bar{\alpha}(s_0)$ at $s_0\in I$ then $f(\alpha(s))=\bar{\alpha}(s)$ and the Frenet trihedrons of $f(\alpha(s))$ and $\bar{\alpha}(s)$ is identically.

Proof. Now, assume that two curves $f(\alpha)=f(\alpha(s))$ and $\bar{\alpha}=\bar{\alpha}(s)$ satisfy the conditions $k(s)=\bar{k}(s)$ and $\tau(s)=\bar{\tau}(s)$, $s\in I$. Let T_0, N_0, B_0 and $\bar{T}_0$, $\bar{N}_0$, $\bar{B}_0$ be the Frenet trihedrons of $f(\alpha(s))$ and $\bar{\alpha}(s)$ at $s_0\in I$, respectively. Since $f(\alpha(s_0))=\bar{\alpha}(s_0)$ then $T_0=\bar{T}$, $N_0=\bar{N}$ and $B_0=\bar{B}$, where $T(s)$, $N(s)$, $B(s)$ and $\bar{T}(s)$, $\bar{N}(s)$, $\bar{B}(s)$ are the Frenet trihedrons of $f(\alpha(s))$ and $\bar{\alpha}(s)$ $\forall s\in I$, respectively. We now observe, by using the Frenet equations at (1), that

$$
\begin{aligned}
&\frac{\mathrm{d}}{\mathrm{d}s}\left\{\left|T(s)-\bar{T}(s)\right|^2+\left|N(s)-\bar{N}(s)\right|^2+\left|B(s)-\bar{B}(s)\right|^2\right\}\\
&=\frac{\mathrm{d}}{\mathrm{d}s}\left\{\left\langle T(s)-\bar{T}(s),T(s)-\bar{T}(s)\right\rangle+\left\langle N(s)-\bar{N}(s),N(s)-\bar{N}(s)\right\rangle+\left\langle B(s)-\bar{B}(s),B(s)-\bar{B}(s)\right\rangle\right\}\\
&=2\left\langle T(s)-\bar{T}(s),T'(s)-\bar{T}'(s)\right\rangle+2\left\langle N(s)-\bar{N}(s),N'(s)-\bar{N}'(s)\right\rangle+2\left\langle B(s)-\bar{B}(s),B'(s)-\bar{B}'(s)\right\rangle\\
&=2\left\langle T(s)-\bar{T}(s),kN(s)-k\bar{N}(s)\right\rangle+2\left\langle N(s)-\bar{N}(s),\varepsilon_1 KT(s)-\varepsilon_2\tau B(s)-\varepsilon_1 K\bar{T}(s)+\varepsilon_2\tau\bar{B}(s)\right\rangle\\
&\quad+2\left\langle B(s)-\bar{B}(s),\varepsilon_3\tau N(s)-\varepsilon_2\tau\bar{N}(s)\right\rangle\\
&=2\left\langle T(s)-\bar{T}(s),kN(s)-k\bar{N}(s)\right\rangle+2\left\langle N(s)-\bar{N}(s),\varepsilon_1 kT(s)\varepsilon_1 k\bar{T}(s)\right\rangle\\
&\quad-2\left\langle N(s)-\bar{N}(s),\varepsilon_1\tau B(s)-\varepsilon_2\tau\bar{B}(s)\right\rangle+2\left\langle B(s)-\bar{B}(s),\varepsilon_3\tau N(s)-\varepsilon_3\tau\bar{N}(s)\right\rangle=0
\end{aligned}
$$

for all $s \in I$. Thus, the above expression is constant, and, since it is zero for $s = s_0$, it is identically zero. It follows that $T(s) = \bar{T}(s)$, $N(s) = \bar{N}(s)$, $B(s) = \bar{B}(s)$ for all $s \in I$.

Since $\frac{d\alpha}{ds} = T(s) = \bar{T}(s) = \frac{d\bar{\alpha}}{ds}$, we obtain $\frac{d}{ds}(\alpha - \bar{\alpha}) = 0$.

Thus $\alpha(s) = \bar{\alpha}(s) + a$, where a is a constant vector. Since $\alpha(s_0) = \bar{\alpha}(s_0)$, we have $a = 0$; hence, $\bar{\alpha}(s)\ \alpha(s) =$ for all $s \in I$.

Corollary 6. Given differentiable functions $k(s) > 0$ and $\tau(s)$, $s \in I$ such that s is the arc length, there exists a regular parameterized spacelike curve under the contraction with the spacelike vector N, $C(\alpha(s))$, in the space-time E_1^3. Also, $k(s)$ is the curvature and $\tau(s)$ is the torsion of $C(\alpha(s))$. Moreover, any other spacelike curve $\bar{\alpha}(s)$ with the spacelike vector $\bar{N}$ satisfying the same conditions and $C(\alpha(s_0)) = \bar{\alpha}(s_0)$ at $s \in I$ then $C(\alpha(s)) = \bar{\alpha}(s)$ and the Frenet trihedrons of $C(\alpha(s))$ and $\bar{\alpha}(s)$ is identically.

Theorem 5. Given differentiable functions $k(s) > 0$, $\tau(s)$ and $\sigma(s), s \in I$ such that s is the arc length, there exists a regular parameterized spacelike curve under the folding with the spacelike vectors N and B_1, $f(\alpha(s))$ in the space-time E_1^4. Also, $k(s)$ is the curvature, $\tau(s)$ is the first torsion, and $\sigma(s)$ is the second torsion of $f(\alpha(s))$. Moreover, any other spacelike curve $\bar{\alpha}(s)$ with the spacelike vectors $\bar{N}$ and $\bar{B}_1$ satisfying the same conditions and $f(\alpha(s_0)) = \bar{\alpha}(s_0)$, where $s_0 \in I$ then $f(\alpha(s)) = \bar{\alpha}(s)$ and the FrenetSerret formulas of $f(\alpha(s))$ and $\bar{\alpha}(s)$ is identically.

Proof. Now assume that two curves $f(\alpha) = f(\alpha(s))$ and $\bar{\alpha} = \bar{\alpha}(s)$ satisfy the conditions $K(s) = \bar{K}(s)$, $\tau(s) = \bar{\tau}(s)$ and $\sigma(s) = \bar{\sigma}(s)$, $s \in I$. Let T_0, N_0, B_{1_0}, B_{2_0} and $\bar{T}_0$, $\bar{N}_0$, $\bar{B}_{1_0}$, $\bar{B}_{2_0}$ be the Frenet-Serret formulas of $f(\alpha(s))$ and $\bar{\alpha}(s)$ at $s_0 \in I$, respectively. Since $f(\alpha(s_0)) = \bar{\alpha}(s_0)$ then $T_0 = \bar{T}_0$, $N_0 = \bar{N}_0$, $B_{1_0} = \bar{B}_{1_0}$, and $B_{2_0} = \bar{B}_{2_0}$. And $T(s)$, $N(s)$, $B_1(s)$,

$B_2(s)$ and $\overline{T}(s)$, $\overline{N}(s)$, $\overline{B}_1(s)$, $\overline{B}_2(s)$ are the Frenet-Serret formulas of $f(\alpha(s))$ and $\overline{\alpha}(s)$ $\forall s \in I$, respectively. We now observe, by using the Frenet equations at (3), that

$$
\begin{aligned}
&\frac{\mathrm{d}}{\mathrm{d}s}\left\{\left|T(s)-\overline{T}(s)\right|^2+\left|N(s)-\overline{N}(s)\right|^2+\left|B_1(s)-\overline{B}_1(s)\right|^2+\left|B_2(s)-\overline{B}_2(s)\right|^2\right\}\\
&=\frac{\mathrm{d}}{\mathrm{d}s}\left\{\left\langle T(s)-\overline{T}(s),T(s)-\overline{T}(s)\right\rangle+\left\langle N(s)-\overline{N}(s),N(s)-\overline{N}(s)\right\rangle\right.\\
&\quad\left.+\left\langle B_1(s)-\overline{B}_1(s),B_1(s)-\overline{B}_1(s)\right\rangle+\left\langle B_2(s)-\overline{B}_2(s),B_2(s)-\overline{B}_2(s)\right\rangle\right\}\\
&=2\left\langle T(s)-\overline{T}(s),T'(s)-\overline{T}'(s)\right\rangle+2\left\langle N(s)-\overline{N}(s),N'(s)-\overline{N}'(s)\right\rangle\\
&\quad+2\left\langle B_1(s)-\overline{B}_1(s),B_1'(s)-\overline{B}_1'(s)\right\rangle+2\left\langle B_2(s)-\overline{B}_2(s),B_2'(s)-\overline{B}_2'(s)\right\rangle\\
&=2\left\langle T(s)-\overline{T}(s),KN(s)-K\overline{N}(s)\right\rangle+2\left\langle N(s)-\overline{N}(s),-KT(s)+\tau B_1(s)+K\overline{T}(s)-\tau\overline{B}_1(s)\right\rangle\\
&\quad+2\left\langle B_1(s)-\overline{B}_1(s),-\tau N(s)-\sigma B_2(s)+\tau\overline{N}(s)+\sigma\overline{B}_2(s)\right\rangle+2\left\langle B_2(s)-\overline{B}_2(s),\sigma B_1(s)-\sigma B_1(s)\right\rangle\\
&=2k\left\langle T(s)-\overline{T}(s),N(s)-\overline{N}(s)\right\rangle-2k\left\langle N(s)-\overline{N}(s),T(s)-\overline{T}(s)\right\rangle+2\tau\left\langle N(s)-\overline{N}(s),B_1(s)-\overline{B}_1(s)\right\rangle\\
&\quad-2\tau\left\langle B_1(s)-\overline{B}_1(s),N(s)-\overline{N}(s)\right\rangle-2\sigma\left\langle B_1(s)-\overline{B}_1(s),B_2(s)-\overline{B}_2(s)\right\rangle\\
&\quad+2\sigma\left\langle B_2(s)-\overline{B}_2(s),B_1(s)-\overline{B}_1(s)\right\rangle=0
\end{aligned}
$$

for all $s \in I$. Thus, the above expression is constant and, since it is zero for $s = s_0$, it is identically zero. It follows that

$$
\begin{aligned}
&T(s)=\overline{T}(s),\ N(s)=\overline{N}(s),\\
&B_1(s)=\overline{B}_1(s),\ B_2(s)=\overline{B}_2(s)
\end{aligned}
$$

for all $s \in I$. Since $\frac{\mathrm{d}\alpha}{\mathrm{d}s}=T(s)=\overline{T}(s)=\frac{\mathrm{d}\overline{\alpha}}{\mathrm{d}s}$. We obtain $\frac{\mathrm{d}}{\mathrm{d}s}\left(\alpha(s)-\overline{\alpha}(s)\right)=0$. Thus $\alpha(s)=\overline{\alpha}(s)+a$, where a is a constant vector. Since $\alpha(s_0)=\overline{\alpha}(s_0)$, we have $a=0$; hence, $\alpha(s)=\overline{\alpha}(s)$ for all $s \in I$.

Theorem 6. Given differentiable functions $k(s)>0$, $\tau(s)$ and $\sigma(s), s \in I$ such that s is the arc length, there exists a regular parameterized spacelike curve under the deformation retract with the spacelike vectors N and B_1, $D(\alpha(s))$, in the space-time E_1^4. Also, $k(s)$ is the curvature, $\tau(s)$ is the first torsion, and $\sigma(s)$ is the second torsion of $D(\alpha(s))$. Moreover, any other spacelike curve $\overline{\alpha}(s)$ with the spacelike vectors $\overline{N}$ and $\overline{B}_1$ satisfying the same conditions and $D(\alpha(s_0))=\bar{\alpha}(s_0)$ where $s_0 \in I$ then $D(\alpha(s))=\bar{\alpha}(s)$ and the Frenet-Serret formulas of $D(\alpha(s))$ and $\bar{\alpha}(s)$ is identically.

Theorem 7. Let $f(\alpha(s))$ be a simple closed hyperplane curve under the folding in E_1^3 with length l, and let A be the area of the region bounded by $f(\alpha(s))$. Then

$$l^2 - 4\pi A \geq 0 \tag{4}$$

and equality holds if and only if $f(\alpha(s))$ is a circle.

Proof. Let E and E' be two parallel lines which do not meet the closed curve $f(\alpha(s))$ and moves them together until they first meet $f(\alpha(s))$. We thus obtain two parallel tangent lines to $f(\alpha(s))$, L and L', so that the curve is entirely contained in the strip bounded by L and L'. Consider a circle S^1 which is tangent to both L and L' and does not meet $f(\alpha(s))$. Let O be the center of S^1 and take a coordinate system with origin at O and the x axis perpendicular to L and L'. Parameterize $f(\alpha(s))$ by arc length, since $f(\alpha(s))$ simple closed a hyperplane curve, then $f(\alpha(s))=(x(s),y(s),0)$, so that it is positively oriented and the tangency points of L and L' are $s=0$ and $S=S_1$, respectively. We can assume that the equation of S^1 is

$$\bar{\alpha}(s)=(\bar{x}(s),\bar{y}(s),0)=(x(s),\bar{y}(s),0), s\in[0,1],$$

where 2r is the distance between L and L'. Denoting by $\bar{A}$ the area bounded by S^1, we have

$$A=\int_0^l x\dot{y}'ds,\ \bar{A}=\pi r^2=-\int_0^l \bar{y}\dot{x}'ds$$

thus

$$A+\pi r^2=\int_0^l (x\dot{y}'-\bar{y}x')\,ds \le \int_0^l \sqrt{(x\dot{y}'-\bar{y}x')^2}\,ds$$
$$\le \int_0^l \sqrt{(x^2+\bar{y}^2)\left((x')^2+(\dot{y}')^2\right)}ds$$
$$=\int_0^l \sqrt{(x^2+\bar{y}^2)}ds = lr \qquad (5)$$

We now notice the fact that the geometric mean of two positive numbers is smaller than or equal to their arithmetic mean, and equality holds if and only if they are equal. It follows that

$$\sqrt{A}\sqrt{\pi r^2} \le \frac{1}{2}\left(A+\pi r^2\right) \le \frac{1}{2}lr \qquad (6)$$

Therefore, $4A\pi r^2 \le l^2 r^2$ then $l^2-4\pi A \ge 0$.

Now, assume that equality holds in Equation (4). Then equality must hold everywhere in Equations (5) and (6). From the equality in Equation (6) it follows that $A=\pi r^2$. Thus, $l=2\pi r$ and r does not depend on the choice of the direction of L. Furthermore, equality in Equation (5) implies that $(x\dot{y}'-\bar{y}x')^2=(x^2-\bar{y}^2)\left((x')^2+(\dot{y}')^2\right)$, or $(xx'+\bar{y}y')^2=0$; that is,

$$\frac{x}{y'}=-\frac{\bar{y}}{x'}$$

$$\frac{\sqrt{x^2+\bar{y}^2}}{\sqrt{(x')^2+(\dot{y}')^2}}=\pm r.\text{ thus, } x=\pm ry'$$

Since r does not depend on the choice of the direction of L, we can interchange x and y in the last relation and obtain $y=\pm rx'$. Thus, $x^2+y^2=r^2\left((x')^2+(\dot{y}')^2\right)=r^2$ and $f(\alpha(s))$ is a circle.

Theorem 8. Let $D(\alpha(s))$ be a simple closed a hyperplane curve under the deformation retract in E_1^3 with length l, and let A be the area of the region bounded by $D(\alpha(s))$. Then $l^2-4\pi A\geq 0$, and equality holds if and only if $r(\alpha(s))$ is a circle.

Corollary 7. Let $C(\alpha(s))$ be a simple closed a hyperplane curve under the contraction in E_1^3 with length l, and let A be the area of the region bounded by $C(\alpha(s))$. Then $l^2-4\pi A\geq 0$, and equality holds if and only if $C(\alpha(s))$ is a circle.

Any n vectors forming a basis for E_1^n will be written $e_1, e_2, \cdots, e_n$, i.e. the basis will be written (e_i). Relative to a basis (e_i), any vector λ in E_1^n is uniquely expressible in the form

$$\lambda = \lambda^1 e_1 + \lambda^2 e_2 + \cdots + \lambda^n e_n$$

$$\lambda = \sum_{i=1}^{n} \lambda^i e_i$$

The numbers λ^i, where $i = 1, 2, \cdots, n$, are called the components of λ relative to the basis (e_i). If μ^i are the components of another vector μ relative to the same basis $\mu = \sum_{i=1}^{n} \mu^i e_i$. Let the vectors $e_{1'}, e_{2'}, \cdots, e_{n'}$ form another basis of E_1^n. Since each vector $e_{i'}$ is uniquely expressible as a linear combination of the vectors (e_i), we have

$$e_{i'} = \sum_{i=1}^{n} \boldsymbol{p}_{i'}^{i} e_i \tag{7}$$

where $(\boldsymbol{p}_{i'}^{i})$ is an $n \times n$ matrix, non-singular because the vectors e_i, are linearly independent. Similarly, the vector $e_{i'}$ is uniquely expressible in the form

$$e_i = \sum_{i'=1}^{n} \boldsymbol{p}_i^{i'} e_{i'} \tag{8}$$

where $(p_i^{i'})$ is a non-singular $n \times n$ matrix. Then

$$e_i = \sum_{i'=1}^{n} p_i^{i'} e_{i'} = \sum_{i',j=1}^{n} p_i^{i'} p_{i'}^{j} e_j$$

The linear independence of the basis vectors implies that $p_i^{i'} p_{i'}^{j} = \delta_i^j$, where δ_i^j called the kroneckel delta, takes the value 1 if $i = j$ and is otherwise zero.

Theorem 9. The components of a vector λ in E_1^n where defined relative to the basis (e_i), and a change of basis will induce a change of components.

Proof. The law of transformation for the components of the vector λ will now be found when the basis is change from (e_i) to $(e_{i'})$ according the Equation (7). If the vector λ has components λ^i relative to the basis (e_i), it is convenient to write $\lambda^{i'}$ for it's components relative to the new basis $(e_{i'})$, related to the former by (7). Then

$$\lambda = \sum_{i=1}^{n} \lambda^i e_i = \sum_{i'=1}^{n} \lambda^{i'} e_{i'} \tag{9}$$

Equations (7), (9) give

$$\sum_{i=1}^{n} \lambda^i e_i = \sum_{i,i'=1}^{n} \lambda^{i'} p_{i'}^{i} e_i$$

From which, since the basis vectors (e_i) are linearly independent, it follows that

$$\lambda^i = p_{i'}^{i} \lambda^{i'} \tag{10}$$

Similarly, substitute in (9) for e_i from (8) to get

$$\sum_{i,i'=1}^{n} \lambda^i p_i^{i'} e_{i'} = \sum_{i'=1}^{n} \lambda^{i'} e_{i'}$$

From which, since the basis vectors $(e_{i'})$ are linearly independent

$$\lambda^{i'} = p_i^{i'} \lambda^i \tag{11}$$

The Equation (11) expresses the new components in terms of the old component, while Equation (10) expresses the old components in terms of the new component.

Theorem 10. Under the retraction, given differentiable functions $k(s)>0$, $\tau(s)$ and $\sigma(s)$, there exists a retraction of regular parameterized timelike curve $\alpha(s)$, $r(\alpha(s))$, such that $k(s)$ is the curvature, $\tau(s)$ and $\sigma(s)$ are, respectively, the first and second torsion of $r(\alpha(s))$. Any other curve $\bar{\alpha}(s)$ satisfying the same conditions, different from $r(\alpha(s))$. By a Poincaré transformation.

Proof. Let us assume that two timelike curves $r(\alpha(s))$ and $\bar{\alpha}(s)$ satisfy the conditions $k(s)=\bar{k}(s)$, $\tau(s)=\bar{\tau}(s)$ and $\sigma(s)=\bar{\sigma}(s)$, with $s \in I$, where I is an open interval of $\mathbb{R}$, and $\bar{k}(s)$, $\bar{\tau}(s)$ and $\bar{\sigma}(s)$ are, respectively, the curvature, first and second torsion of $\bar{\alpha}(s)$. Let $\left\{u_{(A)}\right\}=\left\{T,N,B_1,B_2\right\}$ and

$\left\{\bar{u}_{(A)}\right\}=\left\{\bar{T},\bar{N},\bar{B}_1,\bar{B}_2\right\}$, where $A=1,2,3,4$ be the Serret-Frenet tetrads at $s_0 \in I$ of $r(\alpha(s))$ and $\bar{\alpha}(s)$, respectively. Now, the two Serret-Frenet tetrads of $r(\alpha(s))$ and $\bar{\alpha}(s)$ satisfy the equations

$$\begin{bmatrix} T' \\ N' \\ B_1' \\ B_2' \end{bmatrix} = \begin{bmatrix} 0 & k & 0 & 0 \\ k & 0 & \tau & 0 \\ 0 & -\tau & 0 & \sigma \\ 0 & 0 & -\sigma & 0 \end{bmatrix} \begin{bmatrix} T \\ N \\ B_1 \\ B_2 \end{bmatrix}$$

and

$$\begin{bmatrix} \bar{T}' \\ \bar{N}' \\ \bar{B}_1' \\ \bar{B}_2' \end{bmatrix} = \begin{bmatrix} 0 & k & 0 & 0 \\ k & 0 & \tau & 0 \\ 0 & -\tau & 0 & \sigma \\ 0 & 0 & -\sigma & 0 \end{bmatrix} \begin{bmatrix} \bar{T} \\ \bar{N} \\ \bar{B}_1 \\ \bar{B}_2 \end{bmatrix}$$

This can be written in a more compact form as

$$\frac{\mathrm{d}u_{(A)}}{\mathrm{d}s} = \sum\nolimits_A^{\dots B} a_{AB} u_{(B)} \tag{12}$$

$$\frac{\mathrm{d}\bar{u}_{(A)}}{\mathrm{d}s} = \sum\nolimits_A^{\dots B} \bar{a}_{AB} \bar{u}_{(B)}$$

where $B = 1,2,3,4$, with $\sum\nolimits_A^{\dots B} a_{AB}$ and $\sum\nolimits_A^{\dots B} \bar{a}_{AB}$ denoting the elements of the Serret-Frenet matrix. Clearly, the two tetrads $\{u_{(A)}(s)\}$, $\{\bar{u}_{(A)}(s)\}$ are related by an equation of the type

$$\bar{u}_{(A)}(s) = \wedge_A^{\dots B} b_{AB}(s) u_{(B)}(s) \tag{13}$$

with the elements of the matrix $\wedge(s)$ satisfying the condition

$$\wedge_A^{\dots B} b_{AB}(s_0) = \delta_B^A$$

Since we are assuming that $u_{(A)}^{\alpha}(s_0) = \bar{u}_{(A)}^{\alpha}(s_0)$.

From (12) and (13) we obtain a system of first-order differential equations for the elements of $\wedge(s)$ given by

$$\frac{\mathrm{d}\wedge_{A}^{\cdots B} b_{AB}}{\mathrm{d}s} u_{(B)} + \wedge_{A}^{\cdots B} \sum_{B}^{\cdots D} b_{AB} a_{BD} u_{(D)}$$
$$-\sum_{A}^{\cdots C} \wedge_{C}^{\cdots M} \bar{a}_{AC} b_{CM} u_{(M)} = 0 \tag{14}$$

By assumption, $\sum_{A}^{\ldots B,C,D,G}$ are differentiable functions of the proper parameter s. From the theory of ordinary differential equations, we know that if we are given a set of initial conditions $\wedge_{A}^{\cdots B} b_{AB}(s_0)$ then the above system admits a unique solution

$$\left\{\wedge_{A}^{\cdots B} b_{AB}(s) = \wedge_{A}^{\cdots B} b_{AB}(s)\right\}$$

defined in an open interval $J \subset I$ containing s_0. On the other hand, it is easily seen that $\wedge_{A}^{\cdots B} b_{AB}(s) = \delta_{B}^{A}$, is a solution of (14). Therefore, we conclude that

A. E. El-Ahmady, E. Al-Hesiny $\bar{u}_{(A)}(s) = u_{(B)}(s)$.

REFERENCES

1. E. El-Ahmady and E. Al-Hesiny, "Conditional Retraction of Some Curves in Minkowski 3-Space," International Journal of Applied Mathematics and Statistics, Vol. 32, No. 2, 2012, pp. 39-47.
2. E. El-Ahmady, "The Variation of the Density on Chaotic Spheres in Chaotic Space-Like Minkowski Space Time," Chaos, Solitons and Fractals, Vol. 31, No. 5, 2007, pp. 1272-1278. doi:10.1016/j.chaos.2005.10.112
3. E. El-Ahmady and E. Al-Hesiny, "The Topological Folding of the Hyperbola in Minkowski 3-Space," The International Journal of Nonlinear Science, Vol. 11, No. 4, 2011, pp. 451-458.
4. E. El-Ahmady and E. Al-Hesiny, "Folding and Differential Equations of Some Curves in Minkowski Space," Life Science Journal, Vol. 9, No. 2, 2012, pp. 579-584.
5. E. El-Ahmady and E. Al-Hesiny, "Folding and Deformation Retract of Hyperhelix," Journal of Mathematics and Statistics, Vol. 8, No. 2, 2012, pp. 241-247.
6. E. El-Ahmady, "Retraction of Chaotic Black Hole," The Journal of Fuzzy Mathematics, Vol. 19, No. 4, 2011, pp. 833-838.
7. E. El-Ahmady, "Limits of Fuzzy Retractions of Fuzzy Hyperspheres and Their Foldings," Tamkang Journal of Mathematics, Vol. 37, No. 1, 2006, pp. 47-55.
8. E. El-Ahmady, "The Deformation Retract and Topo Logical Folding of Buchdahi Space," Periodica Mathematica Hungarica, Vol. 28, No. 1, 1994, pp. 19-30.
9. E. El-Ahmady, "Folding and Fundamential Group of Buchdahi Space," Indian Journal of Science and Technology, Vol. 6, No. 1, 2013, pp. 3940-3945.

10. E. El-Ahmady, "On the Fundamental Group and Folding of Klein Bottle," International Journal of Applied Mathematics and Statistics, Vol. 37, No. 6, 2013, pp. 56-64.
11. E. El-Ahmady, "The Geodesic Deformation Retract of Klein Bottle and Its Folding," The International Journal of Nonlinear Science, Vol. 12, No. 3, 2011, pp. 323-330.
12. E. El-Ahmady and E. Al-Hesiny, "On Some Curves in Minkowski 3-Space and Its Deformation Retract," International Journal of Applied Mathematics and Statistics, Vol. 36, No. 6, 2013, pp. 42-53.
13. G. L. Naber, "Topology, Geometry and Gauge Fields, Foundations," Springer-Verlage, New York, Berlin, 2011.
14. M. Reid and B. Szendroi, "Geometry and Topology," Cambridge University Press, Cambridge, New York, 2005.
15. M. Reid and B. Szendroi, "Topology and Geometry," Cambridge University Press, New York, 2005.
16. M. Arkowitz, "Introduction to Homotopy Theory," Springer-Village, New York, 2011.
17. J. Strom, "Modern Classical Homotopy Theory," American Mathematical Society, 2011.
18. J. M. Lee, "Introduction to Topological Manifolds," Springer-Verlage, New York, 2011.
19. R. Lopez, "Differential Geometry of Curves and Surfaces in Lorentz-Minkowski Space," Instituto de Matematica e Estatistica, University of Sao Paulo, Sao Paulo, 2008.

CHAPTER 15

Topological Properties of the Catastrophe Map of a General Equilibrium Production Model with Uncertain States of Nature

Pascal Stiefenhofer

University College London, London, UK

ABSTRACT

This paper shows existence and efficiency of equilibria of a production model with uncertainty, where production is modeled in the demand function of the consumer. Existence and efficiency of equilibria are a direct consequence of the catastrophe map being smooth and proper. Topological properties of the equilibrium set are studied. It is shown that the equilibrium set has the structure of a smooth submanifold of the Euclidean space which is diffeomorphic to the sphere implying connectedness, simple connectedness, and contractibility. The set of economies with discontinuous price systems is shown to be of Lebesgue measure zero.

Keywords: Differential Topology, General Equilibrium, Uncertainty, Production

1. INTRODUCTION

This paper considers the Arrow-Debreu model with a complete set of contingent claims [1] and production. Existence and efficiency of equilibria of this model are shown by [2] in a seminal paper. This paper, however, derives global topological properties of the set of equilibria. The structure of this set has been studied before by Balasko [3] in the context of deterministic pure exchange economics. Such models lack a time structure, and as a consequence do not incorporate uncertainty [3] - [5] (Balasko (Preprint 2011) for example).1

The aim of this paper is to consider a reformulation of the Arrow-Debreu model in terms of an exchange model with production in the utility function. Preliminary results are found in [6] . A version of the decentralized production model is found in [7] . The formulation of the production model considered in this paper allows extending some of the known results about deterministic economies to production economies with uncertainty and production of adjusted demand functions. It is shown that the set of equilibria is a smooth manifold. Its dimension depends on the number of goods available for consumption, the number of uncertain states of nature and the number of consumers. This manifold is also shown to be diffeomorphic to a sphere. This result has deep economic implication. It implies that geodesics can be defined on it. This property is particularly useful when designing economic policies.

The paper is organized in three sections. Section 2 introduces the model. Section 3 establishes the results, and Section 4 is a conclusion.

2. THE LONG RUN PRIVATE OWNERSHIP PRODUCTION MODEL WITH UNCERTAIN STATES OF NATURE

We describe the two period private ownership production model $\mathrm{P}(L)$ introduced in [1] , chapter 7. Uncertainty is defined by a finite set of mutually exclusive and exhaustive states of nature denoted by $s \in \{0,1,\mathrm{L}\ ,S\}$, where s = 0 is the certain event in time period one. In total there are S + 1 states of nature. There are $i \in \{1,\mathrm{L}\ ,m\}$ consumers, $j \in \{1,\mathrm{L}\ ,n\}$ producers, and $k \in \{1,\mathrm{L}\ ,l\}$ physical goods. For all consumers $i \in \{1,\mathrm{L}\ ,m\}$, a consumption bundle is a collection of vectors

$x_i = \left(x_i(0),\ldots,x_i(s),\ldots,x_i(S)\right) \in X_i = \mathbb{R}_{++}^{l(S+1)}$, where consumption in a particular state $s \in \{0,1,\ldots,S\}$ is a vector $x_i(s) = \left(x_i^1(s),\ldots,x_i^l(s)\right) \in \mathbb{R}_{++}^l$. Associated with physical commodities is a set of normalized prices, denoted $\mathbf{S} = \left\{p \in \mathbb{R}_{++}^{l(S+1)} : p^l(s) = 1, \forall s \in \{0,1,\ldots,S\}\right\}$. Consumers are further endowed with a fraction $\theta_{ij} \in [0,1]$ of the profits of each firm. θ_{ij} represents the exogenously determined ownership structure of the private ownership production economy. It satisfies for each $j \in \{1,\ldots,n\}$ and $i \in \{1,\ldots,m\}$, $0 \le \theta_{ij} \le 1$, and $\sum_i \theta_{ij} = 1$. Denote the set of ownership structures

$$\Theta = \left\{\theta_{ij} \in \mathbb{R}_+^{nm} : \sum_i \theta_{ij} = 1, \forall j \in \{1,\ldots,n\}\right\}.$$

Consumers are endowed with a collection of vectors of initial resources

$$\omega_i = \left(\omega_i(0),\ldots,\omega_i(s),\ldots,\omega_i(S)\right) \in \Omega_i = \mathbb{R}_{++}^{l(S+1)},$$

where initial endowments in a particular state $s \in \{0,1,\ldots,S\}$ is a vector $\omega_i(s) = \left(\omega_i^1(s),\ldots,\omega_i^l(s)\right) \in \mathbb{R}_{++}^l$. Consumer $i \in \{1,\ldots,m\}$ is further characterized by a smooth Marschallian demand function

$$f_i : \mathbf{S} \times \mathbb{R}_{++}^{S+1} \to \mathbb{R}_{++}^{l(S+1)},$$

where $f_i(p, w_i)$ is defined for price vector $p \in \mathbf{S}$ and wealth level $w_i \in \mathbb{R}_{++}^{S+1}$ [8], where $w_i = p \cdot \omega_i$.2

Producers are characterized by production sets and their smooth supply functions. The main property of the long run production model is that all activities of the firm are variable. An activity y_j is a collection of vectors $y_j = \left(y_j(0),\ldots,y_j(s),\ldots,y_j(S)\right) \in \mathbb{R}^{l(S+1)}$, where an activity in state $s = 0$ is a vector of inputs $y_j(0) = \left(y_j^1(0),\ldots,y_j^l(0)\right) \in \mathbb{R}_-^l$, and $y_j(s) = \left(y_j^1(s),\ldots,y_j^l(s)\right) \in \mathbb{R}_+^l$ is the associated vector of outputs in state $s \in \{1, \ldots, S\}$. Let $\xi_j : \mathbf{S} \to \mathbf{R}^{l(S+1)}$ denote the smooth supply function of firm $j \in \{1, \ldots, n\}$, where $\xi_j(p)$ is defined on the set of normalized prices. Standard assumptions of smooth production economies introduced in [1] hold for each production set $Y_j \subset \mathbf{R}^{l(S+1)}$. In particular Y_j is convex, $0 \in Y_j$, and ∂Y_j has a strictly positive Gaussian curvature for every $j \in \{1, \ldots, n\}$.

2.1. Equilibrium $\mathbb{P}(\mathcal{L})$

Each consumer $i \in \{1, \ldots, m\}$ chooses a utility maximizing consumption bundle $x_i \in X_i$ at fixed $\omega_i \in \Omega$ and $\theta_{ij} \in \Theta$ satisfying his budget constraints. Each producer $j \in \{1, \ldots, n\}$ chooses profit maximizing net activities $y_j \in Y_j$ at competitive prices $p \in \mathbf{S}$. Let

$$z(p(s),\omega(s),\theta(s)) := \sum_i^m f_i\left(p(s), p(s)\cdot\omega_i(s) + \sum_j^n \theta_{ij}(s)\, p(s)\cdot\xi_j(p(s))\right) - \sum_i^m \omega_i(s) + \sum_j^n \xi_j(p(s)) \tag{1}$$

be the market excess demand function in state $s \in \{0, 1, \ldots, S\}$. Then, market clearance requires demand to equal supply in each market and uncertain state of the world. Hence

$$z(p(s),\omega(s),\theta(s)) = 0, \quad \forall s \in \{0, 1, \ldots, S\}.$$

An equilibrium is a price vector $p \in \mathbf{S}$ which satisfies this equation for a fixed distribution of initial resources and exogenously given ownership structure. An equilibrium pair is an equilibrium price vector $p \in \mathbf{S}$ with associated $\omega \in \Omega$. An equilibrium allocation is an allocation (x, y, θ) associated with an equilibrium price $p \in \mathbf{S}$. The model of the consumer is to solve a constraint optimization problem. This requires a consumer to maximize utility subject to a sequence of $(S+1)$ budget constraints. Hence, each consumer $i \in \{1, \ldots, m\}$

$$\left(x_i(s)\right) \in \arg\max\left\{u_i\left(x(s)\right) : x_i \in B_i(s)\right\}, \forall s \in \{0,1,\ldots,S\},$$

where $u_i : \mathbb{R}^{l(S+1)} \to \mathbb{R}$ is the consumer's smooth3 utility function. The new production adjusted budget set is now defined by

$$B_i(s) := \left\{x_i(s) \in \mathbb{R}^{l}_{++} : p(s) \cdot x_i(s) = p(s) \cdot \omega_i(s) + \sum_{j=1}^{n} \theta_{ij}(s) p(s) \cdot y_j(s)\right\}$$

The model of the producer is to maximize profits. Each producer solves a constraint optimization profit maximization problem. Hence, each $j \in \{1,\ldots,n\}$

$$\left(y_j(s)\right) \in \arg\max\left\{p(s) \cdot y_j(s) : y_j(s) \in Y_j(s)\right\}, \forall s \in \{0,1,\ldots,S\},$$

where the state dependent production set $Y_j(s)$ for all $s \in \{0,1,\ldots,S\}$ satisfies the assumptions of Debreu [1] also stated in the pervious section for the deterministic case.

Definition 1. An equilibrium of the two period private ownership production model with uncertainty $\mathrm{P}(L)$ is a price vector $p \in \mathbf{S}$ at fixed pair $(\omega,\theta) \in \Omega \times \Theta$ if for utility maximizing consumers $i \in \{1,\ldots,m\}$ and profit maximizing producers $j \in \{1,\ldots,n\}$

$$z(p,\omega,\theta) = 0. \tag{2}$$

An equilibrium allocation is a pair $(x,y) \in \mathbb{R}^{l(S+1)m}_{++} \times \mathbb{R}^{l(S+1)n}$ associated with an equilibrium price vector $p \in \mathbf{S}$ for fixed parameters $(\omega,\theta) \in \Omega \times \Theta$. Let , denote the mathematical operation defined by a state by state inner product. There are $l(S+1)$ equilibrium equations less $(S+1)$ equations satisfying Walras' law $p, z(p,\omega) = 0$, hence we have a system of $l(S+1)-(S+1)$ linearly independent equations. This amounts to the number of unknowns, given the number of normalized prices of $(S+1)$.

A study of the qualitative equilibrium structure of the two period private ownership production model with uncertainty amounts to a study of the structure of the solution set of the equilibrium equation $z(p,\omega)=0$. The first result is an equivalence relation between the two period exchange model with uncertainty and the two period production model with uncertainty. The relation between these models follows from the definition of a two period exchange model with production adjusted Marshallian demand functions.

Let $\zeta_i(p(s))=\sum_j \theta_{ij}(s)\xi_j(p(s))$ for any price system $p \in \mathbf{S}$ and uncertain state of the world

$s \in \{0,1,2,\ldots,S\}$. Let $h_i : S \times \mathbf{R}^{S+1}_{++} \to \mathbf{R}^{l(S+1)}_{++}$ defined by $h_i(p,w_i)=f_i(p,w_i+p\cdot\zeta_i(p))-\zeta_i(p)$, where $h_i(s)$ is given by

$$h_i(p(s),w_i(s))=f_i(p(s),w_i(s)+p(s)\cdot\zeta_i(p(s)))-\zeta_i(p(s)),\ \forall s \in \{0,1,2,\ldots,S\} \tag{3}$$

denote the individual demand function of the two period "production adjusted" exchange model $\mathrm{P}(E)$, where for every $i \in \{1,\ldots,m\}$ ownership structure $\theta_{ij}(s)$ is fixed, and total wealth defined by

$$p(s)\cdot\omega_i(s)+p(s)\cdot\zeta_i(p(s))$$

in every $i \in \{1,\ldots,m\}$. Now, let the equilibrium equation of the production model $\mathrm{P}(L)$ be given by

$$\sum_i^m f_i(p(s),p(s)\cdot(\omega_i(s)+\zeta_i(p(s))))=\sum_i^m \omega_i(s)+\sum_j^n \xi_j(p(s)),$$

where

$$f_i(p(s),p(s)\cdot(\omega_i(s)+\zeta_i(p(s)))) \tag{4}$$

denotes the individual demand function. This follows immediately from rewriting the excess demand equation in terms of demand equal to supply. Rewriting $\xi_j(s)$ in terms of ownership $\theta_{ij}(s)$, summing over i, and using $\zeta_i(p(s))=\sum_j \theta_{ij}(s)\xi_j(p(s))$ yields

$$\sum_j^n \xi_j(p(s)) = \sum_j^n \sum_i^m \theta_{ij}(s)\xi_j(p(s)) = \sum_i^m \zeta_i(p(s)).$$

Hence, the equilibrium equation of the production model $P(L)$ writes

$$\sum_i^m f_i\left(p(s), p(s)\cdot\left(\omega_i(s) + \zeta_i(p(s))\right)\right) = \sum_i^m \omega_i(s) + \sum_i^m \zeta_i(p(s)),$$

since $\sum_i^m \theta_{ij}(s) = 1$. This can be rewritten as

$$\sum_i^m f_i\left(p(s), p(s)\cdot\left(\omega_i(s) + \zeta_i(p(s))\right)\right) - \sum_i^m \zeta_i(p(s)) = \sum_i^m \omega_i(s).$$

This is the equilibrium equation of the exchange model with production adjusted demand functions $P(E)$. Hence, by definition of the production adjusted demand function h_i obtain

$$\sum_i^m h_i(p(s), w_i(s)) = \sum_i^m \omega_i(s), \quad \forall s \in \{0,1,\ldots,S\}, \tag{5}$$

since $w_i(s) = p(s)\cdot\omega_i(s) + p(s)\cdot\zeta_i(p(s))$ for all $s \in \{0,1,\ldots,S\}$. This is the equilibrium equation of the production adjusted exchange model $P(E)$ in terms of $h_i(s)$. This concludes the proof of theorem (1).

Theorem 1. For fixed $\theta \in \Theta$, $(p,\omega) \in \boldsymbol{S}\times\Omega$ is an equilibrium of the long run production model with uncertainty $P(L)$ if and only if $(p,\omega) \in \boldsymbol{S}\times\Omega$ is an equilibrium of the two period exchange model with uncertainty and production adjusted demand functions $P(E)$.

We have established a relationship between the production model with a long term time structure and uncertainty $P(L)$, and a pure exchange model with a long term time structure and uncertainty with production adjusted demand functions $P(E)$.

The result suggests that the decentralized production model can be reformulated as a centralized model. It is efficiently applied in establishing many properties about production economies in the next section.

2.2. Equilibrium Structure $\mathbb{E}$ of $\mathbb{P}(E)$

Let $\mathbf{E} \subset \mathbf{S} \times \Omega$ denote the set of equilibrium solutions of the production adjusted exchange model $\mathrm{P}(E)$ or the set of solutions of the long run production model $\mathrm{P}(L)$.4 This set consists of pairs $(p, \omega) \in \mathbf{S} \times \Omega$ satisfying the equilibrium equations $z(p, \omega) = 0$. Formally, for the case of the production model $\mathrm{P}(L)$ we have

$$\mathrm{E} = \left\{ (p, \omega) \in \mathbf{S} \times \Omega : \sum_{i}^{m} f_i \left(p, p \cdot \omega_i + \sum_{j}^{n} \theta_{ij} p \cdot \xi_j(p) \right) - \left(\sum_{i}^{m} \omega_i + \sum_{j}^{n} \xi_j(p) \right) = 0 \right\},$$

and in the case of the production adjusted exchange model $\Pi(E)$, we have

$$\mathrm{E} = \left\{ (p, \omega) \in S \times \Omega : \sum_{i}^{m} h_i(p, w_i) - \sum_{i}^{m} \omega_i = 0 \right\}.$$

Theorem 2. The set E of model $\mathrm{P}(E)$ is a closed subset of the Euclidean space defined by $S \times \Omega$.

Proof. E is defined by pairs $(p, \omega) \in \mathbf{S} \times \Omega$ satisfying the equilibrium Equation (5). E is the preimage of the vector $0 \in \mathbf{R}^{l(S+1)}$ by the smooth mapping $(p, \omega) \text{a} \sum_i h_i(p(s), w_i(s)) - \sum_i \omega_i(s)$ for all $s \in \{0, 1, \mathrm{L}, S\}$, and closed by the closed map lemma closed ([9] , p. 553). The closed map lemma requires the excess demand mapping to be continuous and the domain to be a compact set and the range a Hausdorff space. Recall that the excess demand mapping is differentiable at any order required. This is a consequence of subtracting differentiable aggregate supply mappings from differentiable aggregate demand mappings. Differentiability of demand and supply mappings is in turn a consequence of the assumptions of the model discussed earlier ([1] [8]).

Theorem 3. The set E of model $P(E)$ is a smooth manifold of dimension $(S+1)lm$.

Proof. Consider the mapping $Z:\mathbf{S}\times\Omega$ into $\mathbf{R}^{l(S+1)}$ defined by the smooth mapping

$$(p,\omega)\mapsto \sum_i h_i(p,w_i)-\sum_i \omega_i .$$

By theorem the regular value theorem E is the preimage of $0\in\mathbf{R}^{l(S+1)}$. We need to prove that this mapping does not contain critical points. This follows by showing that the linear tangent map $D_\omega Z$ is onto. The onto property follows directly from the rank property of the Jacobian matrix chosen for any arbitrary individual $i\in\{1,\ldots,m\}$ and state of nature $s\in\{0,1,\ldots,S\}$. By the chain rule, we obtain

$$D_{\omega_i}Z=\begin{pmatrix} \frac{\partial h_i^1(s)}{\partial\omega_i(s)}p^1(s)-1 & \cdots & \frac{\partial h_i^1(s)}{\partial\omega_i(s)}p^{l-1}(s) & \frac{\partial h_i^1(s)}{\partial\omega_i(s)} \\ \vdots & \ddots & \vdots & \vdots \\ \frac{\partial h_i^{l-1}(s)}{\partial\omega_i(s)}p^1(s) & \cdots & \frac{\partial h_i^{l-1}(s)}{\partial\omega_i(s)}p^{l-1}(s)-1 & \frac{\partial h_i^{l-1}(s)}{\partial\omega_i(s)} \end{pmatrix}.$$

By simple algebraic manipulations we obtain the new matrices

$$\begin{pmatrix} \frac{\partial h_i^1(s)}{\partial\omega_i(s)}p^1(s)-1-\frac{\partial h_i^1(s)}{\partial\omega_i(s)}p^1(s) & \ldots & \frac{\partial h_i^1(s)}{\partial\omega_i(s)}p^{l-1}(s) & \frac{\partial h_i^1(s)}{\partial\omega_i(s)} \\ \vdots & \ddots & \vdots & \vdots \\ \frac{\partial h_i^{l-1}(s)}{\partial\omega_i(s)}p^1(s)-\frac{\partial h_i^1(s)}{\partial\omega_i(s)}p^1(s) & \ldots & \frac{\partial h_i^{l-1}(s)}{\partial\omega_i(s)}p^{l-1}(s)-1 & \frac{\partial h_i^{l-1}(s)}{\partial\omega_i(s)} \end{pmatrix},$$

$$\begin{pmatrix} -1 & \cdots & \dfrac{\partial h_i^1(s)}{\partial \omega_i(s)} p^{l-1}(s) - \dfrac{\partial h_i^1(s)}{\partial \omega_i(s)} p^{l-1}(s) & \dfrac{\partial h_i^1(s)}{\partial \omega_i(s)} \\ \vdots & \ddots & \vdots & \vdots \\ 0 & \cdots & \dfrac{\partial h_i^{l-1}(s)}{\partial \omega_i(s)} p^{l-1}(s) - 1 - \dfrac{\partial h_i^1(s)}{\partial \omega_i(s)} p^{l-1}(s) & \dfrac{\partial h_i^{l-1}(s)}{\partial \omega_i(s)} \end{pmatrix}.$$

Finally, we obtain

$$\begin{pmatrix} -1 & & 0 & \dfrac{\partial h_i^1(s)}{\partial w_i(s)} \\ & \ddots & & \vdots \\ 0 & & -1 & \dfrac{\partial h_i^{l-1}(s)}{\partial w_i(s)} \end{pmatrix}$$

from which we extract the information required. Rank $D_{\omega_i} Z$ is equal to $(l-1)$ in every state $s \in \{0,1,\ldots,S\}$. By the regular value theorem E is a smooth manifold. This manifold is parameterized by smooth coordinate functions $\omega = (\omega(0),\ldots,\omega(s),\ldots,\omega(S)) \in \Omega$. From the regular value theorem it also follows that its dimension is equal to the dimension of $\mathrm{S} \times \Omega$ minus $(l-1)(S+1)$, hence

$$\dim(\mathrm{E}) = ((l-1)(S+1) + ml(S+1) - (l-1)(S+1)) = ml(S+1).$$

The following theorem illustrates other economically interesting global properties of the equilibrium manifold. It says that by construction of a diffeomorphism f restricted to the equilibrium manifold E into $S \times \mathbf{R}_{++}^{m(S+1)} \times \mathbf{R}_{++}^{(l-1)(m-1)(S+1)}$ that E is diffeomorphic to the sphere in $\mathbf{R}_{++}^{lm(S+1)}$ implying that the equilibrium manifold is arc-connected, simply connected, and contractible. These properties are particularly useful in applied work such as economic policy equilibrium analysis.5 For example, economic policy is often concerned with finding a path between a current point on E and a desired point on E.

Theorem 4. The smooth equilibrium manifold $\mathbf{E}$ of model $\mathrm{P}(E)$ is diffeomorphic to the sphere of dimension $\mathrm{R}_{++}^{lm(S+1)}$.

Proof. The aim of the proof is to define two smooth mappings between smooth manifolds such that we can apply the theorem (Hirsch [10] , pp. 15-16). Hence, let

$$f:\mathbf{S}\times\Omega\to\mathbf{S}\times\mathrm{R}_{++}^{(S+1)m}\times\mathrm{R}_{++}^{(l-1)(S+1)(m-1)}$$

be smooth mappings defined by

$$\begin{aligned}&f\left(p(s),\omega_1(s),\omega_1(s),\ldots,\omega_m(s)\right)\\&=\left(p(s),(s)\cdot\omega_1(s),p(s)\cdot\omega_2(s),\ldots,p(s)\cdot\omega_m(s),\bar{\omega}_1(s),\bar{\omega}_2(s),\ldots,\bar{\omega}_{m-1}(s)\right),\ \forall s\in\{0,1,\ldots,S\}.\end{aligned}$$

Then, let

$$g:\mathbf{S}\times\mathrm{R}_{++}^{(S+1)m}\times\mathrm{R}_{++}^{(l-1)(S+1)(m-1)}\to\mathbf{S}\times\Omega$$

denote smooth mappings defined by

$$g\left(p(s),\bar{\omega}(s)_1,\omega_1^l(s),\ldots,\bar{\omega}_{m-1}(s),\omega_{m-1}^l(s),\bar{\omega}_m(s)\right),\ \forall s\in\{0,1,\ldots,S\}.$$

Observe that the coordinates for the lth good of the m – 1 consumers in $s\in\{0,1,\ldots,S\}$, $\omega_1^l(s),\ldots,\omega_{m-1}^l(s)$ are defined

$$\omega_i^l(s)=w_i(s)-\left(\sum_{k=1}^{l-1}p^k(s)\cdot\omega_i^l(s)\right),\ \forall i\in\{1,\ldots,m-1\},\ \forall k\in\{1,\ldots,l-1\}. \tag{6}$$

Also observe that the coordinates for the mth consumer of the $l-1$ goods in $s\in\{0,1,\ldots,S\}$, $\omega_m^1(s),\omega_m^2(s),\ldots,\omega_m^{l-1}(s)$ are defined by

$$\omega_m(s)=\sum_{i=1}^{m}h_i\left(p(s),w_i(s)\right)-\sum_{i=1}^{m-1}\omega_i(s). \tag{7}$$

The application of theorem ([10]) requires to show that $E = Im(g)$ and that $f \circ g = Id$. The first part of the proof requires to calculate two inclusions, (i) $Im(g) \subset E$ and (ii) $E \subset Im(g)$. We start by showing the second part first. Now, to show that (i) $Im(g) \subset E$, take any consumption bundle

$$x(s) = \left(p(s), w_1(s), \ldots, w_m(s), \bar{\omega}_1(s), \ldots, \bar{\omega}_{m-1}(s)\right),$$

and compute the inner product of (7) with $p(s)$ $s \in \{0,1,\ldots,S\}$, and apply Walras' law to obtain

$$w_m(s) = p \cdot \omega_m(s), \ \forall s \in \{0,1,\ldots,S\}.$$

From that a reformulation of (7) readily follows in terms of the equilibrium equation

$$\sum_i^m h_i\left(p(s), w_i(s)\right) = \sum_i^m \omega_i(s), \ \forall s \in \{0,1,\ldots,S\},$$

This is the equilibrium Equation (5), hence $Im(g) \subset E$. Next, need to show that (ii) $E \subset Im(g)$. Take $(p,\omega) \in E$. It is then trivial to do the computations proving following equality

$$f(s) \circ g\left(p(s), \omega(s)\right) = \left(p(s), \omega(s)\right), \forall s \in \{0,1,\ldots,S\}$$

from which it readily follows that $E \subset Im(g)$. Clearly we have constructed the two smooth relations such that

$$f(s) \circ g(s) = Id(s), \forall s \in \{0,1,\ldots,S\},$$

where Id is the identity map defined on $\left(S \times \mathbb{R}_{++}^{(S+1)m} \times \mathbb{R}_{++}^{(l-1)(S+1)(m-1)}\right)$.

We have shown that the smooth mapping f restricted to the equilibrium manifold E defines a diffeomorphism between E and the sphere of dimension $\mathbf{R}^{(S+1)lm}$.

It remains to be shown that equilibria in the long run production model with uncertainty always exist. The strategy of the proof is to show that the natural projection mapping $\pi:\mathrm{E}\to\Omega$ is smooth and proper. Existence of long run equilibria of the production model with uncertainty follows immediately from the smoothness lemma (1) and the properness lemma (2) below.

Theorem 5. Equilibria of the two period production model with uncertainty $\mathrm{P}(E)$ always exist.

Lemma 1 (Smoothness) $\pi:\mathrm{E}\to\Omega$ is smooth.

Proof. Recall that E is a smooth submanifold of $\mathbf{S}\times\Omega$. It follows from the definition of a smooth manifold that its natural embedding $\hat{\pi}:\mathrm{E}\to\mathbf{S}\times\Omega$ is itself smooth. The projection mapping $\overline{\pi}:\mathbf{S}\times\Omega\to\mathrm{E}$ being itself smooth, it follows that π the restriction of the natural projection to E as the composition of two smooth mappings $\pi=\overline{\pi}\circ\hat{\pi}$ is therefore smooth.

The next lemma makes use of theorem (see [11] , p. 174).

Lemma 2 (Properness) $\pi:\mathrm{E}\to\Omega$ is proper.

Proof. Pick an arbitrary ω_i for $i\in\{1,\mathrm{L}\ ,m\}$. Let $\omega_i\in K_i$ be an element in a compact set K_i. Now, for every $p\in\mathbf{S}$ and $\omega_i\in K_i$ need to show (i) that $f_i(p,w_i)$ is bounded from below. It follows that

$$u_i(\omega_i)\le u_i(f_i(p,w_i))$$

and by non-satiation have also

$$u_i(\omega_i')\le u_i(\omega_i),$$

which by monotonicity of u_i implies that

$$u_i(\omega_i')\le u_i(f_i(p,w_i)).$$

Clearly, there exists some $x_i' \in \mathbf{R}_{++}^{l(S+1)}$ for every $p \in \mathbf{S}$ and $\omega_i \in K_i$ satisfying

$$x_i' \leq u_i\left(f_i\left(p, w_i\right)\right)$$

by boundedness of indifference mappings from below for every $i \in \{1, \ldots, m\}$. (ii) We now show that for every $p \in \mathbf{S}$ and $\omega_i \in K_i$, $f_i(p, w_i)$ is also bounded from above. For (p, ω_i) have

$$\left(f_i\left(p, w_i\right)\right) = \sum_i \omega_i - \sum_{-i} f_i\left(p, w_i\right)$$

where

$$\sum_i \omega_i - \sum_{-i} f_i\left(p, w_i\right) \leq \sum_i \omega_i - \sum_{-i} x_i'$$

Clearly, $f_i(p, w_i)$, is bounded above by some $x_i'' \in \mathbf{R}_{++}^{l(S+1)}$, since for $(p, \omega) \in \mathrm{E}$, $\sum_{-1} \omega_i$ is bounded from above for every $\omega \in K$. Hence have established upper and lower bounds defining a compact set

$$\left\{x_i' \leq f_i\left(p, w_i\right) \leq x_i''\right\}$$

for every $(p, \omega) \in \pi^{-1}(K)$. Let G be a compact set defined by the preimage of the diffeomorphism $f_i(p, w_i)$ projected onto S. Now, by continuity of $\pi : \mathrm{E} \rightarrow \Omega$, $\pi^{-1}(K)$ is closed in $\mathbf{E}$, which by theorem (2) is a closed subset of $\mathbf{S} \times \Omega$. Closedness of $\pi^{-1}(K)$ follows from closedness of $\pi^{-1}(K) \cap G \times K \subset G \times K$.

The number of equilibria of the long run production model with uncertainty is odd for any regular economy $\omega \in \Omega$. The modulo 2 degree of π is $+1$. See Guillemin and Pollack for example [12].

I now define a subset of points on E at which pairs $(p,\omega)\in\mathrm{E}$ are not regular.

Definition 2. E_c is the set of singular equilibria $(p,\omega)\in\mathrm{E}$ given by the singular points of π.

Proposition 1. E_c is closed.

Proof. A necessary and sufficient condition for an equilibrium pair $(p,\omega)\in\mathrm{E}$ to be singular is that the determinant of the Jacobian matrix of π, denoted by $det(D\pi)$ is equal to zero. Now, the set of critical points E_c defined by the preimage of $0\in det(D\pi)$ is closed by the closed mapping lemma ([9]), since π, $D\pi$, and the coefficients of $det(D\pi)$, are all continuous, from which the result follows.

Definition 3. $\Sigma=\{\omega\in\Omega: \textit{for all } \omega\in\Omega \textit{ singular values of } \pi\}$.

A singular value $\omega\in\Omega$ is the image of π of a singular point $(p,\omega)\in\mathrm{E}_c$ into Ω. The set of regular values is defined by $\mathrm{R}=\{\omega\in\Omega: \text{for all } \omega\in\Omega \text{ regular values of } \pi\}$. It follows that $\mathrm{R}=\Omega\setminus\Sigma$ represents the sets of regular economies. The next proposition states the Σ is closed and of measure zero. This means that the probability of observing an economy with this property is "close" to zero. Hence, its complement R is an open dense set.

Proposition 2. The set of singular economies Σ is closed and of Lebesgue measure zero in Ω.

Proof. The proof follows from the application of Sards's theorem which describes the set of singular values of a smooth mapping having the property of Lebesgue measure zero. Hence know that Σ is a set of Lebesgue measure zero. Closedness of Σ follows from the properness of π.

3. CONCLUSION

Pascal Stiefenhofer This paper discusses local and global equilibrium properties of a production economy with a long-term time structure. Production is modeled in the demand functions of the consumers.

The advantage of this way of modeling production is that it enables us to establish a relationship between production and pure exchange economies. Adding uncertainty to the production model is a further step towards realism.

It is shown that the equilibrium set of all production economies with uncertainty has the structure of a smooth submanifold of the Euclidean space which is diffeomorphic to a sphere. These topological properties are of significant economic importance in terms of economic policy design since they imply connectedness and contractability of the set of solutions. It is also shown that the set of singularities of the catastrophe map is closed, and of Lebesuge measure zero.

The practical implication of this result is that the probability of observing an economy with a discontinuous price system is close to zero.

REFERENCES

1. Debreu, G. (1959) Theory of Value. Wiley, New York.
2. Arrow, K. and Debreu, G. (1954) Existence of an Equilibrium for a Competitive Economy. Econometrics, 22, 265-290.
3. Balasko, Y. (1988) The Equilibrium Manifold: Postmodern Developments in the Theory of General Economic Equilibrium. The MIT Press, Cambridge, Massachusetts.
4. Jouini, E. (1993) The Graph of the Walras Correspondence. The Production Economies Cas. Journal of Mathematical Economics, 22, 139-147.
5. Fuchs, G. (1974) Private Ownership Economies with a Finite Number of Equilibria. Journal of Mathematical Economics, 1, 141-158.
6. Stiefenhofer, P. (2011) Equilibrium Structure of Production Economies with Uncertainty: The Natural Projection Approach. Discussion Papers in Economics, University of York, York, 7.
7. Stiefenhofer, P. (2013) The Catastrophe Map of a Two Period Production Model with Uncertainty. Applied Mathematics, 4, No. 8A.
8. Debreu, G. (1972) Smooth Preferences. Econometrica, 40, 603-615.
9. Lee, J.M. (2004) Introduction to Topological Manifolds. Springer, New York.
10. Hirsch, M. (1972) Differential Topology. Springer Verlag, New York.
11. Lee, J.M. (2000) Introduction to Smooth Manifolds. Springer, New York.
12. Guillemin, V. and Pollack, A. (1974) Differential Topology. Prentice Hall, Upper Saddle River.

NOTES

[1]Discussion paper: the natural projection approach to smooth production economies, 2011.

[3]"Smoothness" follows from the same reasons as in the previous section [8] . It essentially means that all functions are differentiable at any order required.

[5]The construction of a Riemanian metric on the equilibrium manifold would be a very useful result towards economic policy analysis.

CHAPTER 16

On Links between Rough Sets and Digital Topology

El-Sayed A. Abo-Tabl[1,2]

[1]*Mathematics Department, Faculty of Science, Assiut University, Assiut, Egypt*

[2]*Mathematical Department, Faculty of Science and Arts, Muthnab, Qassim University, Qassim, Saudi Arabia*

ABSTRACT

Rough set theory is a powerful tool for dealing with uncertainty, granularity, and incompleteness of knowledge in information systems. In addition, digital topology deals with properties and features of two-dimensional or three-dimensional digital images that correspond to topological properties of objects. So, we try to describe the relationship between rough sets and digital topology. Firstly, we will study the classifications of topologies in rough sets. Secondly, we will use the upper approximation operator to span the digital line, which is the basic building block of the digital spaces.

Keywords: Rough Sets, Upper Approximation Operator, Topology and Digital Topology

1. INTRODUCTION

The theory of rough sets, proposed by Pawlak [1] [2] , is an extension of the set theory for the study of intelligent systems characterized by insufficient and incomplete information. Using the concepts of the lower and the upper approximations from the rough set theory, knowledge hidden in information systems may be unraveled and expressed in the form of decision rules [2] [3] .

The classical rough approximations are based on equivalence relations, but this requirement is not satisfied in some situations. Thus the classical rough approximations have been extended to the similarity relation based rough sets [4] [5] , the tolerance relation based rough sets [6] , the dominance relation based rough sets [7] , the arbitrary binary relation based rough sets [8] -[10] and the covering-based rough sets [11] -[13] .

An interesting and natural research topic in rough set theory is to study rough set theory via topology. Indeed, Polkowski [14] pointed: topological aspects of rough set theory were recognized early in the framework of topology of partitions. Skowron [15] and Wiweger [16] separately discussed this topic for classical rough set theory in 1988. Polkowski [17] constructed and characterized topological spaces from rough sets based on information systems. Kortelainen [18] considered relationships between modified sets, topological spaces and rough sets based on a pre-order. Skowron et al. [6] [19] generalized the classical approximation spaces to tolerance approximation spaces, and discussed the problems of attribute reduction in these spaces (also see [20]). Lashin et al. [21] introduced the topology generated by a subbase, also defined a topological rough membership function by the subbase of the topology. In addition, connections between fuzzy rough set theory and fuzzy topology were also investigated (see [22] -[24]).

General topology has been considered as the entrance to understanding topology science, moreover the base of general topology is the topological space, which has been considered as a representation of universal space in general, and geometric shape in special, also the mathematical analysis concepts. The general topology has become the appropriated frame for every collection connected by relations. Topology is also a mathematical tool to study information systems and rough sets [15] [21] .

General topology has applications in the theory of image processed by exhibiting algorithms, which apply current knowledge of digital topology (i.e., the study of the geometric and topological properties of digital images [25]). The problems that might arise are, for example finding connected components, set boundaries or any other operations which are needed in image processing. The well-known digital Jordan curve theorem is proved by using topological approach [26] -[28] . The theorem is important in the theory of computer graphics.

The basic building block of the digital n-space is the digital line or the so called Khalimsky line [26] [29] -[31] . This is the set of the integers, Z, equipped with the topology K, generated by $S_K = \{\{2n-1, 2n, 2n+1\} : n \in Z\}$ as a subbase. Thus a set U is open in K if and only if whenever $x \in U$ is an even integer, then $x-1, x+1 \in U$. Nowadays, this topology, called the Khalimsky topology, is one of the most important concepts of the digital topology. It has been studied and used by many authors, e.g., [26] -[37] .

The remainder of this paper is organized as follows. In Section 2, we present the fundamental concepts and properties of Pawlak's rough set theory, general topology and digital topology. Section 3 presents the classifications of topologies in rough sets using the properties of the binary relations. In addition, we use the upper approximation operator to obtain the digital line, which is the basic building block of the digital spaces in Section 4 to try to describe the link between rough sets and digital topology. This paper concludes in Section 5.

2. PRELIMINARIES

In this section, we introduce a review of some basic concepts of topological space and Pawlak approximation space.

A topological space is a pair (U, τ) consisting of a set U and family τ of subset of U satisfying the following conditions:

(T1) $\varnothing \in \tau$ and $U \in \tau$.

(T2) τ is closed under arbitrary union.

(T3) τ is closed under finite intersection.

The pair (U, τ) is called a space, the elements of U are called points of the space, the subsets of U belonging to τ are called open sets in the space, and their complement are called closed sets in the space; the family τ of open subsets of U is also called a topology for U [38] .

Definition 2.1. [39] A topology τ on the set U is called an Alexandroff topology if the intersection of arbitrarily many open sets is still open, or equivalently, the union of arbitrarily many closed sets is still closed.

Definition 2.2. [37] Let U be a nonempty set and cl: $P(U) \rightarrow P(U)$ be a function. The space (cl, U) is called Smyth space if the following conditions hold.

1) $cl(\varnothing) = \varnothing$

2) $cl(A \cup B) \subseteq cl(A) \cup cl(B)$

3) if $A \subseteq B$ then $cl(A) \subseteq cl(B)$.

Definition 2.3. For the set of integer Z we have:

(1) In Z the 2-neighbors of x is $N_2(x) = \{x-1, x+1\}$

(2) In Z^2 the 4-neighbors of (x, y) is $N_4((x, y)) = \{(x-1, y), (x+1, y), (x, y-1), (x, y+1)\}$,

(3) In Z^2 the 8-neighbors of (x, y) is

$N_8((x, y)) = \{(x-1, y-1), (x-1, y), (x-1, y+1), (x+1, y-1), (x+1, y), (x+1, y+1), (x, y-1), (x, y+1)\}$.

Definition 2.4. Let X be a nonempty set. Then the following are defined.

1) A relation R on a set X is reflexive if and only if xRx for all x in X.

2) A relation R on a set X is symmetric if and only if xRy implies yRx.

3) A relation R on a set X is transitive if xRy, yRz, then xRz.

A relation R on a set X is called an equivalence relation if it is reflexive, symmetric and transitive.

Definition 2.5. [2] Let R be an equivalence relation on a nonempty set U. For any subset $A \subseteq U$, the lower and upper approximations of A according to R are then defined as

$$\underline{R}(A) = \{x \in U : [x]_R \subseteq A\}$$

$$\overline{R}(A) = \{x \in U : [x]_R \cap A \neq \varnothing\}$$

where $[x]_R$ is called an equivalence class of $x \in U$.

Yao [37] extended the Pawlak's definitions of a rough set to any binary relation R as follows:

Definition 2.6. For the pair (U,R), the set xR is defined as $xR=\{y\in U: xRy\}$ called the right neighborhood of an element $x\in U$.

Definition 2.7. Let R be any binary relation on a nonempty set U. For any subset $A\subseteq U$, the lower and the upper approximations of A according to R are then defined as $\underline{R}(A)=\{x\in U: xR\subseteq A\}$ $\overline{R}(A)=\{x\in U: xR\cap A\neq\varnothing\}$

Obviously, if R is an equivalence relation, $xR=[x]_R$ and these definitions are equivalent to the original Pawlak's definitions.

We list the properties that are of interest in the theory of rough sets, let $A,B\subseteq U$:

L_1. $\overline{R}(A)=\left(\overline{R}(A^C)\right)^C$, where A^C denotes the complementation of A in U.

L_2. $\underline{R}(U)=U$.

L_3. $\underline{R}(A\cap B)=\underline{R}(A)\cap\underline{R}(B)$.

L_4. $\underline{R}(A\cup B)\supseteq\underline{R}(A)\cup\underline{R}(B)$.

L_5. $A\subseteq B\Rightarrow\underline{R}(A)\subseteq\underline{R}(B)$.

L_6. $\underline{R}(\varnothing)=\varnothing$.

L_7. $\underline{R}(A)\subseteq A$.

L_8. $A\subseteq\underline{R}(\overline{R}(A))$.

L_9. $\underline{R}(A)\subseteq\underline{R}(\underline{R}(A))$.

L_{10}. $\overline{R}(A)\subseteq\underline{R}(\overline{R}(A))$.

U_1. $\overline{R}(A)=\left(\overline{R}(A^C)\right)^C$.

U_2. $\overline{R}(\varnothing)=\varnothing$.

U_3. $\overline{R}(A\cup B)=\overline{R}(A)\cup\overline{R}(B)$.

U4. $\overline{R}(A\cap B)\subseteq \overline{R}(A)\cap \overline{R}(B)$.

U5. $A\subseteq B\Rightarrow \overline{R}(A)\subseteq \overline{R}(B)$.

U6. $\overline{R}(U)=U$.

U7. $A\subseteq \overline{R}(A)$.

U8. $\overline{R}(\underline{R}(A))\subseteq A$.

U9. $\overline{R}(\overline{R}(A))\subseteq \overline{R}(A)$.

U10. $\overline{R}(\underline{R}(A))\subseteq \underline{R}(A)$.

Properties L_1 and U_1 state that two approximations are dual to each other. Hence, properties with the same numbers may be regarded as dual properties. Properties L_9, L_{10}, U_9 and U_{10} are expressed in terms of set inclusion. The standard version using set equality can be derived from L_1-L_{10} and U_1 - U_{10}. For example, it follows from L_7 and L_9 that $\underline{R}(A)=\underline{R}(\underline{R}(A))$. It should also be noted that these properties are not independent.

We can find the basic axioms for extended topological spaces in Table 1.

3. TOPOLOGY CLASSIFICATIONS BASED ON ROUGH SETS

In this section, we discuss the classification of topologies in the theory of rough sets.

Proposition 3.1. For any relation R on a nonempty set U and for every $A,B\subseteq U$ the properties L_1 - L_5 and U_1 - U_5 hold according to Definition 2.7.

Proof. See [9] .

Example 3.1. Let $R=\{(a,a),(b,a),(b,c),(b,d),(c,d),(d,c)\}$ be a relation on a nonempty set $U=\{a,b,c,d\}$. Then $aR=\{a\}$, $bR=\{a,c,d\}$, $cR=\{d\}$ and $dR=\{c\}$. Then $\tau=\{A:R(A)=A\}=\{\varnothing,U,\{a\},\{c,d\}\}$ is not a topology on U.

Theorem 3.1. Suppose that $(U,\underline{R},\overline{R})$ is a rough approximation space. If $(U,\underline{R},\overline{R})$ satisfies the properties L_1 - L_5 and U_1 - U_5, then $(U,\overline{R})$ is a Smyth space [37] .

Proposition 3.2. For any reflexive relation R on a nonempty set U and for every $A,B\subseteq U$ the properties L_1 - L_7 and U_1 - U_7 hold according to Definition 2.7.

Proof. See [9] .

Example 3.2. Let $R=\{(a,a),(b,b),(c,c),(d,d),(a,b),(c,d),(d,a)\}$ be a reflexive relation on a nonempty set $U=\{a,b,c,d\}$.

Then $aR=\{a,b\},bR=\{b\},cR=\{c,d\}$ and $dR=\{a,d\}$.

Then $\tau=\{\varnothing,U,\{b\},\{a,b\},\{a,b,d\}\}$ is a topology on U.

Theorem 3.2. If $(U,\underline{R},\overline{R})$ satisfies the properties L_1 - L_7 and U_1 - U_7, then $(U,\overline{R})$ is a pre-topology or Čech closure space.

Proposition 3.3. For any reflexive and symmetric relation R on a nonempty set U and for every $A,B\subseteq U$ the properties L_1 - L_8 and U_1 - U_8 hold according to Definition 2.7.

Proof. See [9] .

Example 3.3. Let $R=\{(a,a),(b,b),(c,c),(d,d),(a,b),(b,a),(b,c),(c,b)\}$ be a reflexive and symmetric relation on a nonempty set $U=\{a,b,c,d\}$. Then $aR=\{a,b\},bR=\{a,b,c\},cR=\{b,c\}$ and $dR=\{d\}$. Then $\tau=\{\varnothing,U,\{d\},\{a,b,c\}\}$ is a topology on U in which each open (closed) set is also closed (open) set.

Theorem 3.3. If $(U,\underline{R},\overline{R})$ satisfies the properties L_1 - L_8 and U_1 - U_8, then $(U,\overline{R})$ is a pre-topology or Čech closure space in which each open (closed) set is also closed (open).

Proposition 3.4. For any reflexive and transitive relation R on a nonempty set U and for every $A,B\subseteq U$ the properties L_1 - L_7, L_9, U_1 - U_8 and U_9 hold according to Definition 2.7.

Proof. See [9] .

Example 3.4. Let $R=\{(a,a),(b,b),(c,c),(d,d),(a,b),(b,d),(a,d),(a,c),(c,d)\}$ be a reflexive and transitive relation on a nonempty set $U=\{a,b,c,d\}$. Then $aR=\{a,b,c,d\}, bR=\{b,d\}, cR=\{c,d\}$ and $dR=\{d\}$. Then $\tau=\{\emptyset,U,\{d\},\{b,d\},\{c,d\},\{b,c,d\}\}$ is an Alexandroff topology on U.

Birkhoff [40] showed that there exists a one-to-one correspondence between the set of all reflexive and transitive binary relations and Alexandroff topologies.

Table 1. The basic axioms for extended topological spaces.

Axioms	Property	Name
K_0	L_2 U_2	Normality Co-normality
K_1	L_3 U_3	Multiplication Addition
K_2	L_5 U_5	Monotone Monotone
K_3	L_7 U_7	Counteraction Extension
K_4	L_8 U_8	Symmetry Symmetry
K_5	L_9 U_9	Idempotent Idempotent
K_6	L_{10} U_{10}	Super-Symmetry Super-Symmetry

Theorem 3.4. If $(U,\underline{R},\overline{R})$ satisfies the properties L_1 - L_7, L_9, U_1 - U_7 and U_9, then $(U,\overline{R})$ is an Alexandroff topological space.

Proposition 3.5. For any equivalence relation R on a nonempty set U and for every $A,B\subseteq U$ the properties L_1 - L_{10} and U_1 - U_{10} hold according to Definition 2.7.

Proof. See [9] .

Example 3.5. Let $R=\{(a,a),(b,b),(c,c),(d,d),(a,b),(b,a),(b,c),(c,b),(a,c),(c,a)\}$ be an equivalence relation on a nonempty set $U=\{a,b,c,d\}$. Then $aR=bR=cR=\{a,b,c\}$ and $dR=\{d\}$. Then $\tau=\{\varnothing,U,\{d\},\{a,b,c\}\}$ is a topology on U in which each open (closed) set is also closed (open) set, and socalled clo-open topology.

Theorem 3.5. If $(U,\underline{R},\overline{R})$ satisfies the properties L_1 - L_{10} and U_1 - U_{10}, then $(U,\overline{R})$ is an Alexandroff topological space in which each open (closed) set is also closed (open) set, and so-called clo-open space.

We summarize these results in Table 2, which indicates the generalized topological spaces depending on the properties of rough sets.

4. CONNECTING DIGITAL TOPOLOGY AND ROUGH SETS

In this section, we try to get the structure of digital topology from the concepts of the theory of rough sets (the upper approximate operator).

Definition 4.1. Let R be any relation on a set of integer Z. We define another relation R' on Z as follows:

$$xR'y \Leftrightarrow x \in \overline{R}(y)$$

on the other hand, we can write the relation R′ in an equivalent form as follows:

$$xR'y \Leftrightarrow x \rightarrow y$$

Theorem 4.1.

1) The relation R on Z is reflexive if and only if R′ is also reflexive.

2) The relation R on Z is transitive if and only if R′ is also transitive.

Proof. 1) Let R be a reflexive relation on $Z \Leftrightarrow x \in R(x) \Leftrightarrow xR'x \Leftrightarrow R'$ is reflexive.

2) Assume that xR′y and yR′z, then $x \in \bar{R}(y)$ and $y \in \bar{R}(z) \Rightarrow xRy$ and yRz, hence by transitivity of R we get xRz, thus $x \in \bar{R}(z)$, that implies xR′z. Therefore R′ is transitive. Conversely, suppose that xRy and yRz, then $x \in \bar{R}(y)$ and $y \in \bar{R}(z)$, that implies xR′y and yR′z, hence by transitivity of R′ we get xR′z, thus $x \in \bar{R}(z)$, so that xRz. Therefore R is transitive.

The following conditions hold on Z using the relation →1) if two points $x, y \in Z$ are 2-neighbors, then either x → y or y → x2) if two points $x, y \in Z$ are not 2-neighbors, then both x → y and y → x not hold.

Lemma 4.1. If $x_1, x_2, x_3 \in Z$ such that $x_2 \in N_2(x_1)$ and $x_3 \in N_2(x_2)$, then we have either $x_1 \to x_2 \leftarrow x_3$ or $x_1 \leftarrow x_2 \to x_3$.

Proof. By condition (1) we have either $x_1 \to x_2$ or $x_2 \to x_1$. Let $x_1 \to x_2$. By condition (1) either $x_2 \to x_3$ or $x_3 \to x_2$. If $x_2 \to x_3$, then by transitivity $x_1 \to x_3$ which contradiction to condition (2). Hence $x_3 \to x_2$.

Then by Lemma 4.1 we can draw the graph of the relation → as follows (Figure 1).

Note that, in the line → ○ ← is equivalent to the open set containing the element "○" and ← □ → is equivalent to the closed set containing the element "□", also the digital line is an Alexandroff topology.

Table 2. Generalized topological spaces defining axioms are indicated by*.

Generalized topologies/Axioms	K_0	K_1	K_2	K_3	K_4	K_5	K_6
Smyth space	*	*	*				
Pre-topology	*	*	*	*			
Pre-topology with every open set is closed	*	*	*	*	*		
Alexandroff topology	*	*	*	*		*	
Clo-open topology	*	*	*	*	*	*	*

Now, if we consider R is a relation on Z^2, then the following conditions hold:

a) If two points $x, y \in Z^2$ are 4-neighbors, then either $x \rightarrow y$ or $y \rightarrow x$,

b) If two points $x, y \in Z^2$ are not 8-neighbors, then both $x \rightarrow y$ and $y \rightarrow x$ not hold.

Lemma 4.2. If x_1, x_2, x_3 which are in the same horizontal or vertical gride line of Z^2 such that $x_2 \in N_4(x_1)$ and $x_3 \in N_4(x_2)$, then we have either $x_1 \rightarrow x_2 \leftarrow x_3$ or $x_1 \leftarrow x_2 \rightarrow x_3$ on a gride line.

Proof. By condition a) we have either $x_1 \rightarrow x_2$ or $x_2 \rightarrow x_1$. Let $x_1 \rightarrow x_2$. Also, by condition a) either $x_2 \rightarrow x_3$ or $x_3 \rightarrow x_2$. If $x_2 \rightarrow x_3$, then by transitivity $x_1 \rightarrow x_3$ which contradiction to condition b). Hence $x_3 \rightarrow x_2$.

In the plane note that, the element "○" is equivalent to the open-open, the element "□" is equivalent to the closed-closed, and the element "●" is called the mixed point and equivalent to the open-closed or the closedopen.

We must note that, there are two topologies on Z^2 which satisfy the conditions 1 and 2, one of them was described in Khalimsky et al. [26] and the other was introduced by Marcus et al. [41] .

Figure 2 describe the topology in [26] . Furthermore, if there is no mixed point, then we have the topology in [41] , which described in Figure 3.

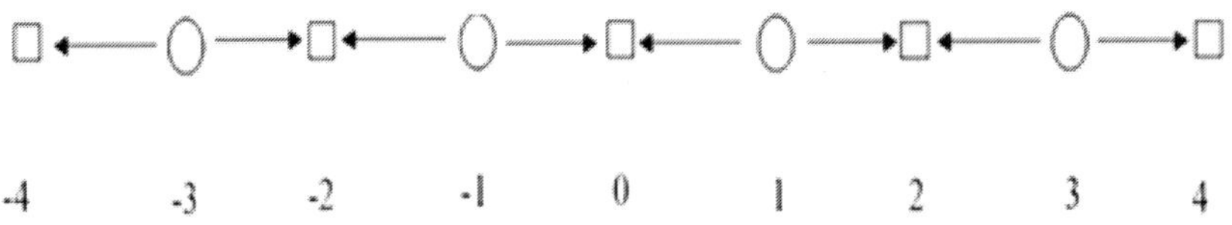

Figure 1. Khalimsky line.

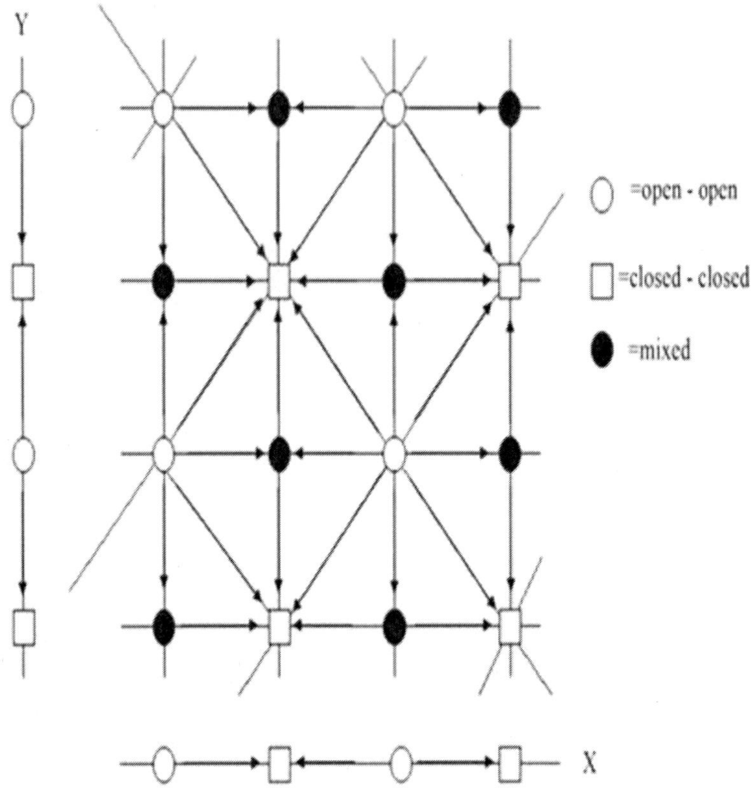

Figure 2. Khalimsky topology in the plane.

5. CONCLUSIONS

In short, topology is a branch of mathematics, whose concepts are fundamental not only to all branches of mathematics, but also in real life applications. Image plays important role in real life.

In the past the process of image analysis took place via various mathematical models with an acceptable amount of error. Digital topology is a new accurate approach.

El-Sayed A. Abo-Tabl, So we use the theory of rough sets to generate the digital spaces, in order to narrow the gap between rough sets and digital topology.

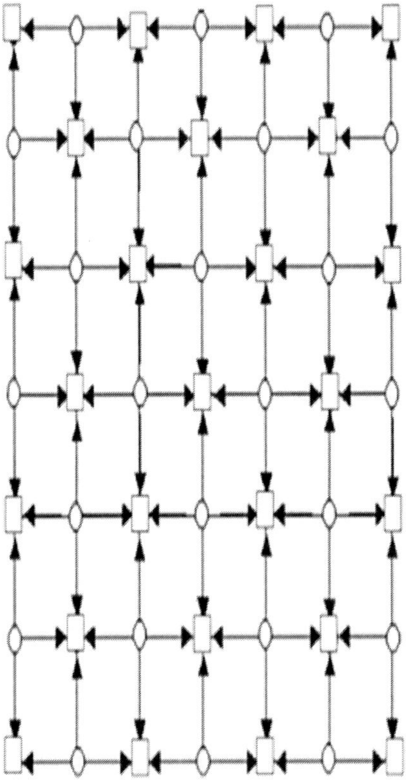

Figure 3. Marcus topology in the plane.

REFERENCES

1. Pawlak, Z. (1982) Rough Sets. International Journal of Computer and Information Science, 11, 341-356.
2. Pawlak, Z. (1991) Rough Sets: Theoretical Aspects of Reasoning about Data. Kluwer Academic Publishers, Boston.
3. Kryszkiewicz, M. (1998) Rough Set Approach to Incomplete Information Systems. Information Sciences, 112, 39-49.
4. Abo-Tabl, E.A. (2013) Rough Sets and Topological Spaces Based on Similarity. International Journal of Machine Learning and Cybernetics, 4, 451-458.
5. Slowinski, R. and Vanderpooten, D. (2000) A Generalized Definition of Rough Approximations Based on Similarity. IEEE Transactions on Knowledge and Data Engineering, 12, 331-336.
6. Skowron, A. and Stepaniuk, J. (1996) Tolerance Approximation Spaces. Fundamenta Informaticae, 27, 245-253.
7. Zhang, H., Ouyang, Y. and Wangc, Z. (2009) Note on Generalized rough Sets Based on Reflexive and Transitive Relations. Information Sciences, 179, 471-473.

8. Liu, G.L. and Zhu, W. (2008) The Algebraic Structures of Generalized Rough Set Theory. Information Sciences, 178, 4105-4113.
9. Yao Y.Y. (1998) Relational Interpretations of Neighborhood Operators and Rough Set Approximation Operators. Information Sciences, 111, 239-259.
10. Zhu, W. (2009) Relationship between Generalized Rough Sets Based on Binary Relation and Covering. Information Sciences, 179, 210-225.
11. Liu, G.L. and Sai, Y. (2009) A Comparison of Two Types of Rough Sets Induced by Coverings. International Journal of Approximate Reasoning, 50, 521-528.
12. Zhu, W. (2007) Topological Approaches to Covering Rough Sets. Information Sciences, 177, 1499-1508.
13. Zhu, W. (2009) Relationship among Basic Concepts in Covering-Based Rough Sets. Information Sciences, 179, 2478- 2486.
14. Polkowski, L. (2002) Rough Sets: Mathematical Foundations. Physica-Verlag, Heidelberg.
15. Skowron, A. (1988) On Topology in Information System. Bulletin of Polish Academic Science and Mathematics, 36, 477-480.
16. Wiweger, A. (1988) On Topological Rough Sets. Bulletin of the Polish Academy of Sciences Mathematics, 37, 51-62.
17. Polkowski, L. (2001) On Fractals Defined in Information Systems via Rough Set Theory. Proceedings of the RSTGC-2001. Bulletin International Rough Set Society, 5, 163-166.
18. Kortelainen J. (1994) On Relationship between Modified Sets, Topological Spaces and Rough Sets. Fuzzy Sets and Systems, 61, 91-95.
19. Skowron, A., Swiniarski, R. and Synak, P. (2005) Approximation Spaces and Information Granulation. Transactions on Rough Sets III: Lecture Notes in Computer Science, 3400, 175-189.
20. Jarvinen, J. and Kortelainen, J. (2007) A Unifying Study between Model-Like Operators, Topologies, and Fuzzy Sets. Fuzzy Sets and Systems, 158, 1217-1225.
21. Lashin, E., Kozae, A., Khadra, A.A. and Medhat, T. (2005) Rough Set Theory for Topological Spaces. International Journal of Approximate Reasoning, 40, 35-43.
22. Li, T.J., Yeung, Y. and Zhang, W.X. (2008) Generalized Fuzzy Rough Approximation Operators Based on Fuzzy Covering. International Journal of Approximate Reasoning, 48, 836-856.
23. Qin, K. and Pei, Z. (2005) On the Topological Properties of Fuzzy Rough Sets. Fuzzy Sets and Systems, 151, 601-613.
24. Srivastava, A.K. and Tiwari, S.P. (2003) On Relationships among Fuzzy Approximation Operators, Fuzzy Topology, and Fuzzy Automata. Fuzzy Sets and Systems, 138, 197-204.
25. Rosenfeld, A. (1979) Digital topology. The American Mathematical Monthly, 86, 621-630.
26. Khalimsky, E.D., Kopperman, R. and Meyer, P.R. (1990) Computer Graphics and Connected Topologies on Finite Ordered Sets. Topology and Its Applications, 36, 1-17. [Citation Time(s):5]
27. Kiselman, C.O. (2000) Digital Jordan Curve Theorems. Discrete Geometry for Computer Imagery (DGCI), 1953, 46- 56.
28. Slapal, J. (2013) Topological Structuring of the Digital Plane. Discrete Mathematics and Theoretical Computer Science, 15, 165-176.
29. Khalimsky, E.D., Kopperman, R. and Meyer, P.R. (1990) Boundaries in Digital Planes. Journal of Applied Mathematics and Stochastic Analysis, 3, 27-55.
30. Kong, T.Y., Kopperman, R. and Meyer, P.R. (1991) A Topological Approach to Digital Topology. The American Mathematical Monthly, 98, 901-917.
31. Kopperman, R., Meyer, P.R. and Wilson, R.G. (1991) A Jordan Surface Theorem for Three-Dimensional Digital Spaces. Discrete and Computational Geometry, 6, 155-161.
32. Bai, Y., Han, X. and Prince, J.L. (2009) Digital Topology on Adaptive Octree Grids. Journal of Mathematical Imaging and Vision, 34, 165-184.
33. Eckhardt, U. and Latecki, L.J. (2003) Topologies for the Digital Spaces Z^2 and Z^3. Computer Vision and Image Understanding, 90, 295-312.
34. Melin, E. (2007) Digital Surfaces and Boundaries in Khalimsky Spaces. Journal of Mathematical Imaging and Vision, 28, 169-177.

35. Slapal, J. (2003) Closure Operations for Digital Topology. Theoretical Computer Science, 305, 457-471.
36. Slapal, J. (2006) Digital Jordan Curves. Topology and Its Application, 153, 3255-3264.
37. Smyth, M.B. (1995) Semi-Metrics, Closure Spaces and Digital Topology. Theoretical Computer Science, 151, 257-276.
38. Sierpenski, W. and Krieger, C. (1956) General Topology. University of Toronto, Toronto.
39. Alexandroff, P. (1937) Diskrete Raume. Matematicheskii Sbornik, 2, 501-519.
40. Birkhoff, G. (1937) Rings of Sets. Duke Mathematical Journal, 3, 383-548.
41. Marcus, D., Wyse, F., et al. (1970) A Special Topology for the Integers (Problem 5712). The American Mathematical Monthly, 77, 1119.

CHAPTER 17

A Unified Theory (I) for Neighborhood Systems and Basic Concepts on Fuzzifying Topological Spaces

Osama Rashed Sayed

Department of Mathematics, Faculty of Science, Assiut University, Assiut, Egypt

ABSTRACT

This paper considers fuzzifying topologies, a special case of I-fuzzy topologies (bifuzzy topologies), introduced by Ying [25]. It investigates topological notions defined by means of Δ -open sets when these are planted into the framework of Ying's fuzzifying topological spaces (by Łukasiewicz logic in [0, 1]). In this paper we introduce some sorts of operations, called general fuzzifying operations from $P(X)$ to $\Im(X)$, where (X,τ) is a fuzzifying topological space. By making use of them we contract neighborhood structures, derived sets, closure operations and interior operations.

Keywords: Łukasiewicz Logic; Semantics; Fuzzifying Topology

1. INTRODUCTION

In the last few years fuzzy topology, as an important research field in fuzzy set theory, has been developed into a quite mature discipline [1-6]. In contrast to classical topology, fuzzy topology is endowed with richer structure, to a certain extent, which is manifested with different ways to generalize certain classical concepts. So far, according to Ref. [2], the kind of topologies defined by Chang [7] and Goguen [8] is called the topologies of fuzzy subsets, and further is naturally called L-topological spaces if a lattice L of membership values has been chosen. Loosely speaking, a topology of fuzzy subsets (resp. an L-topological space) is a family τ of fuzzy subsets (resp. L-fuzzy subsets) of nonempty set X, and τ satisfies the basic conditions of classical topologies [9]. On the other hand, Höhle in [10] proposed the terminology L-fuzzy topology to be an L-valued mapping on the traditional powerset $P(X)$ of X. The authors in [4,5,11,12] defined an L-fuzzy topology to be an L-valued mapping on the L-powerset L^X of X. In 1952, Rosser and Turquette [13] proposed emphatically the following problem: If there are many-valued theories beyond the level of predicates calculus, then what are the detail of such theories? As an attempt to give a partial answer to this problem in the case of point set topology, Ying in 1991 [14,15] used a semantical method of continuousvalued logic to develop systematically fuzzifying topology. Briefly speaking, a fuzzifying topology on a set X assigns each crisp subset of X to a certain degree of being open, other than being definitely open or not. In factfuzzifying topologies are a special case of the L-fuzzy topologies in [11,12] since all the t-norms on I are included as a special class of tensor products in these paper. Ying uses one particular tensor product, namely Łukasiewicz conjunction. Thus his fuzzifying topologies are a special class of all the I-fuzzy topologies considered in the categorical frameworks [11,12]. Roughly speaking, the semantical analysis approach transforms formal statements of interest, which are usually expressed as implication formulas in logical language, into some inequalities in the truth value set by truth valuation rules, and then these inequalities are demonstrated in an algebraic way and the semantic validity of conclusions is thus established. So far, there has been significant research on fuzzifying topologies [16-21]. In 1979, several characterizations of compactness are unified by the operation introduced by Kasahara [22]. Also, he studied the concept of α-continuity (where α is an operation) and defined some types of spaces by using this operation. In 1981, the concept of other type of continuity which generalizes the α-continuity [22] was introduced by Jankoviĉ [23]. In 1983, Abd El-Monsef, et al. [24] introduced an operation $\tilde{\alpha}$ on the family $\Im$ of all closed sets in the topological space (X,τ) which is dual to the operation α. In 1991, Kerre et al. [25] introduced an extension of the concept of an operation on the class of all fuzzy sets on X endowed with Chang fuzzy topology [7]. It was shown that a lot of characterizations and properties of many concepts and stronger forms can be unified by using this notion. In 1991, Kandil et al. [26] applied the concept of the operation defined in [25] to unify and generalize several characterizations and properties of a lot of already existing weaker and stronger forms of fuzzy continuity. A basic structure of this paper is as follows: First, in Section 2 we offer some definition and results which will be needed in this paper. In Section 3 the concepts of fuzzifying Δ-open sets, CΔ-open sets, Δ-closed sets and CΔ-closed sets are introduced and some of their properties are discussed. In Section 4 the fuzzifying Δ- and CΔ-neighborhood systems are presented and a fuzzifying topology induced by CΔ- neighborhood system is introduced. In Section 5 the concepts of fuzzifying Δ- and CΔ-derived sets, Δ- and CΔ-closure operations and Δ- and CΔ-interior operations were established and some of their properties are studied. Finally, in Section 6, we summarize the main results obtained and raise some related problems for further study. Thus we fill a gap in the existing literature on fuzzifying topology.

NOTE: All corollaries in this paper are results in [14- 21].

2. PRELIMINARIES

We present the fuzzy logical and corresponding set theoretical notations [14,15] since we need them in this paper.

For any formula φ, the symbol $[\varphi]$ means the truth value of φ, where the set of truth values is the unit interval [0, 1]. We write $\vDash \varphi$ if $[\varphi]=1$ for any interpretation. Also, $\Im(X)$ is the family of all fuzzy sets in X. The truth valuation rules for primary fuzzy logical formulae and corresponding set theoretical notations are:

a) $[\alpha]=\alpha(\alpha\in[0,1])$;

b) $[\varphi\wedge\psi]=\min([\varphi],[\psi])$;

c) $[\varphi\to\psi]=\min(1,1-[\varphi]+[\psi])$.

1) If $\tilde{A}\in\Im(X),\left[x\in\tilde{A}\right]:=\tilde{A}(x)$.

2) If X is the universe of discourse, then

$$[\forall x \varphi(x)] := \inf_{x \in X} [\varphi(x)].$$

In addition the truth valuation rules for derived formulae are:

1) $[\neg\varphi] := [\varphi \to 0] = 1 - [\varphi]$;

2) $[\varphi \vee \psi] := [\neg(\neg\varphi \wedge \neg\psi)] = \max([\varphi],[\psi])$;

3) $[\varphi \leftrightarrow \psi] := [(\varphi \to \psi) \wedge (\psi \to \varphi)]$;

4) $[\varphi \otimes \psi] := [\neg(\varphi \to \neg\psi)] = \max(0, [\varphi] + [\psi] - 1)$;

5) $[\exists x \varphi(x)] := [\neg \forall x \neg \varphi(x)] := \sup_{x \in X} [\varphi(x)]$;

6) If $\tilde{A}, \tilde{B} \in \Im(X)$, then

a) $[\tilde{A} \subseteq \tilde{B}] := [\forall x (x \in \tilde{A} \to x \in \tilde{B})]$

$= \inf_{x \in X} \min(1, 1 - \tilde{A}(x) + \tilde{B}(x))$;

b) $[\tilde{A} \equiv \tilde{B}] := [\tilde{A} \subseteq \tilde{B}] \wedge [\tilde{B} \subseteq \tilde{A}]$;

c) $[\tilde{A} \doteq \tilde{B}] := [\tilde{A} \subseteq \tilde{B}] \otimes [\tilde{B} \subseteq \tilde{A}]$.

We give now the following definitions and results in fuzzifying topology [14-21] which are used in the sequel.

Definition 2.1 [14]. *Let X be a universe of discourse, and* $\tau \in \Im(P(X))$ *satisfy the following conditions:*

1) $\tau(X) = \tau(\varnothing) = 1$;

2) for any $A, B \in P(X), \tau(A \cap B) \ge \tau(A) \wedge \tau(B)$;

3) for any

$$\{A_\lambda \in P(X) : \lambda \in \Lambda\}, \tau\left(\bigcup_{\lambda \in \Lambda} A_\lambda\right) \ge \bigwedge_{\lambda \in \Lambda} \tau(A_\lambda).$$

Then τ is a fuzzifying topology and (X, τ) is a fuzzifying topological space.

Note: In the rest of this paper (X,τ) (or briefly X) is always fuzzifying topological space.

Definition 2.2 [14] *The family of all fuzzifying closed sets, denoted by* $F\in\Im(P(X))$, *is defined as* $A\in F:=(X-A)\in\tau$, *where* $X-A$ *is the complement of A.*

Definition 2.3 [14] *The neighborhood system* $N_x\in\Im(P(X))$ *of* $x\in X$ *is defined as*

$$N_x(A)=\sup_{x\in B\subseteq A}\tau(B).$$

Definition 2.4 [15] *The interior* A° *or* $Int(A)$ *of* $A\subseteq X$ *is defined as* $Int(A)(x)=N_x(A)$.

Definition 2.5 (*Lemma 5.2.* [14]). *The closure* $Cl(A)$ *or* $\overline{A}$ *of A is defined as* $Cl(A)(x)=1-N_x(X-A)$. *In Theorem 5.3 [14], Ying proved that the closure* $Cl:P(X)\to\Im(X)$*sssis a fuzzifying closure operator since its extension* $Cl:\Im(X)\to\Im(X)$, $Cl(\tilde{A})=\bigcup_{\alpha\in[0,1]}\alpha Cl(\tilde{A}_\alpha)$, $\tilde{A}\in\Im(X)$, *where* $\tilde{A}_\alpha=\{\alpha:\tilde{A}(x)\geq\alpha\}$ *is the* α *-cut of A and* $\alpha\tilde{A}=\alpha\wedge\tilde{A}(x)$ *satisfies the following Kuratowski closure axioms:*

1) $\vDash Cl(\varnothing)=\varnothing$;

2) for any $\tilde{A}\in\Im(X)$, $\vDash \tilde{A}\subseteq Cl(\tilde{A})$;

3) for any $\tilde{A},\tilde{B}\in\Im(X)$, $\vDash Cl(\tilde{A}\cup\tilde{B})=Cl(\tilde{A})\cup Cl(\tilde{B})$;

4) for any $\tilde{A}\in\Im(X)$, $\vDash Cl(Cl(\tilde{A}))\subseteq Cl(\tilde{A})$.

Definition 2.6 [18] *For any* $\tilde{A}\in\Im(X)$,

$$\vDash Int(\tilde{A})\equiv X-Cl(X-\tilde{A}).$$

Theorem 2.1 [18] *For any* $\tilde{A},\tilde{B}\in\Im(X)$,

1) $Int(X) \equiv X$;

2) $Int(\tilde{A}) \subseteq \tilde{A}$;

3) $Int(\tilde{A} \cap \tilde{B}) = Int(\tilde{A}) \cap Int(\tilde{B})$;

4) $Int(Int(\tilde{A})) \supseteq Int(\tilde{A})$.

Theorem 2.2 [18] *For any* $\tilde{A}, \tilde{B} \in \Im(X)$, *if* $[\tilde{A} \subseteq \tilde{B}] = 1$, *then*

1) $Cl(\tilde{A}) \subseteq Cl(\tilde{B})$;

2) $Int(\tilde{A}) \subseteq Int(\tilde{B})$;

3) $Int(Cl(Int(\tilde{A}))) \subseteq Int(Cl(Int(\tilde{B})))$;

4) $Cl(Int(\tilde{A})) \subseteq Cl(Int(\tilde{B}))$;

5) $Int(Cl(\tilde{A})) \subseteq Int(Cl(\tilde{B}))$;

6) $Cl(Int(Cl(\tilde{A}))) \subseteq Cl(Int(Cl(\tilde{B})))$.

Theorem 2.3 *For any* $\tilde{A} \in \Im(X)$

1) $X - Int(Cl(Int(\tilde{A}))) \equiv Cl(Int(Cl(X - \tilde{A})))$ [19];

2) $X - Cl(Int(\tilde{A})) \equiv Int(Cl(X - \tilde{A}))$ [18];

3) $X - Cl(Int(\tilde{A})) \equiv Cl(Int(X - \tilde{A}))$ [18];

4) $X - Cl(Int(Cl(\tilde{A}))) \equiv Int(Cl(Int(X - \tilde{A})))$ [16].

Theorem 2.4

1) 惘$X \equiv Int(Cl(Int(X)))$; $\phi \equiv Int(Cl(Int(\phi)))$;

2) 惘$X \equiv Cl(Int(X))$; $\phi \equiv Cl(Int(\phi))$;

3) $X \equiv Int(Cl(X)); \quad \phi \equiv Int(Cl(\phi));$

4) $X \equiv Cl(Int(Cl(X))); \quad \phi \equiv Cl(Int(Cl(\phi))).$

Theorem 2.5 *For any* $A \in P(X)$

1) $Int(A) \subseteq Int(Cl(Int(A)))$ [19];

2) $Int(Cl(Int(A))) \subseteq Int(Cl(A));$

3) $Int(Cl(Int(A))) \subseteq Cl(Int(A));$

4) $Cl(Int(A)) \subseteq Cl(Int(Cl(A)));$

5) $Int(Cl(A)) \subseteq Cl(Int(Cl(A))).$

Theorem 2.6 *For any* $A \in P(X)$

1) $Int(A) \subseteq Cl(Int(A));$

2) $Int(A) \subseteq Int(Cl(A))$ [17];

3) $Int(A) \subseteq Cl(Int(Cl(A)))$ [16].

Definition 2.7 *Let X be a non-empty set.*

1) By the symbol $\Im(X)^{P(X)}$ we denote the set of all functions from $P(X)$ into $\Im(X)$. Each member of $\Im(X)^{P(X)}$ will be called a general fuzzifying operation.

2) Let $\Delta, \nabla \in \Im(X)^{P(X)}$.

a) We say that $\Delta \leq \nabla$, if $\Delta(A) \leq \nabla(A)$ for each $A \subseteq X$.

b) We say that Δ and ∇ are dual if

$X - \Delta(A) \equiv \nabla(X - A);$ equivalently

$X - \nabla(A) \equiv \Delta(X - A)$ for each $A \subseteq X$.

3) A general fuzzifying operation $\Delta \in \Im(X)^{P(X)}$ is said to be monotone if ‘ $A \subseteq B \rightarrow \Delta(A) \subseteq \Delta(B)$.

4) A general fuzzifying operation $\Delta \in \Im(X)^{P(X)}$ is said to be of type $O1$ if ‘ $X \subseteq \Delta(X)$; equivalently ‘ $\nabla(\phi) \subseteq \phi$, where Δ and ∇ are dual.

5) A general fuzzifying operation $\Delta \in \Im(X)^{P(X)}$ is said to be of type $O2$ if $\Delta(A \cap B) = \Delta(A) \wedge \Delta(B)$ for any $A, B \subseteq X$.

Example 2.1

1) From Theorem 2.4 we have O°, $O^{-}, O^{\circ-\circ}, O^{-\circ}, O^{\circ-}$, $O^{-\circ} \cup O^{\circ-}$, and $O^{-\circ-}$ are of type $O1$ and each member of them is monotone from Theorem 2.2.

2) The fuzzifying operations $O^{\circ}, O^{-}, O^{\circ-\circ}, O^{-\circ}, O^{\circ-}$, $O^{-\circ} \cup O^{\circ-}$, and $O^{-\circ-}$ and the fuzzifying operations $O^{-}, O^{\circ}, O^{-\circ-}, O^{\circ-}, O^{-\circ}$, $O^{-\circ} \cup O^{\circ-}$, and $O^{\circ-\circ}$ are dual respectively (see Theorem 2.3).

3) From Theorem 2.1 (3), O° is of type $O2$.

4) From Theorem 2.5, one can easily deduce that:

a) $O^{\circ} \leq O^{\circ-\circ}$;

b) $O^{\circ-\circ} \leq O^{-\circ}$;

c) $O^{\circ-\circ} \leq O^{\circ-}$;

d) $O^{\circ-\circ} \leq O^{-\circ} \cup O^{\circ-}$;

e) $O^{\circ-} \leq O^{-\circ-}$;

f) $O^{-\circ} \leq O^{-\circ-}$;

g) $O^{-\circ} \cup O^{\circ-} \leq O^{-\circ-}$.

Note: In the rest of this paper always $\Delta, \nabla \in \Im(X)^{P(X)}$.

3. FUZZIFYING OPEN SETS

In this section the concepts of fuzzifying Δ-open sets, CΔ-open sets, Δ-closed sets and CΔ-closed sets are introduced and some of their properties are discussed.

Definition 3.1 *1) The family of all fuzzifying Δ-open sets, denoted by* $\tau_\Delta \in \Im(P(X))$, *is defined as follows:*

$$A \in \tau_\Delta := \forall x \left(x \in A \rightarrow x \in \Delta(A) \right),$$

i.e.,

$$\tau_\Delta(A) = \inf_{x \in A} \Delta(A)(x);$$

2) The family of all fuzzifying CΔ-open sets, denoted by $\tau_{C\Delta} \in \Im(P(X))$, is defined as follows:

$$A \in \tau_{C\Delta} := \forall x \left(x \in A \cap \Delta(A) \rightarrow x \in A^{\circ} \right),$$

i.e.,

$$\tau_{C\Delta}(A) = \inf_{x \in A} \min\left(1, 1 - \Delta(A)(x) + A^{\circ}(x)\right);$$

3) The family of all fuzzifying Δ (resp. CΔ-closed sets, denoted by $\hat{o}_\Delta$ (resp. $\hat{o}_{C\Delta}$) $\in \Im(P(X))$, is defined as follows:

$$A \in \text{越}_\Delta \left(\text{resp. } {}_{C\Delta}\right) := X - A \in \tau_\Delta \left(\text{resp. } \tau_{C\Delta}\right)$$

Definition 3.2 *1) If* $\Delta = ()^{\circ}$, *(resp.* $()^{\circ-\circ}, ()^{\circ-}, ()^{-\circ}$, $()^{-\circ} \cup ()^{\circ-}, ()^{-\circ-}$*), the notion of fuzzifying Δ-open sets coincides with the notion of fuzzifying open (resp.* α *-open, semi-open, pre-open,* γ *-open,* β *-open) sets and will be denoted by* τ *(resp.* $\tau_\alpha, \tau_S, \tau_P, \tau_\gamma, \tau_\beta$*);*

2) If $\Delta = ()^{\circ - \circ}$ (resp. $()^{\circ -}, ()^{- \circ}, ()^{- \circ} \cup ()^{\circ -}, ()^{- \circ -}$), the notion of fuzzifying CΔ-open sets coincides with the notion of fuzzifying $c\alpha$ (resp. csemi, cpre, $c\gamma$, $c\beta$)-open sets and will be denoted by $\tau_{C\alpha}$ (resp. τ_{CS}, τ_{CP}, $\tau_{C\gamma}, \tau_{C\beta}$);

3) If $\Delta = ()^{\circ}$, (resp. $()^{\circ - \circ}, ()^{\circ -}, ()^{- \circ}, ()^{- \circ} \cup ()^{\circ -}, ()^{- \circ -}$), the notion of fuzzifying Δ-closed sets coincides with the notion of fuzzifying closed (resp. α-closed, semiclosed, pre-closed, γ-closed, β-closed) sets and will be denoted by ô (resp. 趑$_\alpha$, $_S$,趑$_P$, $_\gamma$, $_\beta$);

4) If $\Delta = ()^{\circ - \circ}$ (resp. $()^{\circ -}, ()^{- \circ}, ()^{- \circ} \cup ()^{\circ -}, ()^{- \circ -}$), the notion of fuzzifying CΔ-closed sets coincides with the notion of fuzzifying $c\alpha$ (resp. csemi, cpre, $c\gamma$, $c\beta$)-closed sets and will be denoted by ô$_{C\alpha}$ (resp. 趑$_{CS}$, $_{CP}$,趑$_{C\gamma}$, $_{C\beta}$);

Theorem 3.1 *1) If Δ is of type O1, then*

a) $\models \phi \in \tau_\Delta$; $\models X \in \tau_\Delta$;

b) $\models \phi \in \tau_{C\Delta}$; $\models X \in \tau_{C\Delta}$;

2) If Δ is monotone, then a) for any $\{A_\lambda : \lambda \in \Lambda\} \subseteq P(X)$,

$$\tau_\Delta\left(\bigcup_{\lambda \in \Lambda} A_\lambda\right) \geq \bigwedge_{\lambda \in \Lambda} \tau_\Delta(A_\lambda).$$

b) for any $A, B \in P(X)$,

$$\tau_{C\Delta}(A \cap B) \geq \tau_{C\Delta}(A) \wedge \tau_{C\Delta}(B)$$

Proof. 1) a) $\tau_\Delta(X) = [X \subseteq \Delta(X)] = 1$;

$\tau_\Delta(\phi) = [\phi \subseteq \Delta(\phi)] = 1.$

b) $\tau_{C\Delta}(X) = [X \cap \Delta(X) \subseteq X^\circ] = [X \subseteq X] = 1$;

$\tau_{C\Delta}(\phi) = [\phi \cap \Delta(\phi) \subseteq \phi^\circ] = [\phi \subseteq \phi] = 1.$

2) a) Since Δ is monotone, then

$$\begin{aligned}\tau_\Delta\left(\bigcup_{\lambda\in\Lambda} A_\lambda\right) &= \inf_{x\in\bigcup_{\lambda\in\Lambda} A_\lambda} \Delta\left(\bigcup_{\lambda\in\Lambda} A_\lambda\right)(x)\\ &= \inf_{\lambda\in\Lambda}\inf_{x\in A_\lambda} \Delta\left(\bigcup_{\lambda\in\Lambda} A_\lambda\right)(x)\\ &\geq \inf_{\lambda\in\Lambda}\inf_{x\in A_\lambda} \Delta(A_\lambda)(x) = \bigwedge_{\lambda\in\Lambda}\tau_\Delta(A).\end{aligned}$$

b) For any $\{A_\lambda : \lambda\in\Lambda\}\subseteq\Im(X)$,

$$\text{'}\ \forall\lambda\left(\lambda\in\Lambda\to\tilde{A}_\lambda\subseteq\tilde{B}_\lambda\right)\to\bigcap_{\lambda\in\Lambda}\tilde{A}_\lambda\subseteq\bigcap_{\lambda\in\Lambda}\tilde{B}_\lambda$$

(see Lemma 1.1 (1) [14]). Since Δ is monotone, then from Theorem 2.1 (3) we have

$$\begin{aligned}&\tau_{C\Delta}(A)\wedge\tau_{C\Delta}(B)\\ &=\left[A\cap\Delta(A)\subseteq A^\circ\wedge B\cap\Delta(B)\subseteq B^\circ\right]\\ &\leq\left[(A\cap B)\cap(\Delta(A)\cap\Delta(B))\subseteq\left(A^\circ\cap B^\circ\right)\right]\\ &=\left[(A\cap B)\cap(\Delta(A)\cap\Delta(B))\subseteq(A\cap B)^\circ\right]\\ &\leq\left[(A\cap B)\cap\Delta(A\cap B)\subseteq(A\cap B)^\circ\right]\\ &=\tau_{C\Delta}(A\cap B).\end{aligned}$$

Corollary 3.1

1) a) $\text{'}\ X\in\tau\left(\text{resp. }\tau_\alpha,\tau_S,\tau_P,\tau_\gamma,\tau_\beta\right);$

$$\text{'}\ \phi\in\tau\left(\text{resp. }\tau_\alpha,\tau_S,\tau_P,\tau_\gamma,\tau_\beta\right);$$

b) for any $\{A_\lambda : \lambda\in\Lambda\}\subseteq P(X)$,

$$\tau\left(\text{resp. } \tau_{\alpha}, \tau_{S}, \tau_{P}, \tau_{\gamma}, \tau_{\beta}\right)\left(\bigcup_{\lambda \in \Lambda} A_{\lambda}\right)$$
$$\geq \bigwedge_{\lambda \in \Lambda} \tau\left(\text{resp. } \tau_{\alpha}, \tau_{S}, \tau_{P}, \tau_{\gamma}, \tau_{\beta}\right)\left(A_{\lambda}\right).$$

2) a) ‘ $X \in \tau_{C\alpha}\left(\text{resp. } \tau_{CS}, \tau_{CP}, \tau_{C\gamma}, \tau_{C\beta}\right);$

‘ $\phi \in \tau_{C\alpha}\left(\text{resp. } \tau_{CS}, \tau_{CP}, \tau_{C\gamma}, \tau_{C\beta}\right);$

b) for any $A, B \in P(X)$,

$$\tau_{C\alpha}\left(\text{resp. } \tau_{CS}, \tau_{CP}, \tau_{C\gamma}, \tau_{C\beta}\right)(A \cap B)$$
$$\geq \tau_{C\alpha}\left(\text{resp. } \tau_{CS}, \tau_{CP}, \tau_{C\gamma}, \tau_{C\beta}\right)(A)$$
$$\wedge \tau_{C\alpha}\left(\text{resp. } \tau_{CS}, \tau_{CP}, \tau_{C\gamma}, \tau_{C\beta}\right)(B).$$

Theorem 3.2

1) If Δ is of type 01, then

a) 戶$\phi \in {}_{\Delta}$; 戶$X \in {}_{\Delta}$;

b) 戶$\phi \in {}_{C\Delta}$; 戶$X \in {}_{C\Delta}$;

2) If Δ is monotone, then a) for any $\{A_{\lambda} : \lambda \in \Lambda\} \subseteq P(X)$,

$$趑_{\Delta}\left(\bigcap_{\lambda \in \Lambda} A_{\lambda}\right) \geq \bigwedge_{\lambda \in \Lambda} {}_{\Delta}\left(A_{\lambda}\right).$$

b) for any $A, B \in P(X)$,

趑$_{C\Delta}(A \cap B) \geq {}_{C\Delta}(A) \wedge ?_{C\Delta}(B).$

Proof. It is immediate from Theorem 3.1.

Corollary 3.2 1) a) 戶$X \in$ (resp. 趑$_{\alpha}$, $_{S}$, 趑$_{P}$, $_{\gamma}$, $?_{\beta}$);

戶$\phi \in$ (resp. 趑$_{\alpha}$, $_{S}$, 趑$_{P}$, $_{\gamma}$, $?_{\beta}$);

b) 戶$X \in {}_{C\alpha}$ (resp. 趑$_{CS}$, $_{CP}$, 趑$_{C\gamma}$, $_{C\beta}$);

$\text{戸}\phi \in {}_{C\alpha}\left(\text{resp. 趑}_{CS},\ {}_{CP}, \text{趑}_{C\gamma},\ {}_{C\beta}\right);$

2) a) for any $\{A_\lambda : \lambda \in \Lambda\} \subseteq P(X),$

$$\text{趑}\left(\text{resp. } {}_{\alpha}, \text{趑}_{S},\ {}_{P}, \text{趑}_{\gamma},\ {}_{\beta}\right)\left(\bigcap_{\lambda\in\Lambda} A_\lambda\right)$$
$$\geq \bigwedge_{\lambda\in\Lambda}\left(\text{resp. 趑}_{\alpha},\ {}_{S}, \text{趑}_{P},\ {}_{\gamma}, ?\ {}_{\beta}\right)(A_\lambda).$$

b) for any $A, B \in P(X),$

$$\text{趑}_{C\alpha}\left(\text{resp. } {}_{CS}, \text{趑}_{CP},\ {}_{C\gamma},\ {}_{C\beta}\right)(A \cap B)$$
$$\geq \text{趑}_{C\alpha}\left(\text{resp. } {}_{CS}, \text{趑}_{CP},\ {}_{C\gamma},\ {}_{C\beta}\right)(A)$$
$$\wedge \text{趑}_{C\alpha}\left(\text{resp. } {}_{CS}, \text{趑}_{CP},\ {}_{C\gamma},\ {}_{C\beta}\right)(A)$$

Theorem 3.3 1) $\text{戸}A \in {}_{\Delta} \leftrightarrow \forall x\left(x \in \nabla(A) \to x \in A\right);$

2) $\text{戸}A \in {}_{C\Delta} \leftrightarrow \forall x\left(x \in \bar{A} \to x \in A \cup \nabla(A)\right)$, where ∇ is the dual of Δ

Proof.

1) $\left[\forall x\left(x \in \nabla(A) \to x \in A\right)\right]$

$$= \left[\forall x\left(x \in X - A \to x \in X - \nabla(A)\right)\right]$$
$$= \left[\forall x\left(x \in X - A \to x \in \Delta(X - A)\right)\right]$$
$$= \tau_\Delta(X - A) = [A \in \hat{o}_{\Delta}].$$

2) $\left[\forall x\left(x \in \bar{A} \to x \in A \cup \nabla(A)\right)\right]$

$$= \left[\forall x\left(x \in \left(X - \left(A \cup \left(\nabla(A)\right)\right)\right) \to x \in \left(X - \bar{A}\right)\right)\right]$$
$$= \left[\forall x\left(x \in (X - A) \cap \left(X - \nabla(A)\right) \to x \in (X - A)^{\circ}\right)\right]$$
$$= \left[\forall x\left(x \in (X - A) \cap \left(\Delta(X - A)\right) \to x \in (X - A)^{\circ}\right)\right]$$
$$= \tau_{C\Delta}(X - A) = [A \in \hat{o}_{C\Delta}].$$

Corollary 3.3

1) $\text{戶}A \in \left(\text{resp. } \text{趑}_{\alpha},\ _{S}, \text{趑}_{P},\ _{\gamma}, ?_{\beta}\right)$

$\leftrightarrow \forall x\Big(x \in A^{-}\left(\text{resp. } A^{-\circ-}, A^{-\circ}, A^{\circ-},\right.$

$\left. A^{-\circ} \cup A^{\circ-}, A^{\circ-\circ} \rightarrow x \in A\right);$

2) $\text{戶}A \in\ _{C\alpha}\left(\text{resp. } \text{趑}_{CS},\ _{CP}, \text{趑}_{C\gamma},\ _{C\beta}\right)$

$\leftrightarrow \forall x\Big(x \in \overline{A} \rightarrow x \in A \cup A^{-\circ-}$

$\left(\text{resp. } A^{-\circ}, A^{\circ-}, A^{-\circ} \cup A^{\circ-}, A^{\circ-\circ}\right)\Big).$

Theorem 3.4 Let $\Delta_1, \Delta_2 \in \Im(X)^{P(X)}$.

1) If $\Delta_1 \leq \Delta_2$, a) $‘\ \tau_{\Delta_1} \subseteq \tau_{\Delta_2}$; b) $\text{戶}\ _{\Delta_1} \subseteq\ _{\Delta_2}$; c) $‘\ \tau_{C\Delta_1} \subseteq \tau_{C\Delta_2}$; d) $\text{戶}\ _{C\Delta_1} \subseteq\ _{C\Delta_2}$;

2) a) $‘\ \tau \subseteq \tau_{C\Delta}$; b) $\text{戶}\ \subseteq\ _{C\Delta}$.

Proof.

1) a) $\tau_{\Delta_1}(A) = [A \subseteq \Delta_1(A)] \leq [A \subseteq \Delta_2(A)] = \tau_{\Delta_2}(A);$

b) From a) above, we have

$\text{趑}_{\Delta_1}(A) = \tau_{\Delta_1}(X - A) \leq \tau_{\Delta_2}(X - A) =\ _{\Delta_2}(A);$

c) $\tau_{C\Delta_1}(A) = [A \cap \Delta_1(A) \subseteq A^{\circ}] \geq [A \cap \Delta_2(A) \subseteq A^{\circ}]$

$= \tau_{C\Delta_2}(A);$

d) From (c) above, we have $\text{趑}_{C\Delta_1}(A) = \tau_{C\Delta_1}(X - A) \leq \tau_{C\Delta_2}(X - A) =\ _{\Delta_2}(A).$

2) a) $\tau(A) = [A \subseteq A^{\circ}] \leq [A \cap \Delta(A) \subseteq A^{\circ}] = \tau_{C\Delta}(A);$

b) From a) above, we have

$\text{趑}(A) = \tau(X - A) \leq \tau_{C\Delta}(X - A) =\ _{C\Delta}(A)$

Corollary 3.4 1) a) i) $‘\ \tau \subseteq \tau_{\alpha}$; ii) $‘\ \tau_{\alpha} \subseteq \tau_{S}$; iii) $‘\ \tau_{\alpha} \subseteq \tau_{P}$; iv) $‘\ \tau_{S} \subseteq \tau_{\gamma}$; (v) $‘\ \tau_{P} \subseteq \tau_{\gamma}$; (vi) $‘\ \tau_{\gamma} \subseteq \tau_{\beta}$.

b) i) 戶 $\subseteq$ $_{\alpha}$; ii) 戶 $_{\alpha} \subseteq ?_{S}$; iii) 戶 $_{\alpha} \subseteq ?_{P}$; iv) 戶 $_{S} \subseteq$ $_{\gamma}$; v) 戶 $_{P} \subseteq ?_{\gamma}$; vi) 戶 $_{\gamma} \subseteq ?_{\beta}$.

c) i) ‘ $\tau_{CS} \subseteq \tau_{C\alpha}$; ii) ‘ $\tau_{CP} \subseteq \tau_{C\alpha}$; iii) ‘ $\tau_{C\gamma} \subseteq \tau_{CS}$; iv) ‘ $\tau_{C\gamma} \subseteq \tau_{CP}$; v) ‘ $\tau_{C\beta} \subseteq \tau_{C\gamma}$.

d) i) 戶 $_{CS} \subseteq ?_{C\alpha}$; ii) 戶 $_{CP} \subseteq ?_{C\alpha}$; iii) 戶 $_{C\gamma} \subseteq$ $_{CS}$; iv) 戶 $_{C\gamma} \subseteq$ $_{CP}$; v) 戶 $_{C\beta} \subseteq$ $_{C\gamma}$.

2) a) ‘ $\tau \subseteq \tau_{C\alpha}\left(\text{resp. } \tau_{CS}, \tau_{CP}, \tau_{C\gamma}, \tau_{C\beta}\right)$;

b) 戶 $\subseteq$ 趛$_{C\alpha}$ (resp. $_{CS}$, 趛$_{CP}$, $_{C\gamma}$, $?_{C\beta}$).

Corollary 3.5

1) a) ‘ $\tau \subseteq \tau_{S}$; b) ‘ $\tau \subseteq \tau_{P}$; c) ‘ $\tau \subseteq \tau_{\beta}$; d) ‘ $\tau \subseteq \tau_{\gamma}$;

2) a) 戶 $\subseteq$ $_{S}$; b) 戶 $\subseteq$ $_{P}$; c) 戶 $\subseteq$ $_{\beta}$; d) 戶 $\subseteq$ $_{\gamma}$.

Theorem 3.5 Let ‘ $A^{\circ} \subseteq \Delta(A)$ for each $A \subseteq X$.

1) ‘ $A \in \tau_{C\Delta} \leftrightarrow A^{\circ} \equiv A \cap \Delta(A)$;

2) 戶$A \in$ $_{C\Delta} \leftrightarrow \overline{A} \equiv A \cup \nabla(A)$.

Proof. 1) Since ‘ $A^{\circ} \subseteq \Delta(A)$ and ‘ $A^{\circ} \subseteq A$, then ‘ $A^{\circ} \subseteq A \cap \Delta(A)$. Thus

$$\begin{aligned}\tau_{C\Delta}(A) &= \left[A^{\circ} \subseteq A \cap \Delta(A)\right] \wedge \left[A \cap \Delta(A) \subseteq A^{\circ}\right] \\ &= \left[A^{\circ} \equiv A \cap \Delta(A)\right].\end{aligned}$$

2) Since for every $A \subseteq X$, ‘ $A^{\circ} \subseteq \Delta(A)$, then one can deduce that ‘ $\nabla(A) \subseteq \overline{A}$ for every $A \subseteq X$. Also, since ‘ $A \subseteq \overline{A}$ one can have that ‘ $A \cup \nabla(A) \subseteq \overline{A}$. From Theorem 3.3 (2) we have

$$\begin{aligned}\hat{o}_{C\Delta}(A) &= \left[\overline{A} \subseteq A \cup \nabla(A)\right] \\ &= \left[\overline{A} \subseteq A \cup \nabla(A)\right] \wedge \left[A \cup \nabla(A) \subseteq \overline{A}\right] \\ &= \left[\overline{A} \equiv A \cup \nabla(A)\right].\end{aligned}$$

Corollary 3.6 *For any* $A \subseteq X$

1) ‘ $A \in \tau_{C\alpha}\left(\text{resp. } \tau_{CS}, \tau_{CP}, \tau_{C\gamma}, \tau_{C\beta}\right)$

$$\leftrightarrow A^{\circ} \equiv A \cap A^{\circ - \circ} \left(\text{resp. } A^{\circ -}, A^{- \circ}, A^{\circ -} \cup A^{- \circ}, A^{- \circ -}\right);$$

2) $\text{戶} A \in \ _{C\alpha} \left(\text{resp. 趑}_{CS}, \ _{CP}, \text{趑}_{C\gamma}, \ _{C\beta}\right)$

$$\leftrightarrow \overline{A} \equiv A \cup A^{- \circ -} \left(\text{resp. } A^{- \circ}, A^{\circ -}, A^{- \circ} \cup A^{\circ -}, A^{\circ - \circ}\right).$$

Remark 3.1 *In crisp setting, i.e., if the underlying fuzzifying topology is the ordinary topology, one can have*

$$` \ A \in \tau \leftrightarrow \left(A \in \tau_{\alpha} \wedge A \in \tau_{C\alpha}\right)$$
$$\left(\text{resp.} \left(A \in \tau_{S} \wedge A \in \tau_{CS}\right),\right.$$
$$\left(A \in \tau_{P} \wedge A \in \tau_{CP}\right), \left(A \in \tau_{\gamma} \wedge A \in \tau_{C\gamma}\right),$$
$$\left.\left(A \in \tau_{\beta} \wedge A \in \tau_{C\beta}\right)\right).(*)$$

Of course the implication "®" in (*) is either the Łukaciewicz's implication or the Boolean's implication since these implications are identical in crisp setting. But in fuzzifying setting the statement (*) may not be true as illustrated by the following example.

Example 3.1 *Let* $X = \{a, b, c\}$ *and* τ *be a fuzzifying topology on X defined as follows:'*

$$\tau(X) = \tau(\phi) = \tau(\{a\}) = \tau(\{a, c\}) = 1;$$

$$\tau(\{b\}) = \tau(\{a, b\}) = 0 \text{ and } \tau(\{c\}) = \tau(\{b, c\}) = \frac{1}{4}.$$

From the definitions of the interior and the closure of a subset of X and the definitions of the interior and the closure of a fuzzy subset of X and some calculations we have:

$$\tau_{\alpha}(\{a, b\}) = \tau_{S}(\{a, b\}) = \tau_{P}(\{a, b\})$$
$$= \tau_{\gamma}(\{a, b\}) = \tau_{\beta}(\{a, b\}) = \frac{3}{4}$$

and

$$\tau_{C\alpha}(\{a,b\}) = \tau_{CS}(\{a,b\}) = \tau_{CP}(\{a,b\})$$
$$= \tau_{C\gamma}(\{a,b\}) = \tau_{C\beta}(\{a,b\}) = \frac{1}{4}.$$

So,

$$\tau_{\alpha}(\{a,b\}) \wedge \tau_{C\alpha}(\{a,b\})$$
$$= \tau_{S}(\{a,b\}) \wedge \tau_{CS}(\{a,b\})$$
$$= \tau_{P}(\{a,b\}) \wedge \tau_{CP}(\{a,b\})$$
$$= \tau_{\gamma}(\{a,b\}) \wedge \tau_{C\gamma}(\{a,b\})$$
$$= \tau_{\beta}(\{a,b\}) \wedge \tau_{C\beta}(\{a,b\})$$
$$= \frac{3}{4} \wedge \frac{1}{4} = \frac{1}{4} \neq 0 = \tau(\{a,b\}).$$

Theorem 3.6 *Let* ' $A^{\circ} \subseteq \Delta(A)$ *for each* $A \subseteq X$. *Then*

1) ' $A \in \tau \rightarrow (A \in \tau_{\Delta} \wedge A \in \tau_{C\Delta})$;

2) ' $(A \in \tau_{\Delta} \otimes A \in \tau_{C\Delta}) \rightarrow A \in \tau$.

Proof. 1) It is obtained from Theorem 2.4 (1)(a) and (2)(a).

2) $[A \in \tau_{\Delta} \otimes A \in \tau_{C\Delta}] = \max(0, \tau_{\Delta}(A) + \tau_{C\Delta}(A) - 1)$

$$= \max\Big(0, \inf_{x \in A} \Delta(A)(x)$$
$$+ \inf_{x \in A} \min\big(1, 1 - \Delta(A)(x) + A^{\circ}(x)\big) - 1\Big)$$
$$= \max\Big(0, \inf_{x \in A} \Delta(A)(x)$$
$$+ \inf_{x \in A} \big(1 - \Delta(A)(x) + A^{\circ}(x)\big) - 1\Big)$$
$$\leq \max\Big(0, \inf_{x \in A} \big(\Delta(A)(x) + 1 - \Delta(A)(x) + A^{\circ}(x) - 1\big)\Big)$$
$$= \inf_{x \in A} A^{\circ}(x) = [A \in \tau].$$

Corollary 3.7

1) ‘ $A \in \tau$

$\to (A \in \tau_\alpha \wedge A \in \tau_{C\alpha})(\text{resp.}(A \in \tau_S \wedge A \in \tau_{CS}),$

$(A \in \tau_P \wedge A \in \tau_{CP}),(A \in \tau_\gamma \wedge A \in \tau_{C\gamma});$

$(A \in \tau_\beta \wedge A \in \tau_{C\beta}));$

2) ‘ $(A \in \tau_\alpha \otimes A \in \tau_{C\alpha})(\text{resp.}(A \in \tau_S \otimes A \in \tau_{CS}),$

$(A \in \tau_P \otimes A \in \tau_{CP}),(A \in \tau_\gamma \otimes A \in \tau_{C\gamma});$

$(A \in \tau_\beta \otimes A \in \tau_{C\beta})) \to A \in \tau.$

Theorem 3.7 *Let* ‘ $A^\circ \subseteq \Delta(A)$ *for each* $A \subseteq X$. *If for every* $A \subseteq X$, ‘ $A \in \tau_\Delta$ *or* ‘ $A \in \tau_{C\Delta}$, *then*

1) ‘ $A \in \tau \leftrightarrow (A \in \tau_\Delta \wedge A \in \tau_{C\Delta});$

2) ‘ $(A \in \tau_\Delta \otimes A \in \tau_{C\Delta}) \leftrightarrow A \in \tau.$

Proof. Using Theorem 3.6 it remains to prove the following

1) Suppose that ‘ $A \in \tau_\Delta$. Then for each $x \in A$ we have $\Delta(A)(x) = 1$. So

$$\begin{aligned}[A \in \tau_\Delta] \wedge [A \in \tau_{C\Delta}] &= [A \in \tau_{C\Delta}] \\ &= \inf_{x \in A} \min(1, 1 - \Delta(A)(x) + A^\circ(x)) \\ &= \inf_{x \in A} A^\circ(x) = \tau(A).\end{aligned}$$

Now, suppose that ‘ $A \in \tau_{C\Delta}$. Since $[A^\circ \subseteq \Delta(A)] = 1$, then

$$\begin{aligned}1 &= \min(1, 1 - \Delta(A)(x) + A^\circ(x)) \\ &= 1 - \Delta(A)(x) + A^\circ(x).\end{aligned}$$

For each $x \in A$ we have $\Delta(A)(x) = A^\circ(x)$. Thus,

$$[A \in \tau_\Delta] \wedge [A \in \tau_{C\Delta}] = [A \in \tau_\Delta] = \tau(A).$$

2) The proof is similar to 1).

Corollary 3.8 *If for every* $A \subseteq X$, ‘ $A \in \tau_\alpha$ *or* ‘ $A \in \tau_{C\alpha}$ *(resp.* ‘ $A \in \tau_S$ *or* ‘ $A \in \tau_{CS}$, ‘ $A \in \tau_P$ *or* ‘ $A \in \tau_{CP}$, ‘ $A \in \tau_\gamma$ *or* 惯$A \in \tau_{C\gamma}$, $A \in \tau_\beta$ *or* ‘ $A \in \tau_{C\beta}$), *then*

1) ‘ $A \in \tau$

$\leftrightarrow (A \in \tau_\alpha \wedge A \in \tau_{C\alpha})(\text{resp.}(A \in \tau_S \wedge A \in \tau_{CS}),$

$(A \in \tau_P \wedge A \in \tau_{CP}),(A \in \tau_\gamma \wedge A \in \tau_{C\gamma});$

$(A \in \tau_\beta \wedge A \in \tau_{C\beta}));$

2) ‘ $A \in \tau$

$\leftrightarrow (A \in \tau_\alpha \otimes A \in \tau_{C\alpha})(\text{resp.}(A \in \tau_S \otimes A \in \tau_{CS}),$

$(A \in \tau_P \otimes A \in \tau_{CP}),(A \in \tau_\gamma \otimes A \in \tau_{C\gamma});$

$(A \in \tau_\beta \otimes A \in \tau_{C\beta})).$

Theorem 3.8 *Let* Δ *be a monotone. Then*

‘ $A \in \tau_\Delta \leftrightarrow \forall x(x \in A \rightarrow \exists B(B \in \tau_\Delta \wedge x \in B \subseteq A)).$

Proof.

$$\left[\forall x(x \in A \rightarrow \exists B(B \in \tau_\Delta \wedge x \in B \subseteq A))\right]$$
$$= \inf_{x \in A} \sup_{x \in B \subseteq A} \tau_\Delta(B).$$

First, we have $\inf_{x \in A} \sup_{x \in B \subseteq A} \tau_\Delta(B) \geq \tau_\Delta(A).$ On the other hand, Suppose that $\beta_x = \{B : x \in B \subseteq A\}$. Then for any $f \in \prod_{x \in A} \beta_x$, we have $\bigcup_{x \in A} f(x) = A.$ Since Δ is monotone, then by Theorem 3.1 (2)(a) we have

$\tau_\Delta(A) = \tau_\Delta\left(\bigcup_{x \in A} f(x)\right) \geq \inf_{x \in A} \tau_\Delta(f(x)).$ By completely distributive law we have

$$\tau_\Delta(A) \geq \sup_{f \in \prod_{x \in A} \beta_x} \inf_{x \in A} \tau_\Delta(f(x)) = \inf_{x \in A} \sup_{x \in B \subseteq A} \tau_\Delta(B).$$

Corollary 3.9

$$\vDash A \in \tau(\text{resp. } \tau_\alpha, \tau_S, \tau_P, \tau_\gamma, \tau_\beta) \leftrightarrow \forall x(x \in A$$
$$\rightarrow \exists B(B \in \tau(\text{resp. } \tau_\alpha, \tau_S, \tau_P, \tau_\gamma, \tau_\beta) \wedge x \in B \subseteq A).$$

Remark 3.2 *The following are valid in crisp setting:*

1) $\vDash A \in \tau_\alpha \wedge B \in \tau_\alpha \rightarrow A \cap B \in \tau_\alpha;$

2) $\vDash A \in \tau \wedge B \in \tau_\gamma \rightarrow A \cap B \in \tau_\gamma;$

3) $\vDash A \in \tau_\alpha \wedge B \in \tau_\gamma \rightarrow A \cap B \in \tau_\gamma,$

but in fuzzifying setting these statement may not be true by the following example.

Example 3.2 *From Example 3.1 we have*

1) $\tau_\alpha(\{a,b\}) \wedge \tau_\alpha(\{b,c\}) = \frac{1}{4} \nleq 0$

$= \tau_\alpha(\{b\}) = \tau_\alpha(\{a,b\} \cap \{b,c\}).$

2) $\tau(\{b,c\}) \wedge \tau_\gamma(\{a,b\}) = \frac{1}{4} \nleq 0$

$= \tau_\gamma(\{b\}) = \tau_\gamma(\{a,b\} \cap \{b,c\}).$

3) $\tau_\alpha(\{a,b\}) \wedge \tau_\gamma(\{b,c\}) = \frac{1}{4} \nleq 0$

$= \tau_\gamma(\{b\}) = \tau_\gamma(\{a,b\} \cap \{b,c\}).$

4. FUZZIFYING NEIGHBORHOOD STRUCTURE OF A POINT

In this section the concepts of Δ-neighborhood system and CΔ-neighborhood system of a point are presented and a fuzzifying topology induced by CΔ-neighborhood system is obtained.

Definition 4.1 *The fuzzifying Δ-(resp. CΔ-) neighborhood system of* $x \in X$, *denoted by* N_x^{Δ} *(resp.* $N_x^{C\Delta}$ *)* $\in \Im(P(X))$, *is defined as follows:*

$$A \in N_x^{\Delta} := \exists B\left(B \in \tau_{\Delta} \wedge x \in B \subseteq A\right)$$
$$\left(\text{resp. } A \in N_x^{C\Delta} := \exists B\left(B \in \tau_{C\Delta} \wedge x \in B \subseteq A\right)\right),$$

i.e.,

$$N_x^{\Delta}(A) = \sup_{x \in B \subseteq A} \tau_{\Delta}(B)\left(\text{resp. } N_x^{C\Delta}(A) = \sup_{x \in B \subseteq A} \tau_{C\Delta}(B)\right).$$

Definition 4.2 *Let* $x \in X$.

1) If $\Delta = ()^{\circ}$, (resp. $()^{\circ-\circ}, ()^{\circ-}, ()^{-\circ}, ()^{-\circ} \cup ()^{\circ-}, ()^{-\circ-}$), the fuzzifying Δ-neighborhood system of x coincides with the fuzzifying (resp. fuzzifying α-, fuzzifying semi-, fuzzifying pre-, fuzzifying γ-, fuzzifying β-) neighborhood system of x and will be denoted by N_x (resp. $N_x^{\alpha}, N_x^{S}, N_x^{P}, N_x^{\gamma}, N_x^{\beta}$);

2) If $\Delta = ()^{\circ-\circ}$ (resp. $()^{\circ-}, ()^{-\circ}, ()^{-\circ} \cup ()^{\circ-}, ()^{-\circ-}$), the fuzzifying CΔ-neighborhood system of x coincides with the fuzzifying $c\alpha$- (resp. fuzzifying csemi-, fuzzifying cpre-, fuzzifying $c\gamma$-, fuzzifying $c\beta$-) neighborhood system of x and will be denoted by $N_x^{C\alpha}$ (resp. $N_x^{CS}, N_x^{CP}, N_x^{C\gamma}, N_x^{C\beta}$).

Theorem 4.1 *Let Δ be a monotone.*

1) $\vDash A \in \tau_{\Delta} \leftrightarrow \forall x\left(x \in A \to \exists B\left(B \in N_x^{\Delta} \wedge B \subseteq A\right)\right)$;

2) $\vDash \forall x\left(x \in A \to A \in N_x^{\Delta}\right) \leftrightarrow A \in \tau_{\Delta}$.

Proof. Using Theorem 3.8 we have

$$\left[\forall x\left(x \in A \to \exists B\left(B \in N_x^{\Delta} \wedge B \subseteq A\right)\right)\right]$$
$$= \inf_{x \in A} \sup_{B \subseteq A} N_x^{\Delta}(B) = \inf_{x \in A} \sup_{B \subseteq A} \sup_{x \in C \subseteq B} \tau_{\Delta}(C)$$
$$= \inf_{x \in A} \sup_{x \in C \subseteq A} \tau_{\Delta}(C) = \left[A \in \tau_{\Delta}\right].$$

2) From 1) the proof is immediate.

Corollary 4.1 *Let* (X,τ) *be a fuzzifying topological space.*

1) $\tau(A)\left(\text{resp. } \tau_{\alpha}(A), \tau_S(A), \tau_P(A), \tau_{\gamma}(A), \tau_{\beta}(A)\right)$

$$= \inf_{x \in A} \sup_{x \in C \subseteq A} \tau(C)\left(\text{resp. } \tau_{\alpha}(C), \tau_S(C), \tau_P(C), \tau_{\gamma}(C), \tau_{\beta}(C)\right);$$

1) $\tau(A)\left(\text{resp. } \tau_{\alpha}(A), \tau_S(A), \tau_P(A), \tau_{\gamma}(A), \tau_{\beta}(A)\right)$

$$= \inf_{x \in A} N_x(A)\left(\text{resp. } N_x^{\alpha}(A), N_x^{S}(A), N_x^{P}(A), N_x^{\gamma}(A), N_x^{\beta}(A)\right);$$

Theorem 4.2 *The mapping* $N^{\Delta}: X \to \Im(P(X))$, $x \mapsto N_x^{\Delta}$, *has the following properties:*

1) If Δ is of type $O1$, then N_x^{Δ} is normal for any $x \in X$;

2) For any $x, A,$ ‘ $A \in N_x^{\Delta} \to x \in A$;

3) For any $x, A, B,$ ‘ $A \subseteq B \to \left(A \in N_x^{\Delta} \to B \in N_x^{\Delta}\right)$;

4) If Δ is monotone, then for any $x, A,$

$$\vDash A \in N_x^{\Delta} \to \exists H\left(H \in N_x^{\Delta} \wedge H \subseteq A \wedge \forall y\left(y \in H \to H \in N_y^{\Delta}\right)\right);$$

5) If Δ is of type $O2$, then

$$\vDash A \in N_x^{\Delta} \wedge B \in N_x^{\Delta} \to A \cap B \in N_x^{\Delta}.$$

Proof. 1) Since Δ is of type $O1$, then

$$N_x^{\Delta}(X) = \sup_{x \in B \subseteq X} \tau_{\Delta}(B) \geq \tau_{\Delta}(X) = [X \subseteq \Delta(X)] = 1.$$

2) If $[A \in N_x^{\Delta}] = 0$, then the results holds. Suppose $[A \in N_x^{\Delta}] = \sup_{x \in H \subseteq A} \tau_{\Delta}(H) > 0$. Then there exists $H_{\circ} \in P(X)$ such that $x \in H_{\circ} \subseteq A$. Now, we have $[x \in A] = 1$. Thus $[A \in N_x^{\Delta}] \leq [x \in A]$ holds always.

3) If $[A \subseteq B] = 0$, then the result holds. Now, suppose that $[A \subseteq B] = 1$. Then we have

$$[A \in N_x^{\Delta}] = \sup_{x \in C \subseteq A} \tau_{\Delta}(C) \leq \sup_{x \in C \subseteq B} \tau_{\Delta}(C) = [B \in N_x^{\Delta}].$$

4) Since Δ is monotone, then from Theorem 4.1 a) we have

$$\begin{aligned}
&\left[\exists H\left(H \in N_x^{\Delta} \wedge H \subseteq A \wedge \forall y\left(y \in H \to H \in N_y^{\Delta}\right)\right)\right] \\
&= \sup_{H \subseteq A}\left(N_x^{\Delta}(H) \wedge \inf_{y \in H} N_y^{\Delta}(H)\right) = \sup_{H \subseteq A}\left(N_x^{\Delta}(H) \wedge \tau_{\Delta}(H)\right) \\
&= \sup_{H \subseteq A} \tau_{\Delta}(H) \geq \sup_{x \in H \subseteq A} \tau_{\Delta}(H) = [A \in N_x^{\Delta}].
\end{aligned}$$

5) $N_x^{\Delta}(A \cap B) = \sup_{x \in C \subseteq A \cap B} \tau_{\Delta}(C)$

$$= \sup_{x\in C_1\subseteq A, x\in C_2\subseteq B} \tau_\Delta(C_1\cap C_2)$$
$$\geq \sup_{x\in C_1\subseteq A, x\in C_2\subseteq B} (\tau_\Delta(C_1)\wedge\tau_\Delta(C_2))$$
$$= \sup_{x\in C_1\subseteq A} \tau_\Delta(C_1)\wedge \sup_{x\in C_2\subseteq B} \tau_\Delta(C_2)$$
$$= N_x^\Delta(A)\wedge N_x^\Delta(B).$$

Corollary 4.2 *The mapping N*

$\left(\text{resp. } N^\alpha, N^S, N^P, N^\gamma, N^\beta\right): X \to \Im(P(X))$,

$x \mapsto N_x\left(\text{resp. } N_x^\alpha, N_x^S, N_x^P, N_x^\gamma, N_x^\beta\right)$ *has the following properties:*

1) For any $x \in X$, $\left(\text{resp. } N_x^\alpha, N_x^S, N_x^P, N_x^\gamma, N_x^\beta\right)$ N_x is normal;

2) For any $x, A,$

$$\text{`}\ A\in N_x\left(\text{resp. } N_x^\alpha, N_x^S, N_x^P, N_x^\gamma, N_x^\beta\right)\to x\in A;$$

3) For any $x, A, B,$

$$\text{`}\ A\subseteq B\to\Big(A\in N_x\left(\text{resp. } N_x^\alpha, N_x^S, N_x^P, N_x^\gamma, N_x^\beta\right)$$
$$\to B\in N_x\left(\text{resp. } N_x^\alpha, N_x^S, N_x^P, N_x^\gamma, N_x^\beta\right)\Big);$$

4) For any $x, A,$

$$\text{`}\ A\in N_x\left(\text{resp. } N_x^\alpha, N_x^S, N_x^P, N_x^\gamma, N_x^\beta\right)$$
$$\to \exists H\Big(H\in N_x\left(\text{resp. } N_x^\alpha, N_x^S, N_x^P, N_x^\gamma, N_x^\beta\right)\wedge H$$
$$\subseteq A\wedge\forall y\Big(y\in H\to H\in N_x\left(\text{resp. } N_x^\alpha, N_x^S, N_x^P, N_x^\gamma, N_x^\beta\right)\Big)\Big);$$

5) For any $x, A, B,$

$$\text{`}\ A\in N_x\wedge B\in N_x\to A\cap B\in N_x.$$

Theorem 4.3 *The mapping* $N^{C\Delta}: X \to \Im(P(X)),\ x \mapsto N_x^{C\Delta}$, *has the following properties:*

1) If Δ is of type $O1$, then $N_x^{C\Delta}$ is normal for any $x \in X$;

2) For any $x, A,$ ‘ $A \in N_x^{C\Delta} \to x \in A$;

3) For any $x, A, B,$

$$‘\ A \subseteq B \to \left(A \in N_x^{C\Delta} \to B \in N_x^{C\Delta}\right);$$

4) If Δ is monotone, then or any $x, A, B,$

$$‘\ A \in N_x^{C\Delta} \wedge B \in N_x^{C\Delta} \to A \cap B \in N_x^{C\Delta}.$$

Conversely, if a mapping $N^{C\Delta}$ satisfies 1), 3) and 4), then it assigns a fuzzifying topology on X, denoted by $\tau_{N^{C\Delta}} \in \Im(P(X))$, is defined as follows:

$$A \in \tau_{N^{C\Delta}} := \forall x\left(x \in A \to A \in N_x^{C\Delta}\right).$$

PROOF. Since $N_x^{C\Delta}$ is normal and satisfies properties (2) and (3) in Theorem 4.2, then $\tau_{N^{C\Delta}}$ is a fuzzifying topology on X. The rest of the proof is similar to the proof of Theorem 4.2.

Corollary 4.3 *The mapping* $N^{C\alpha}$

$$\left(\text{resp. } N^{CS}, N^{CP}, N^{C\gamma}, N^{C\beta}\right): X \to \Im(P(X)), x \mapsto N_x^{C\alpha}$$
$$\left(\text{resp. } N_x^{CS}, N_x^{CP}, N_x^{C\gamma}, N_x^{C\beta}\right),$$

has the following properties:

1) $N_x^{C\alpha}\left(\text{resp. } N_x^{CS}, N_x^{CP}, N_x^{C\gamma}, N_x^{C\beta}\right)$ is normal for any $x \in X$;

2) For any $x, A,$

$$\vDash A \in N_x^{C\alpha}\left(\text{resp. } N_x^{CS}, N_x^{CP}, N_x^{C\gamma}, N_x^{C\beta}\right) \rightarrow x \in A;$$

3) For any $x, A, B,$

$$\vDash A \subseteq B \rightarrow \left(A \in N_x^{C\alpha}\left(\text{resp. } N_x^{CS}, N_x^{CP}, N_x^{C\gamma}, N_x^{C\beta}\right) \rightarrow B \in N_x^{C\alpha}\left(\text{resp. } N_x^{CS}, N_x^{CP}, N_x^{C\gamma}, N_x^{C\beta}\right)\right);$$

4) If Δ is monotone, then or any $x, A, B,$

$$\vDash A \in N_x^{C\alpha}\left(\text{resp. } N_x^{CS}, N_x^{CP}, N_x^{C\gamma}, N_x^{C\beta}\right) \wedge B \in N_x^{C\alpha}\left(\text{resp. } N_x^{CS}, N_x^{CP}, N_x^{C\gamma}, N_x^{C\beta}\right) \rightarrow A \cap B \in N_x^{C\alpha}\left(\text{resp. } N_x^{CS}, N_x^{CP}, N_x^{C\gamma}, N_x^{C\beta}\right).$$

5) The mapping $N^{C\alpha}\left(\text{resp. } N^{CS}, N^{CP}, N^{C\gamma}, N^{C\beta}\right)$ assigns a fuzzifying topology on X, denoted by $\tau_{N^{C\alpha}}\ \left(\text{resp. } \tau_{N^{CS}}, \tau_{N^{CP}}, \tau_{N^{C\gamma}}, \tau_{N^{C\beta}}\right) \in \Im(P(X))$, is defined as follows:

$$A \in \tau_{N^{C\alpha}}\left(\text{resp. } \tau_{N^{CS}}, \tau_{N^{CP}}, \tau_{N^{C\gamma}}, \tau_{N^{C\beta}}\right) := \forall x\left(x \in A \rightarrow A \in N_x^{C\alpha}\left(\text{resp. } N_x^{CS}, N_x^{CP}, N_x^{C\gamma}, N_x^{C\beta}\right)\right).$$

Theorem 4.4 *If Δ is of type* $O1$ *and monotone, then* $\vDash \tau_{C\Delta} \subseteq \tau_{N^{C\Delta}}$.

Proof. Let $B \subseteq X$. Then

$$\tau_{N^{C\Delta}}(B) = \inf_{x \in B} N_x^{C\Delta}(B) = \inf_{x \in B} \sup_{x \in A \subseteq B} \tau_{C\Delta}(A) \geq \tau_{C\Delta}(B).$$

Corollary 4.4

1) $\vDash \tau_{C\alpha} \subseteq \tau_{N^{C\alpha}}$;

2) $\vDash \tau_{CS} \subseteq \tau_{N^{CS}}$;

3) $\tau_{CP} \subseteq \tau_{N^{CP}}$;

4) $\tau_{C\gamma} \subseteq \tau_{N^{C\gamma}}$;

5) $\tau_{C\beta} \subseteq \tau_{N^{C\beta}}$.

5. CLOSURE AND INTERIOR OPERATIONS IN FUZZIFYING TOPOLOGY

The purpose of this section is to establish the concepts of fuzzifying Δ- and CΔ-derived sets, fuzzifying Δ- and CΔ-closure operation and fuzzifying Δ- and CΔ-interior operation and study some of their properties.

Definition 5.1 *The fuzzifying Δ- (resp. CΔ-) derived set* $d_\Delta(A)$ *(resp.* $d_{C\Delta}(A)$*)* $\in \Im(X)$ *of* $A \subseteq X$ *is defined as follows:*

$$x \in d_\Delta(A) := \forall B\left(B \in N_x^\Delta \rightarrow B \cap \left(A - \{x\}\right) \neq \phi\right)$$

$$\left(\text{resp. } x \in d_{C\Delta}(A) := \forall B\left(B \in N_x^{C\Delta} \rightarrow B \cap \left(A - \{x\}\right) \neq \phi\right)\right)$$ i.e.,

$$d_\Delta(A)(x) = \inf_{B \cap (A - \{x\}) = \phi} \left(1 - N_x^\Delta(B)\right)$$

$$\left(\text{resp. } d_{C\Delta}(A)(x) = \inf_{B \cap (A - \{x\}) = \phi} \left(1 - N_x^{C\Delta}(B)\right)\right).$$

Definition 5.2 *For* $A \subseteq X$.

1) If $\Delta = 0^{\circ}$, (resp. $0^{\circ-\circ}, 0^{\circ-}, 0^{-\circ}, 0^{-\circ} \cup 0^{\circ-}, 0^{-\circ-}$), the notion of fuzzifying Δ-derived set of $A \subseteq X$ coincides with the notion of fuzzifying derived (resp. α- derived, semi-derived, pre-derived, γ-derived, β-derived) set and will be denoted by

$$A'\left(\text{resp. } d_\alpha(A), d_S(A), d_P(A), d_\gamma(A), d_\beta(A)\right);$$

2) If $\Delta = ()^{\circ - \circ}$ (resp. $()^{\circ -}, ()^{- \circ}, ()^{- \circ} \cup ()^{\circ -}, ()^{- \circ -}$), the notion of fuzzifying cΔ-derived set of $A \subseteq X$ coincides with the notion of fuzzifying derived (resp. $c\alpha$-derived, csemi-derived, cpre-derived, $c\gamma$-derived, $c\beta$-derived) set and will be denoted by

$$d_{C\alpha}(A)\left(\text{resp. } d_{CS}(A), d_{CP}(A), d_{C\gamma}(A), d_{C\beta}(A)\right).$$

Theorem 5.1 *For every* $A \subseteq X$ *we have*

1) $d_{\Delta}(A)(x) = 1 - N_x^{\Delta}\left((X - A) \cup \{x\}\right);$

2) $d_{C\Delta}(A)(x) = 1 - N_x^{C\Delta}\left((X - A) \cup \{x\}\right).$

Proof. 1) Using Theorem 4.3 3) we have

$$\begin{aligned} d_{\Delta}(A)(x) &= \inf_{B \cap (A - \{x\}) = \phi} \left(1 - N_x^{\Delta}(B)\right) \\ &= 1 - \sup_{B \cap (A - \{x\}) = \phi} N_x^{\Delta}(B) \\ &= 1 - \sup_{B \subseteq (X - A) \cup \{x\}} N_x^{\Delta}(B) \\ &= 1 - N_x^{\Delta}\left((X - A) \cup \{x\}\right). \end{aligned}$$

2) It is similar to the proof. of 1).

Corollary 5.1

1) a) $A'(x) = 1 - N_x\left((X - A) \cup \{x\}\right);$

b) $d_{\alpha}(A)(x) = 1 - N_x^{\alpha}\left((X - A) \cup \{x\}\right);$

c) $d_S(A)(x) = 1 - N_x^S\left((X - A) \cup \{x\}\right);$

d) $d_P(A)(x) = 1 - N_x^P\left((X - A) \cup \{x\}\right);$

e) $d_{\gamma}(A)(x) = 1 - N_x^{\gamma}\left((X - A) \cup \{x\}\right);$

f) $d_{\beta}(A)(x) = 1 - N_x^{\beta}\left((X - A) \cup \{x\}\right);$

2) a) $d_{C\alpha}(A)(x) = 1 - N_x^{C\alpha}((X - A) \cup \{x\});$

b) $d_{CS}(A)(x) = 1 - N_x^{CS}((X - A) \cup \{x\});$

c) $d_{CP}(A)(x) = 1 - N_x^{CP}((X - A) \cup \{x\});$

d) $d_{C\gamma}(A)(x) = 1 - N_x^{C\gamma}((X - A) \cup \{x\});$

e) $d_{C\beta}(A)(x) = 1 - N_x^{C\beta}((X - A) \cup \{x\});$

Theorem 5.2 *For every $A \subseteq X$ we have*

1) If Δ is monotone, then $\models A \in {}_{\Delta} \leftrightarrow d_{\Delta}(A) \subseteq A;$

2) If Δ is of type $O1$ and monotone, then

$$\models A \in {}_{N^{C\Delta}} \leftrightarrow d_{C\Delta}(A) \subseteq A$$

.

Proof. Using Theorem 4.1 (2) we have 1) $[d_{\Delta}(A) \subseteq A] = \inf_{x \in X - A} (1 - d_{\Delta}(A)(x))$

$$= \inf_{x \in X - A} N_x^{\Delta}((X - A) \cup \{x\})$$
$$= \inf_{x \in X - A} N_x^{\Delta}(X - A) = \tau_{\Delta}(X - A) = \hat{o}_{\Delta}(A).$$

2) It is similar to 1).

Corollary 5.2

1) a) $\models A \in \quad \leftrightarrow A' \subseteq A;$

b) $\models A \in {}_{\alpha} \leftrightarrow d_{\alpha}(A) \subseteq A;$

c) $\models A \in {}_{S} \leftrightarrow d_{S}(A) \subseteq A;$

d) $\models A \in {}_{P} \leftrightarrow d_{P}(A) \subseteq A;$

e) $\models A \in {}_{\gamma} \leftrightarrow d_{\gamma}(A) \subseteq A;$

f) $\models A \in {}_{\beta} \leftrightarrow d_{\beta}(A) \subseteq A;$

2) a) $\models A \in {}_{N^{C\alpha}} \leftrightarrow d_{C\alpha}(A) \subseteq A;$

b) $\models A \in {}_{N^{CS}} \leftrightarrow d_{CS}(A) \subseteq A;$

c) $\models A \in {}_{N^{CP}} \leftrightarrow d_{CP}(A) \subseteq A;$

d) $\models A \in {}_{N^{C\gamma}} \leftrightarrow d_{C\gamma}(A) \subseteq A;$

e) $\models A \in {}_{N^{C\beta}} \leftrightarrow d_{C\beta}(A) \subseteq A;$

Definition 5.3 *The fuzzifying Δ-(resp. CΔ-) closure* $Cl_{\Delta}(A)$ *(resp.* $Cl_{C\Delta}(A)$*)* $\in \Im(X)$ *of* $A \subseteq X$ *is defined as follows:*

$$x \in Cl_{\Delta}(A) := \forall B((B \supseteq A) \wedge (B \in \hat{o}_{\Delta}) \rightarrow x \in B)$$

$$\left(\text{resp. } x \in Cl_{C\Delta}(A) := \forall B((B \supseteq A) \wedge (B \in \hat{o}_{C\Delta}) \rightarrow x \in B)\right)$$ i.e.,

$$Cl_{\Delta}(A)(x) = \inf_{x \notin B \supseteq A} (1 - \hat{o}_{\Delta}(B))$$

$$\left(\text{resp. } Cl_{C\Delta}(A)(x) = \inf_{x \notin B \supseteq A} (1 - \hat{o}_{C\Delta}(B))\right).$$

Definition 5.4 *1) If* $\Delta = ()^{\circ}$, *(resp.* $()^{-\circ} \cup ()^{\circ -}, ()^{-\circ -}$) $()^{\circ - \circ}, ()^{\circ -}, ()^{- \circ}$, *the notion of fuzzifying Δ-closure of* $A \subseteq X$ *coincides with the notion of fuzzifying closure (resp.* α *-closure, semi-closure, pre-closure,* γ *-closure,* β *-closure) operation and will be denoted by*

$$\overline{A}\left(\text{resp. } Cl_{\alpha}(A), Cl_{S}(A), Cl_{P}(A), Cl_{\gamma}(A), Cl_{\beta}(A)\right);$$

2) If $\Delta = ()^{\circ - \circ}$ (resp. $()^{\circ -}, ()^{-\circ}, ()^{-\circ} \cup ()^{\circ -}, ()^{-\circ -}$), the notion of fuzzifying CΔ-closure of $A \subseteq X$ coincides with the notion of fuzzifying $c\alpha$ (resp. csemi, cpre, $c\gamma$, $c\beta$)-closure operation and will be denoted by

$$Cl_{C\alpha}(A)\left(\text{resp. } Cl_{CS}(A), Cl_{CP}(A), Cl_{C\gamma}(A), Cl_{C\beta}(A)\right).$$

Theorem 5.3 *For every* $A \subseteq X$ *we have*

1) $Cl_{\Delta}(A)(x) = 1 - N_x^{\Delta}(X - A);$

2) $Cl_{C\Delta}(A)(x) = 1 - N_x^{C\Delta}(X - A);$

Proof.

1) $Cl_{\Delta}(A)(x) = \inf_{x \notin B \supseteq A}\left(1 - \hat{o}_{\Delta}(B)\right)$

$= \inf_{x \in X - B \subseteq X - A}\left(1 - \tau_{\Delta}(X - B)\right)$

$= 1 - \sup_{x \in X - B \subseteq X - A} \tau_{\Delta}(X - B)$

$= 1 - N_x^{\Delta}(X - A).$

2) It is similar to the proof of (1).

Corollary 5.3

1) a) $\overline{A}(x) = 1 - N_x(X - A);$

b) $Cl_{\alpha}(A)(x) = 1 - N_x^{\alpha}(X - A);$

c) $Cl_S(A)(x) = 1 - N_x^S(X - A);$

d) $Cl_P(A)(x) = 1 - N_x^P(X - A);$

e) $Cl_{\gamma}(A)(x) = 1 - N_x^{\gamma}(X - A);$

f) $Cl_{\beta}(A)(x) = 1 - N_x^{\beta}(X - A);$

2) a) $Cl_{C\alpha}(A)(x) = 1 - N_x^{C\alpha}(X - A);$

b) $Cl_{CS}(A)(x) = 1 - N_x^{CS}(X - A);$

c) $Cl_{CP}(A)(x) = 1 - N_x^{CP}(X - A);$

d) $Cl_{C\gamma}(A)(x) = 1 - N_x^{C\gamma}(X - A);$

e) $Cl_{C\beta}(A)(x) = 1 - N_x^{C\beta}(X - A);$

Theorem 5.4 *For every* $A \subseteq X$ *and* $x \in X$ *we have*

1) If Δ is of type 01, then a) ‘ $Cl_{\Delta}(\phi) \equiv \phi;$

b) ‘ $Cl_{C\Delta}(\phi) \equiv \phi;$

2) a) ‘ $A \subseteq Cl_{\Delta}(A);$

b) ‘ $A \subseteq Cl_{C\Delta}(A);$

3) a) ‘ $Cl_{\Delta}(A) \equiv d_{\Delta}(A) \cup A;$

b) ‘ $Cl_{C\Delta}(A) \equiv d_{C\Delta}(A) \cup A;$

4) a) ‘ $x \in Cl_{\Delta}(A) \leftrightarrow \forall B\left(B \in N_x^{\Delta} \to A \cap B \neq \phi\right);$

b) ‘ $x \in Cl_{C\Delta}(A) \leftrightarrow \forall B\left(B \in N_x^{C\Delta} \to A \cap B \neq \phi\right);$

5) a) $\models A \equiv Cl_{\Delta}(A) \leftrightarrow A \in \quad {}_{\Delta};$

b) If Δ is of type 01 and monotone, then

$\models A \equiv Cl_{C\Delta}(A) \leftrightarrow A \in \quad {}_{N^{C\Delta}};$

Proof. We prove only a) of each statements since b) is similar.

1) a) $Cl_{\Delta}(\phi)(x) = 1 - N_x(X - \phi)$

$= 1 - N_x(X) = 1 - 1 = 0.$

2) a) It is clear that $N_x(A) = 0$ for any $A \subseteq X$ and $x \in X$ in case of $x \notin A$. Now, suppose that $x \in A$. Then $Cl_{\Delta}(A)(x) = 1 - N_x^{\Delta}(X - A) = 1$. Therefore

‘ $A \subseteq Cl_{\Delta}(A).$

3) a) Using Theorem 5.1 (1) and (2) above, we have

$$\left(d_\Delta(A)\cup A\right)(x)=\max\left(1-N_x^\Delta\left((X-A)\cup\{x\}\right),A(x)\right)$$
$$=Cl_\Delta(A)(x).$$

4) a) $\left[\forall B\left(B\in N_x^{C\Delta}\rightarrow A\cap B\neq\phi\right)\right]$

$$=\inf_{B\subseteq X-A}\left(1-N_x^\Delta(B)\right)=1-\sup_{B\subseteq X-A}N_x^\Delta(B)$$
$$=1-N_x^\Delta(X-A)=Cl_\Delta(A)(x).$$

5) a) Since $\left[A\subseteq d_\Delta(A)\cup A\right]=1,$ then from Theorem 5.2 (1) and (3) (a) above we have

$$\hat{o}_\Delta(A)=\left[d_\Delta(A)\subseteq A\right]=\left[d_\Delta(A)\cup A\subseteq A\right]$$
$$=\left[A\subseteq d_\Delta(A)\cup A\right]\wedge\left[d_\Delta(A)\cup A\subseteq A\right]$$
$$=\left[d_\Delta(A)\cup A\equiv A\right]=\left[Cl_\Delta(A)\equiv A\right],$$

because $\left[\tilde{B}\subseteq A\right]=\left[\tilde{B}\cup A\equiv A\right]$ for any $\tilde{B}\in\Im(X).$

Corollary 5.4

1) a) ‘ $\bar{\phi}\equiv\phi$;

b) ‘ $Cl_\alpha(\phi)\equiv\phi$;

c) ‘ $Cl_S(\phi)\equiv\phi$;

d) ‘ $Cl_P(\phi)\equiv\phi$;

e) ‘ $Cl_\gamma(\phi)\equiv\phi$;

f) ‘ $Cl_\beta(\phi)\equiv\phi$;

g) ‘ $Cl_{C\alpha}(\phi)\equiv\phi$;

h) ‘ $Cl_{CS}(\phi)\equiv\phi$;

i) ‘ $Cl_{CP}(\phi)\equiv\phi$;

j) ‘ $Cl_{C\gamma}(\phi)\equiv\phi$;

k) ‘ $Cl_{C\beta}(\phi) \equiv \phi$;

2) for any $A \subseteq X$,

a) ‘ $A \subseteq \overline{A}$;

b) ‘ $A \subseteq Cl_{\alpha}(A)$;

c) ‘ $A \subseteq Cl_{S}(A)$;

d) ‘ $A \subseteq Cl_{P}(A)$;

e) ‘ $A \subseteq Cl_{\gamma}(A)$;

f) ‘ $A \subseteq Cl_{\beta}(A)$;

g) ‘ $A \subseteq Cl_{C\alpha}(A)$;

h) ‘ $A \subseteq Cl_{CS}(A)$;

i) ‘ $A \subseteq Cl_{CP}(A)$;

j) ‘ $A \subseteq Cl_{C\gamma}(A)$;

k) ‘ $A \subseteq Cl_{C\beta}(A)$.

3) for any $A \subseteq X$,

a) ‘ $\overline{A} \equiv A' \cup A$;

b) ‘ $Cl_{\alpha}(A) \equiv d_{\alpha}(A) \cup A$;

c) ‘ $Cl_{S}(A) \equiv d_{S}(A) \cup A$;

d) ‘ $Cl_{P}(A) \equiv d_{P}(A) \cup A$;

e) ‘ $Cl_{\gamma}(A) \equiv d_{\gamma}(A) \cup A$;

f) ‘ $Cl_{\beta}(A) \equiv d_{\beta}(A) \cup A$;

g) ‘ $Cl_{C\alpha}(A) \equiv d_{C\alpha}(A) \cup A$;

h) ‘ $Cl_{CS}(A) \equiv d_{CS}(A) \cup A$;

i) ‘ $Cl_{CP}(A) \equiv d_{CP}(A) \cup A$;

j) ‘ $Cl_{C\gamma}(A) \equiv d_{C\gamma}(A) \cup A$;

k) ‘ $Cl_{C\beta}(A) \equiv d_{C\beta}(A) \cup A$.

4) for any $A \subseteq X$ and $x \in X$

a) ‘ $\overline{A} \leftrightarrow \forall B(B \in N_x \to A \cap B \neq \phi)$;

b) ‘ $x \in Cl_\alpha(A) \leftrightarrow \forall B\left(B \in N_x^\alpha \to A \cap B \neq \phi\right)$;

c) ‘ $x \in Cl_S(A) \leftrightarrow \forall B\left(B \in N_x^S \to A \cap B \neq \phi\right)$;

d) ‘ $x \in Cl_P(A) \leftrightarrow \forall B\left(B \in N_x^P \to A \cap B \neq \phi\right)$;

e) ‘ $x \in Cl_\gamma(A) \leftrightarrow \forall B\left(B \in N_x^\gamma \to A \cap B \neq \phi\right)$;

f) ‘ $x \in Cl_\beta(A) \leftrightarrow \forall B\left(B \in N_x^\beta \to A \cap B \neq \phi\right)$;

g) ‘ $x \in Cl_{C\alpha}(A) \leftrightarrow \forall B\left(B \in N_x^{C\alpha} \to A \cap B \neq \phi\right)$;

h) ‘ $x \in Cl_{CS}(A) \leftrightarrow \forall B\left(B \in N_x^{CS} \to A \cap B \neq \phi\right)$;

i) ‘ $x \in Cl_{CP}(A) \leftrightarrow \forall B\left(B \in N_x^{CP} \to A \cap B \neq \phi\right)$;

j) ‘ $x \in Cl_{C\gamma}(A) \leftrightarrow \forall B\left(B \in N_x^{C\gamma} \to A \cap B \neq \phi\right)$;

k) ‘ $x \in Cl_{C\beta}(A) \leftrightarrow \forall B\left(B \in N_x^{C\beta} \to A \cap B \neq \phi\right)$.

5) for any $A \subseteq X$

a) $\vDash A \equiv \overline{A} \leftrightarrow A \in$;

b) $\vDash A \equiv Cl_\alpha(A) \leftrightarrow A \in {}_{\alpha}$;

c) $\vDash A \equiv Cl_S(A) \leftrightarrow A \in {}_{S}$;

d) $\vDash A \equiv Cl_P(A) \leftrightarrow A \in {}_{P}$;

e) $\vDash A \equiv Cl_\gamma(A) \leftrightarrow A \in {}_{\gamma}$;

f) $\models A \equiv Cl_{\beta}(A) \leftrightarrow A \in \ {}_{\beta}$;

g) $\models A \equiv Cl_{C\alpha}(A) \leftrightarrow A \in \ N^{C\alpha}$;

h) $\models A \equiv Cl_{CS}(A) \leftrightarrow A \in \ N^{CS}$;

i) $\models A \equiv Cl_{CP}(A) \leftrightarrow A \in \ N^{CP}$;

j) $\models A \equiv Cl_{C\gamma}(A) \leftrightarrow A \in \ N^{C\gamma}$;

k) $\models A \equiv Cl_{C\beta}(A) \leftrightarrow A \in \ N^{C\beta}$.

Theorem 5.5 *For every* $A, B \subseteq X$ *we have*

1) $\models B \doteq Cl_{\Delta}(A) \leftrightarrow B \in \ {}_{\Delta}$;

2) $\models B \doteq Cl_{C\Delta}(A) \leftrightarrow B \in \ N^{C\Delta}$.

Proof. 1) Suppose $[A \subseteq B] = 0$. Then $[B \doteq Cl_{\Delta}(A)] = 0$ and the results holds. Now, suppose $[A \subseteq B] = 1$. Then $[B \subseteq Cl_{\Delta}(A)] = 1 - \sup_{x \in B-A} N_x^{\Delta}(X-A)$ and $[Cl_{\Delta}(A) \subseteq B] = \inf_{x \in X-B} N_x^{\Delta}(X-A)$. So

$$[B \doteq Cl_{\Delta}(A)]$$
$$= \max\left(0, \inf_{x \in X-B} N_x^{\Delta}(X-A) - \sup_{x \in B-A} N_x^{\Delta}(X-A)\right).$$

Let $[B \doteq Cl_{\Delta}(A)] > t$. Then $\inf_{x \in X-B} N_x^{\Delta}(X-A) > t + \sup_{x \in B-A} N_x^{\Delta}(X-A)$. For any $x \in X - B$, $\sup_{x \in C \subseteq X-A} \tau_{\Delta}(C) > t + \sup_{x \in B-A} N_x^{\Delta}(X-A)$, i.e., there exists $C_x \subseteq X$ such that $x \in C_x \subseteq X - A$ and $\tau_{\Delta}(C_x) > t + \sup_{x \in B-A} N_x^{\Delta}(X-A)$. Now we want to prove $C_x \subseteq X - B$. If not, then there exist $x' \in B - A$ with $x' \in C_x$. Hence

$$\sup_{x \in B-A} N_x^{\Delta}(X-A) \geq N_{x'}^{\Delta}(X-A) \geq \tau_{\Delta}(C_x)$$
$$> t + \sup_{x \in B-A} N_x^{\Delta}(X-A)$$

and this is a contradiction. Therefore

$$\hat{o}_{\Delta}(B) = \tau_{\Delta}(X-B) = \inf_{x\in X-B} N_x^{\Delta}(X-B)$$
$$\geq \inf_{x\in X-B} \tau_{\Delta}(C_x) > t + \sup_{x\in B-A} N_x^{\Delta}(X-A) > t.$$

Since t is arbitrary, it holds that $[B \doteq Cl_{\Delta}(A)] \leq \hat{o}_{\Delta}(B)$

2) The proof is similar to 1).

Corollary 5.5 *For any A and B.*

1) a) $\models B \doteq \overline{A} \leftrightarrow B \in$;

b) $\models B \doteq Cl_{\alpha}(A) \leftrightarrow B \in \ _{\alpha}$;

c) $\models B \doteq Cl_{S}(A) \leftrightarrow B \in \ _{S}$;

d) $\models B \doteq Cl_{P}(A) \leftrightarrow B \in \ _{P}$;

e) $\models B \doteq Cl_{\gamma}(A) \leftrightarrow B \in \ _{\gamma}$;

e) $\models B \doteq Cl_{\beta}(A) \leftrightarrow B \in \ _{\beta}$;

2) a) $\models B \doteq Cl_{C\alpha}(A) \leftrightarrow B \in \ _{N^{C\alpha}}$;

b) $\models B \doteq Cl_{CS}(A) \leftrightarrow B \in \ _{N^{CS}}$;

c) $\models B \doteq Cl_{CP}(A) \leftrightarrow B \in \ _{N^{CP}}$;

d) $\models B \doteq Cl_{C\gamma}(A) \leftrightarrow B \in \ _{N^{C\gamma}}$;

e) $\models B \doteq Cl_{C\beta}(A) \leftrightarrow B \in \ _{N^{C\beta}}$.

Definition 5.5 *The fuzzifying Δ-(resp. CΔ-) interior* $Int_{\Delta}(A)$ *(resp.* $Int_{C\Delta}(A)$*)* $\in \Im(X)$ *of* $A \subseteq X$ *is defined as follows:*

$$x \in Int_{\Delta}(A) := N_x^{\Delta}(A) \left(\text{resp. } x \in Int_{C\Delta}(A) := N_x^{C\Delta}(A)\right).$$

Definition 5.6 1) *If* $\Delta = ()^{\circ}$, *(resp.* $()^{-\circ} \cup ()^{\circ -}, ()^{-\circ -})\ ()^{\circ - \circ}, ()^{\circ -}, ()^{-\circ}$, *the notion of fuzzifying* Δ*-interior of* $A \subseteq X$ *coincides with the notion of fuzzifying interior (resp.* α*-interior, semi-interior, pre-interior,* γ*-interior,* β*-interior) operation and will be denoted by*

$$A^{\circ}\left(\text{resp. } Int_{\alpha}(A), Int_{S}(A), Int_{P}(A), Int_{\gamma}(A), Int_{\beta}(A)\right);$$

2) If $\Delta = ()^{\circ - \circ}$ (resp. $()^{\circ -}, ()^{-\circ}, ()^{-\circ} \cup ()^{\circ -}, ()^{-\circ -}$), the notion of fuzzifying $C\Delta$-interior of $A \subseteq X$ coincides with the notion of fuzzifying $c\alpha$ (resp. csemi, cpre, $c\gamma$, $c\beta$)-interior operation and will be denoted by

$$Int_{C\alpha}(A)\left(\text{resp. } Int_{CS}(A), Int_{CP}(A), Int_{C\gamma}(A), Int_{C\beta}(A)\right).$$

Theorem 5.6 *For every* x, A, B *we have*

1) a) ‘ $B \doteq Int_{\Delta}(A) \to B \in \tau_{\Delta}$;

b) if Δ is of type 01 and monotone, then

$$‘\ B \doteq Int_{C\Delta}(A) \to B \in \tau_{N^{C\Delta}};$$

2) a) ‘ $B \in \tau_{\Delta} \wedge B \subseteq A \to B \subseteq Int_{\Delta}(A)$;

b) if Δ is of type 01 and monotone, then

$$‘\ B \in \tau_{N^{C\Delta}} \wedge B \subseteq A \to B \subseteq Int_{\Delta}(A);$$

3) a) ‘ $A \equiv Int_{\Delta}(A) \leftrightarrow A \in \tau_{\Delta}$;

b) if Δ is of type 01 and monotone, then

$$‘\ A \equiv Int_{C\Delta}(A) \leftrightarrow A \in \tau_{N^{C\Delta}};$$

4) a) ‘ $x \in Int_{\Delta}(A) \leftrightarrow x \in A \wedge x \in (X - d_{\Delta}(X - A));$

b) ‘ $x \in Int_{C\Delta}(A) \leftrightarrow x \in A \wedge x \in (X - d_{C\Delta}(X - A));$

5) a) ‘ $Int_{\Delta}(A) \equiv X - Cl_{\Delta}(X - A);$

b) ‘ $Int_{C\Delta}(A) \equiv X - Cl_{C\Delta}(X - A);$

Proof. We prove only a) of each statements since b) is similar.

First, we prove 2) a) If $B \not\subseteq A,$ then $[B \in \tau_{\Delta} \wedge B \subseteq A] = 0.$ Now, suppose that $B \subseteq A.$ Then we have

$$
\begin{aligned}
&[B \subseteq Int_{\Delta}(A)] = \inf_{x \in B} Int_{\Delta}(A)(x) = \inf_{x \in B} N_x^{\Delta}(A) \\
&\geq \inf_{x \in B} N_x^{\Delta}(B) = \tau_{\Delta}(B) = [B \in \tau_{\Delta} \wedge B \subseteq A].
\end{aligned}
$$

3) a) $[A \equiv Int_{\Delta}(A)]$

$$
\begin{aligned}
&= \min\left(\inf_{x \in A} Int_{\Delta}(A)(x), \inf_{x \in A}(1 - Int_{\Delta}(A)(x))\right) \\
&= \inf_{x \in A} Int_{\Delta}(A)(x) = \inf_{x \in A} N_x^{\Delta}(A) \\
&= \tau_{\Delta}(A) = [A \in \tau_{\Delta}].
\end{aligned}
$$

4) a) If $x \notin A,$

$$
[x \in Int_{\Delta}(A)] = 0 = [x \in A \wedge x \in (X - d_{\Delta}(X - A))].
$$

If $x \in A,$

$$
\begin{aligned}
[x \in d_{\Delta}(X - A)] &= 1 - N_x^{\Delta}(A \cup \{x\}) \\
&= 1 - N_x^{\Delta}(A) = 1 - Int_{\Delta}(A)(x).
\end{aligned}
$$

Thus $\left[x\in A\wedge x\in\left(X-d_{\Delta}\left(X-A\right)\right)\right]=Int_{\Delta}\left(A\right)\left(x\right).$

5) a) Follows from Theorem 5.3 (1).

Finally, we prove 1) a). From 5) a) and Theorem 5.5 1) we have

$$\begin{aligned}\left[B\doteq Int_{\Delta}\left(A\right)\right]&=\left[X-B\doteq Cl_{\Delta}\left(X-A\right)\right]\\&\leq\left[\left(X-B\right)\in\hat{o}_{\Delta}\right]=\tau_{\Delta}\left(B\right).\end{aligned}$$

Corollary 5.6 *For any* $x, A, B,$

1) a) ‘ $B\doteq A^{\circ}\to B\in\tau$;

b) ‘ $B\doteq Int_{\alpha}\left(A\right)\to B\in\tau_{\alpha}$;

c) ‘ $B\doteq Int_{S}\left(A\right)\to B\in\tau_{S}$;

d) ‘ $B\doteq Int_{P}\left(A\right)\to B\in\tau_{P}$;

e) ‘ $B\doteq Int_{\gamma}\left(A\right)\to B\in\tau_{\gamma}$;

f) ‘ $B\doteq Int_{\beta}\left(A\right)\to B\in\tau_{\beta}$;

g) ‘ $B\doteq Int_{C\alpha}\left(A\right)\to B\in\tau_{N^{C\alpha}}$;

h) ‘ $B\doteq Int_{CS}\left(A\right)\to B\in\tau_{N^{CS}}$;

i) ‘ $B\doteq Int_{CP}\left(A\right)\to B\in\tau_{N^{CP}}$;

j) ‘ $B\doteq Int_{C\gamma}\left(A\right)\to B\in\tau_{N^{C\gamma}}$;

k) ‘ $B\doteq Int_{C\beta}\left(A\right)\to B\in\tau_{N^{C\beta}}$;

2) a) ‘ $B\in\tau\wedge B\subseteq A\to B\subseteq A^{\circ}$;

b) ‘ $B\in\tau_{\alpha}\wedge B\subseteq A\to B\subseteq Int_{\alpha}\left(A\right)$;

c) ‘ $B \in \tau_S \wedge B \subseteq A \to B \subseteq Int_S(A)$;

d) ‘ $B \in \tau_P \wedge B \subseteq A \to B \subseteq Int_P(A)$;

e) ‘ $B \in \tau_\gamma \wedge B \subseteq A \to B \subseteq Int_\gamma(A)$;

f) ‘ $B \in \tau_\beta \wedge B \subseteq A \to B \subseteq Int_\beta(A)$;

g) ‘ $B \in \tau_{N^{C\alpha}} \wedge B \subseteq A \to B \subseteq Int_\alpha(A)$;

h) ‘ $B \in \tau_{N^{CS}} \wedge B \subseteq A \to B \subseteq Int_S(A)$;

i) ‘ $B \in \tau_{N^{CP}} \wedge B \subseteq A \to B \subseteq Int_P(A)$;

j) ‘ $B \in \tau_{N^{C\gamma}} \wedge B \subseteq A \to B \subseteq Int_\gamma(A)$;

k) ‘ $B \in \tau_{N^{C\beta}} \wedge B \subseteq A \to B \subseteq Int_\beta(A)$;

3) a) ‘ $A \equiv A^\circ \leftrightarrow A \in \tau$;

b) ‘ $A \equiv Int_\alpha(A) \leftrightarrow A \in \tau_\alpha$;

c) ‘ $A \equiv Int_S(A) \leftrightarrow A \in \tau_S$;

d) ‘ $A \equiv Int_P(A) \leftrightarrow A \in \tau_P$;

e) ‘ $A \equiv Int_\gamma(A) \leftrightarrow A \in \tau_\gamma$;

f) ‘ $A \equiv Int_\beta(A) \leftrightarrow A \in \tau_\beta$;

g) ‘ $A \equiv Int_{C\alpha}(A) \leftrightarrow A \in \tau_{N^{C\alpha}}$;

h) ‘ $A \equiv Int_{CS}(A) \leftrightarrow A \in \tau_{N^{CS}}$;

i) ‘ $A \equiv Int_{CP}(A) \leftrightarrow A \in \tau_{N^{CP}}$;

j) $\vDash A \equiv Int_{C\gamma}(A) \leftrightarrow A \in \tau_{N^{C\gamma}}$;

k) $\vDash A \equiv Int_{C\beta}(A) \leftrightarrow A \in \tau_{N^{C\beta}}$;

4) a) $\vDash x \in A^{\circ} \leftrightarrow x \in A \wedge x \in \left(X - (X - A)'\right)$;

b) $\vDash x \in Int_{\alpha}(A) \leftrightarrow x \in A \wedge x \in \left(X - d_{\alpha}(X - A)\right)$;

c) $\vDash x \in Int_{S}(A) \leftrightarrow x \in A \wedge x \in \left(X - d_{S}(X - A)\right)$;

d) $\vDash x \in Int_{P}(A) \leftrightarrow x \in A \wedge x \in \left(X - d_{P}(X - A)\right)$;

e) $\vDash x \in Int_{\gamma}(A) \leftrightarrow x \in A \wedge x \in \left(X - d_{\gamma}(X - A)\right)$;

f) $\vDash x \in Int_{\beta}(A) \leftrightarrow x \in A \wedge x \in \left(X - d_{\beta}(X - A)\right)$;

g) $\vDash x \in Int_{C\alpha}(A) \leftrightarrow x \in A \wedge x \in \left(X - d_{C\alpha}(X - A)\right)$;

h) $\vDash x \in Int_{CS}(A) \leftrightarrow x \in A \wedge x \in \left(X - d_{CS}(X - A)\right)$;

i) $\vDash x \in Int_{CP}(A) \leftrightarrow x \in A \wedge x \in \left(X - d_{CP}(X - A)\right)$;

j) $\vDash x \in Int_{C\gamma}(A) \leftrightarrow x \in A \wedge x \in \left(X - d_{C\gamma}(X - A)\right)$;

k) $\vDash x \in Int_{C\beta}(A) \leftrightarrow x \in A \wedge x \in \left(X - d_{C\beta}(X - A)\right)$;

5) a) $\vDash A^{\circ} \equiv X - \overline{X - A}$;

b) $\vDash Int_{\alpha}(A) \equiv X - Cl_{\alpha}(X - A)$;

c) $\vDash Int_{S}(A) \equiv X - Cl_{S}(X - A)$;

d) $\vDash Int_{P}(A) \equiv X - Cl_{P}(X - A)$;

e) $\vDash Int_{\gamma}(A) \equiv X - Cl_{\gamma}(X - A)$;

f) ‘ $Int_{\beta}(A) \equiv X - Cl_{\beta}(X - A)$;

g) ‘ $Int_{C\alpha}(A) \equiv X - Cl_{C\alpha}(X - A)$;

h) ‘ $Int_{CS}(A) \equiv X - Cl_{CS}(X - A)$;

i) ‘ $Int_{CP}(A) \equiv X - Cl_{CP}(X - A)$;

j) ‘ $Int_{C\gamma}(A) \equiv X - Cl_{C\gamma}(X - A)$;

k) ‘ $Int_{C\beta}(A) \equiv X - Cl_{C\beta}(X - A)$.

6. CONCLUSIONS

The present paper investigates topological notions when these are planted into the framework of Ying's fuzzifying topological spaces (in semantic method of continuous valued-logic).

It continue various investigations into fuzzy topology in a legitimate way and extend some fundamental results in general topology to fuzzifying topology. An important virtue of our approach (in which we follow Ying) is that we define topological notions as fuzzy predicates (by formulae of Łukasiewicz fuzzy logic) and prove the validity of fuzzy implications (or equivalences). Unlike the (more wide-spread) style of defining notions in fuzzy mathematics as crisp predicates of fuzzy sets, fuzzy predicates of fuzzy sets provide a more genuine fuzzification; furthermore the theorems in the form of valid fuzzy implications are more general than the corresponding theorems on crisp predicates of fuzzy sets.

The main contributions of the paper are to study some sorts of operations, called general fuzzifying operations. There are some problems for further study:

1) Apply the general fuzzifying operation to convergence theory, continuity, separation axioms etc.

Osama Rashed Sayed (2, L) topologies.

REFERENCES

1. U. Höhle, "Many Valued Topology and Its Applications," Kluwer Academic Publishers, Dordrecht, 2001.
2. U. Höhle and S. E. Rodabaugh, "Mathematics of Fuzzy Sets: Logic, Topology, and Measure Theory," Handbook of Fuzzy Sets Series, Vol. 3, Kluwer Academic Publishers, Dordrecht, 1999.
3. U. Höhle, S. E. Rodabaugh and A. Ŝostak, "Special Issue on Fuzzy Topology," Fuzzy Sets and Systems, Vol. 73, 1995, pp. 1-183.
4. T. Kubiak, "On Fuzzy Topologies," Ph.D. Thesis, Adam Mickiewicz University, Poznan, 1985.
5. Y. M. Liu and M. K. Luo, "Fuzzy Topology," World Scientific, Singapore, 1998.
6. G. J. Wang, "Theory of L-Fuzzy Topological Spaces," Shaanxi Normal University Press, Xi'an, 1988 (in Chinese).
7. C. L. Chang, "Fuzzy Topological Spaces," Journal of Mathematical Analysis and Applications, Vol. 24, No. 1, 1968, pp. 182-190.
8. J. A. Goguen, "The Fuzzy Tychonoff Theorem," Journal of Mathematical Analysis and Applications, Vol. 43, No. 3, 1973, pp. 182-190.
9. J. L. Kelley, "General Topology," Van Nostrand, New York, 1955.
10. U. Höhle, "Uppersemicontinuous Fuzzy Sets and Applications," Journal of Mathematical Analysis and Applications, Vol. 78, No. 2, 1980, pp. 659-673.
11. U. Höhle and A. Ŝostak, "Axiomatic Foundations of Fixed-Basis Fuzzy Topology," In: U. Höhle and S. E. Rodabaugh, Eds., Mathematics of Fuzzy Sets: Logic, Topology, and Measure Theory, Vol. 3, Kluwer Academic Publishers, Dordrecht, 1999, pp. 123-272.
12. S. E. Rodabaugh, "Categorical Foundations of VariableBasis Fuzzy Topology," In: U. Höhle and S. E. Rodabaugh, Eds., Mathematics of Fuzzy Sets: Logic, Topology, and Measure Theory, Vol. 3, Kluwer Academic Publishers, Dordrecht, 1999, pp. 273-388.
13. J. B. Rosser and A. R. Turquette, "Many-Valued Logics," North-Holland, Amsterdam, 1952.
14. M. S. Ying, "A New Approach for Fuzzy Topology (I)," Fuzzy Sets and Systems, Vol. 39, No. 3, 1991, pp. 303- 321.
15. M. S. Ying, "A New Approach for Fuzzy Topology (II)," Fuzzy Sets and Systems, Vol. 47, No. 2, 1992, pp. 221- 223.
16. K. M. Abd El-Hakeim, F. M. Zeyada and O. R. Sayed, "β-Continuity and D(c, β)-Continuity in Fuzzifying Topology," The Journal of Fuzzy Mathematics, Vol. 7, No. 3, 1999, pp. 547-558.
17. K. M. Abd El-Hakeim, F. M. Zeyada and O. R. Sayed, "Pre-Continuity and D(c, P)-Continuity in Fuzzifying Topology," Fuzzy Sets and Systems, Vol. 119, No. 3, 2001, pp. 459-471.
18. F. H. Khedr, F. M. Zeyada and O. R. Sayed, "Fuzzy Semi-Continuity and Fuzzy Csemi-Continuity in Fuzzifying Topology," The Journal of Fuzzy Mathematics, Vol. 7, No. 1, 1999, pp. 105-124.
19. F. H. Khedr, F. M. Zeyada and O. R. Sayed, "α-Continuity and $c\alpha$-Continuity in Fuzzifying Topology," Fuzzy Sets and Systems, Vol. 116, No. 3, 2000, pp. 325- 337.
20. T. Noiri and O. R. Sayed, "Fuzzy $c\gamma$ Open Sets and Fuzzy $c\gamma$-Continuity in Fuzzifying Topology," International Journal of Mathematics and Mathematical Sciences, Vol. 31, No. 1, 2002, pp. 51-63.
21. T. Noiri and O. R. Sayed, "Fuzzy γ Open Sets and Fuzzy γ-Continuity in Fuzzifying Topology," Scientiae Mathematicae Japonicae, Vol. 55, No. 2, 2002, pp. 255- 263.
22. S. Kasahara, "Operation-Compact Spaces," Mathematica Japonica, 24, No. 1, 1979, pp. 97-105.
23. D. S. Jankovic', "Properties of $\alpha\beta$-Continuous Functions," The Proceedings of Fifth Prague Topological Symposium, 1981.

24. M. E. Abd El-Monsef, F. M. Zeyada and A. S. Mashhour, "Operations on Topologies and Its Applications on Some Types of Covering," Annales de la Société Scientifique de Bruxelles, Vol. 79, 1983, pp. 155-172.
25. E. E. Kerre, A. A. Nouh and A. Kandil, "Operations and the Class of Fuzzy Sets on a Universe Endowed with a Fuzzy Topology," Proceedings of IFSA, Vol. 109-113, Brussels, 1991.
26. Kandil, E. E. Kerre and A. A. Nouh, "Operations and Mappings on Fuzzy Topological Spaces," Annales de la Société Scientifique de Bruxelles, Vol. 105, No. 4, 1991, pp. 167-188.

CHAPTER 18

An Alternative Manifold for Cosmology Using Seifert Fibered and Hyperbolic Spaces

Maria E. Mejía[1], Reinaldo R. Rosa[2]

[1]*Departamento de Matemáticas, Universidad de Guadalajara, Guadalajara, México*

[2]*Laboratório Associado de Computação e Matemática Aplicada, Instituto Nacional de Pesquisas Espaciais, São José dos Campos, Brasil*

ABSTRACT

We propose a model with 3-dimensional spatial sections, constructed from hyperbolic cusp space glued to Seifert manifolds which are in this case homology spheres. The topological part of this research is based on Thurston's conjecture which states that any 3-dimensional manifold has a canonical decomposition into parts, each of which has a particular geometric structure. In our case, each part is either a Seifert fibered or a cusp hyperbolic space. In our construction we remove tubular neighbourhoods of singular orbits in areas of Seifert fibered manifolds using a splice operation and replace each with a cusp hyperbolic space. We thus achieve elimination of all singularities, which appear in the standard-like cosmological models, replacing them by "a torus to infinity". From this construction, we propose an alternative manifold for cosmology with finite volume and without Friedmann-like singularities. This manifold was used for calculating coupling constants. Obtaining in this way a theoretical explanation for fundamental forces is at least in the sense of the hierarchy.

Keywords: Topology, Cosmology, Thurston's Theory, Singularity-Free

1. INTRODUCTION

In his field equations, Einstein just provided a compact mathematical tool that made it possible to develop a theory to describe the general configuration of matter-energy and space-time for the Universe. Nevertheless, the questions related to the global shape of space and, in particular, its finite or infinite extension, cannot be fully answered by the General Relativity (a local physical theory). That is why, the use of topology (a global mathematical theory) is necessary to construct a new manifold which could answer some of those questions. Accordingly, many topological "alternatives" of three-dimensional spaces can thus be used to build such cosmological universe models.

The investigation of topological changes may bring a new approach to studying the early cosmology of the universe [1] [2] . The problem of changing topology has been discussed from different points of view in both classical and quantum gravity [3] [4] . In addition, Gauge theory [5] provides an alternative to solve some problems that general relativity can not.

Efremov [6] proposed a manifold which admits 3-dimensional sections of Euclidean space-time cobordisms describing topology changes. In this work we take the Seifert Fibered Spaces represented by homologic spheres and we improve the idea of Efremov [6] [7] and Mitskievich [8] [9] adding hyperbolic cusp spaces. We describe the process of construction of a cosmological model which was used to calculate the coupling constants for fundamental forces as shown in [10] .

We like to point out that our Model of Space-Time does not have singular orbits (those are obtained by factorization with respect to the action group S^1) because they have been replaced by an hyperbolic manifold. Also, comparing the coupling constants obtained in [6] with the new ones, we can see that the new model has a better fit with those obtained by experiments (see [10]).

Other works which can be a motivation for using hyperbolic manifolds in physics are: [11] -[14] . However, compared to the work presented here, their approaches are in a very different way because they are limited to a decomposition of Seifert Fibered Spaces and do not use Thurston's theory for the hyperbolic part.

First we make the decomposition of Seifert Fibered Spaces in prime spaces. This process includes a collection of cuts of a 3-dimensional space M^3 in two parts along 2-dimensional spheres such that the space M^3 is separated into two parts and neither of these is a 3-dimensional disc. This operation is the inverse of the connected sum. The result of gluing two of 3-dimensional disc to each component of the boundary is to obtain a simpler 3-manifold every time we do the process. Kneser [15] proved that such process is terminated after a finite number of steps. We call prime components of M^3 as each of the parts that were obtained as a result of this decomposition process. They are solely defined except by homeomorphisms, see [16] .

After that, we involve the decomposition of cuts along a torus. The reverse operation to splice described in [17] , has been discovered by Johanson [18] , Jaco and Shalen [19] . This decomposition along the torus $(T_i^2 \subset M^3)$ gives a set of 3-manifolds of which some are Seifert Fibered manifolds and others allow hyperbolic structures. Previous works [6] and [9] , considered a special case. Namely, a 3-dimensional space contains only Seifert fibered spaces represented by homological spheres as parts.

Here, we extend those ideas using Thurston's conjecture (now proved) [20] -[22] and the work of Jaco [19] [23] and Johanson [18] . This allows glueing a suitable hyperbolic space in Seifert fibered space and taking the Seyfert hyperbolic invariants inside of a hyperbolic space.

2. INTRODUCING SEIFERT MANIFOLDS AND HYPERBOLIC SPACES

There are several definitions in literature about Seifert Manifolds, here we introduce our own definition of Seifert Manifolds. For hyperbolic spaces we will use Thurston's theory. Then we will use all those concepts for the construction of a new Manifold representing the cosmological Universe.

Seifert Manifolds

Definition 1. A Seifert Fibered Manifold (SFM) is an orientable three-manifold M , M is the collection $\{C_\alpha\}$ of pairwise disjoint simple closed curves, called fibers, such that for each α there is a closed neighbourhood V of $C_\alpha \in M$ which is a solid torus and there is a covering map $p: D^2 \times S^1 \to V$ such that:

1) p maps each $x \times S^1, (x \in D^2)$ to C_β.

2) $p^{-1}(C_\alpha)$ are connected.

3) The covering translation group is generated by $\tau_{a,b}$ where

$$\tau_{a,b}\left(re^{i\theta},e^{i\phi}\right)=\left(re^{i\left(\theta+\frac{2a\pi}{b}\right)},e^{i\left(\phi+\frac{2a\pi}{b}\right)}\right).$$

Definition 2. If M is a 3-compact manifold and M has a Seifert fibration, there is a smooth map $f:M\to S$, S surface with the following property. Each $p\in S$ has a closed neighbourhood U such that, the restriction f to $f^{-1}(U)$ is a diffeomorphism from Seifert fibration to a solid torus. To be more precise, there are relatively prime integers p and q, with $0\le q<p$ and the diffeomorphisms $g:U\to D^2$ and $h:f^{-1}(U)\to S^1\times D^2$ such that the diagram

$$\begin{array}{ccc} f^{-1}(U) & \underset{h}{\rightarrow} & S^1\times D^2 \\ \downarrow f_{|f^{-1}(U)} & & \downarrow f_{p,q} \\ U & \underset{g}{\rightarrow} & D^2 \end{array}$$

is commutative.

Applying the previous definitions. Consider the trivial fibered cylinder $D^2\times I$, with fibers1 $\{x\}\times I$. Identifying cylinder caps and making a rotation $\frac{2\pi p}{q}$, $(p,q\in\mathbb{Q})$ to a cylinder we obtain a fibered torus, see Figure 1.

The reader can also see [23] .

See Appendix for other useful definitions.

3. HYPERBOLIC SPACES

In his notes, Thurston [21] describes different models for hyperbolic spaces. However we only take here the upper-half space model. We believe that this particular hyperbolic space has the necessary geometry in order to fulfil the requirements of the physical space. It has itself differents approaches, then in physics we want to be one part flat. In this space inasmuch we can find the ratio of the Euclidean distance. Also has the useful characteristic of having points at infinity.

The upper half-space model (taked from [21]).

This is closely related to the Poincaré disc model and the southern hemisphere, but it is often more convenient for computation or for constructing graphics. To obtain the upper-half space model, rotate the sphere S^n in $\mathbb{R}^{n+1}$ so that the southern hemisphere lies in the half-space $x_n \geq 0$ is $\mathbb{R}^{n+1}$. Now, the stereographic projection from the top of S^n (which is now on the equator) sends the southern hemisphere to the upper half-space in $\mathbb{R}^{n+1}$ As shown in Figure 2.

A hyperbolic line, in the upper half-space, is a circle perpendicular to the bounding plane $\mathbb{R}^{n+1} \subset \mathbb{R}^n$.

The hyperbolic metric is:

$$ds^2 = \left(\frac{1}{x_n}\right)^2 dx^2.$$

This representation of a hyperbolic space is suitable in physics because we have two things. A manifold with finite volume [24] and the Euclidean image of a hyperbolic object moving toward $\mathbb{R}^{n-1}$ has size precisely proportional to the Euclidean distance from $\mathbb{R}^{n-1}$.

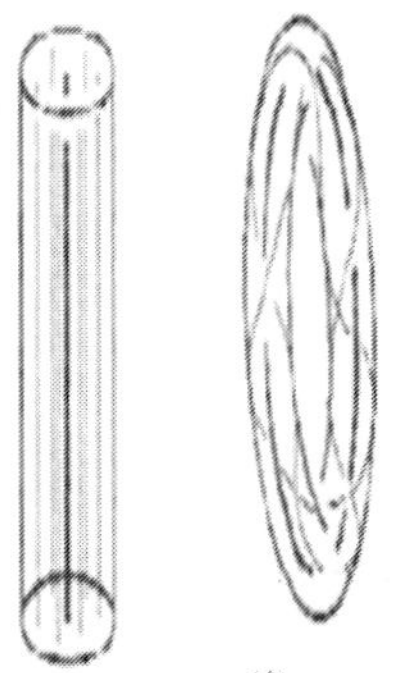

1(a) 1(b)

Figure 1. (a) Represents the cilinder with fibres and 1(b) is the fibered torus that will appear in our construction.

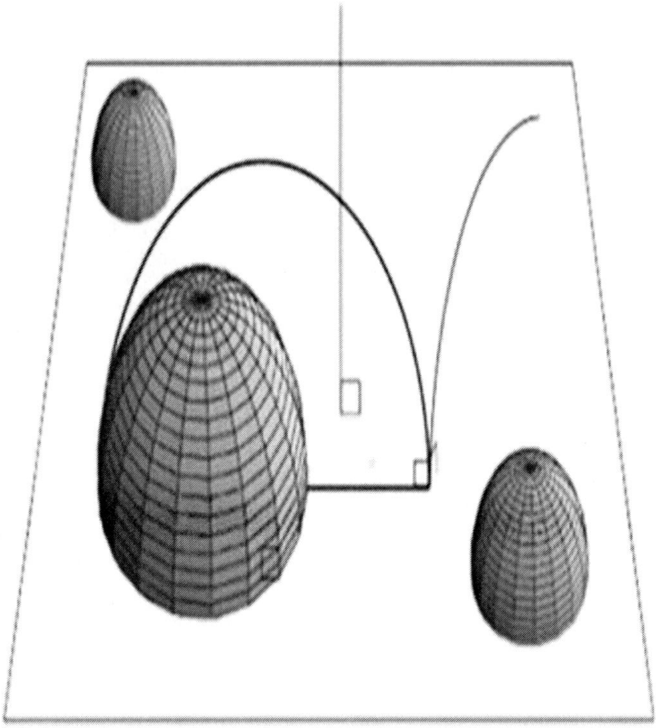

Figure 2. Upper half space model.

Here we present the triangulation of the Seifert spheres and hyperbolic spaces in order to preserve the orientation of manifolds. Those are used in [10] in order to calculate a discrete volume. After this we use the Eisenbud ideas for manifold decompositions and gluing. See [17] for a complete description of links, splice operation and their characteristics [25] . Here we only need highlight the following theorem.

Theorem 1. Jaco-Shalen, Johanson Torus Decom-position.

If M is an oriented, compact, irreducible 3-manifold, then there is a finite collection of incompressible disjoint torus $T_1,\cdots,T_n$ in M such that splitting M along the union of those torus, produces manifolds M_i that are either Seifert or atoroidal (every incompressible torus in M_i is isotopic to a component toroidal of ∂M_i). Moreover, a minimal collection of such torus T_i is unique under isotopy.

This theorem together with the standardization of Thurston's theorem states: there is a collection of incompressible tori $W \subset M$, unique under isotopy ambient, which cut M in Seifert fibered manifolds and hyperbolic manifolds of finite volume. See [26] .

In the next sections we will do the triangulation of Seifert fibered spaces as homology spheres as well as the hyperbolic space. But first, we see some definitions and constructions that will be helpful.

4. SFH-SPHERES, DERIVED AND PRIMARY SEQUENCES

Definition 3. We consider homology Brieskorn spheres $\Sigma(a_1,a_2,a_3)$, which are a special case of the SFHspheres (Seifert fibered homology spheres), which contain only three exceptional orbits. These SFH-spheres are the intersection of the surface

$$z_1^{a_1} + z_2^{a_2} + z_3^{a_3} = 0, \qquad (z_i \in \mathbb{C}_i) \tag{1}$$

in a 3-complex dimensional space with the 5-dimensional unit sphere

$$|z_1|^2 + |z_2|^2 + |z_3|^2 = 1. \tag{2}$$

If a_1, a_2, a_3 are relatively prime and if $a_i > 1$ then there is a unique Seifert fibration with Seiferts invariant not normalized

$$(a_1,b_1);(a_2,b_2);(a_3,b_3)$$

subject to

$$\sum_{i=1}^{3} b_i\sigma_i = 1 \tag{3}$$

where $\sigma_i = a / a_i$ and $a = a_1 a_2 a_3$.

It is possible to represent this relationship by the Euler number of the Seifert fibration of SFH-sphere $\Sigma(\underline{a}) := \Sigma(a_1, a_2, a_3)$ [27].

$$e(\Sigma(\underline{a})) = \sum_{i=1}^{3} b_i / a_i = 1/a \tag{4}$$

which is a topological invariant of $\Sigma(\underline{a})$. The SFH-sphere $\Sigma(\underline{a})$ may also be described as a cyclic cover $a_3 - \text{folds}$ of S^3, branched along the torus knot of type $K(a_1, a_2)$ [28].

4.1. SFH-Spheres Derived

Given a SFH-sphere $\Sigma(\underline{a}) := \Sigma(a_1, a_2, a_3)$ we define the derivative (in the sense that a SFH-sphere is obtained from another SFH-sphere) of a SFH-sphere, as follows:

$$\Sigma^{(1)}(\underline{a}) := \Sigma^{(1)}(a_1, a_2, a_3) := \Sigma(a_1, \sigma_1, a+1) \equiv \Sigma\left(a_1^{(1)}, a_2^{(1)}, a_3^{(1)}\right) = \Sigma^{(1)}\left(\underline{a}^{(1)}\right), \tag{5}$$

and the Seifert invariants are:

$$\left(a_i^{(1)}, b_i^{(1)}\right) = \{(a_1, b_1), (\sigma_1, \delta_1), (a+1, -1)\}, \tag{6}$$

where $\delta_1 = b_2 a_3 + b_3 a_2$, $\sigma_1 = \frac{a}{a_1}$, $a = a_1 a_2 a_3$.

It is easy to prove that the manifold in (5) with invariants (6) is a homology sphere because its Euler number satisfy (4). We have:

$$e\left(\Sigma^{(1)}(\underline{a})\right) = \sum_{i=1}^{3} \frac{b_i^{(1)}}{a_i^{(1)}} = \frac{1}{a^{(1)}},$$

where $a^{(1)} = a_1^{(1)} a_2^{(1)} a_3^{(1)}$ and $a_1^{(1)}, a_2^{(1)}, a_3^{(1)}$ are coprime.

Since any SFH-sphere has unique Seifert fibration, the meaning of derivative of the sphere coincides with the derived homology structure of the Seifert fibered space (i.e. acquires characteristics of SFH-spheres).

Definition 4. By induction, we define the derivative $\Sigma^{(\ell)}(\underline{a}) = \Sigma\left(a_1^{(\ell)}, a_2^{(\ell)}, a_3^{(\ell)}\right)$, of $\Sigma(\underline{a})$ for any order ℓ with a recurrence relation,

$$a^{(\ell)} = a^{(\ell-1)}\left(a^{(\ell-1)} + 1\right). \tag{7}$$

In particular (7) holds for the product of three Seifert invariants $a^{(\ell)} = a_1^{(\ell)} a_2^{(\ell)} a_3^{(\ell)}$.

Definition 5. We define here a sequence of SFH-spheres that we call a primary sequence. Let p_i be a prime number which is the i-th in the set of positive integers $\mathbb{N}$, e.g. $p_1 = 2, p_2 = 3, \cdots, p_9 = 23, p_{10} = 29, \cdots$.

The primary sequence of the SFH-sphere is defined as:

$$\left\{\Sigma\left(q_i, p_{i+1}, p_{i+2}\right) \middle| i \in \mathbb{N}\right\} \tag{8}$$

where $q_i = p_1 \cdots p_i$.

Included (for our model) in this sequence, as its first two terms, ordinary three-dimensional spheres S^3 with Seifert fibration determined by mappings:

$$h_{pq} : S^3 \to S^2$$

which are defined as

$$h_{pq}\left(z_1, z_2\right) = z_1^p \,/\, z_2^q, \quad p, q \in \mathbb{N}. \tag{9}$$

Recall

$$S^3 = \left\{\left(z_1, z_2\right) \in \mathbb{C}^2 \middle| \left|z_1\right|^2 + \left|z_2\right|^2 = 1\right\}, \tag{10}$$

and

$$z_1^p / z_2^q \in \mathbb{C} \cup \infty \cong S^2. \tag{11}$$

The mapping h_{pq}, describes the sphere S^3 with the structure of the Seifert fibration onto an orbifold $S^2(p,q)$, isomorphic to the sphere S^2 with two cone points characterized by the angles $\frac{2\pi}{p}$ and $\frac{2\pi}{q}$.

We denote these 3-dimensional spheres (SF-spheres) as $\Sigma(1,1,2)$, $p=1, q=2$ and $\Sigma(1,2,3)$, $p=2, q=3.$ The Seifert fibration structure in each of these SH-spheres is determined by the numbers p and q but are not unique. Formally, we say that they are special cases of the SFH-spheres. In the notation we use an additional number in the first place of the notation corresponding to an arbitrary regular orbit. This allows us to find the derivative of the above Seifert fibrations $\Sigma(1,1,2)$, $\Sigma(1,2,3)$ and for all elements of the primary sequence using (5).

Now we will form the family of manifolds which belongs to the first-primary SFH-spheres and its derivatives:

Definition 6. Using (5) we define the derivative

$$\left\{\Sigma^{(\ell)}(q_{i-1}, p_i, p_{i+1}) \middle| i \in \overline{0,8}, l \in \overline{0,4}\right\}, \tag{12}$$

where $\overline{0,n}$ is an integer range from 0 to n. The spheres $\Sigma(1,1,2)$ and $\Sigma(1,2,3)$ that are included in this family are $i=0,1$ respectively $(q_{-1}=q_0=p_0=1)$.

Example: As an example of such derived sequence we present the sequence of the Poincaré homology sphere $\Sigma(p_1, p_2, p_3)=\Sigma(2,3,5)$.

$$\Sigma^{(1)}(2,3,5)=\Sigma(2,15,31)$$

$$\Sigma^{(2)}(2,3,5)=\Sigma(2,465,931)$$

$$\Sigma^{(3)}(2,3,5)=\Sigma(2,432915,865831)$$

$$\Sigma^{(4)}(2,3,5)=\Sigma(2,374831227365,749662454731)$$

4.2. SFH-Spheres Triangulation

Definition 7. Any smooth compact 3-dimensional manifold X admits a finite closed partition $\tau=\{T_1,\cdots,T_N\}$ which comprises closed 3-dimensional simplices (tetrahedra) T_i such that $dim(T_i \cap T_j)<3,\ \forall i,j=1,\cdots,N, i\neq j$. We will call this partition a c-triangulation (or canonical triangulation).

The reader can see [8] .

We first take S^2 and extract 3 discs that are tubular neighbourhoods of singular orbits, dividing $S^2(a_1,a_2,a_3)$ on 6 discs: 3 correspond to singular orbits and 3 for regular orbits. See Figure 3.

In Figure 4 we can see an antecedent of a graph.

We start our construction with a singular orbit S_i. We will see how to represent a regular orbit h in a boundary neighbourhood of a singular orbit

$$\partial TS_i = T^2.$$

Regular orbits are represented by

$$h=\sigma_i m + a_i l$$

where $\sigma_i = \frac{a}{a_i}.$

Figure 5 shows a T^2 triangulation represented as a rectangle with edges identified correspondingly to $\sigma_i=3$ and $a_i=2$ (as a particular case, but it can be generalized to any σ_i and a_i).

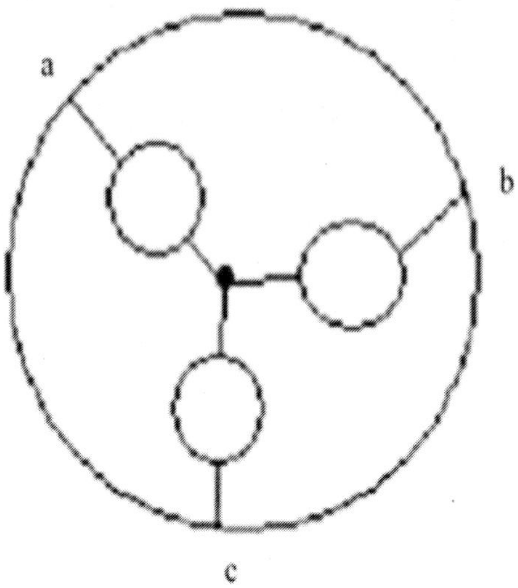

Figure 3. The orbits before removal.

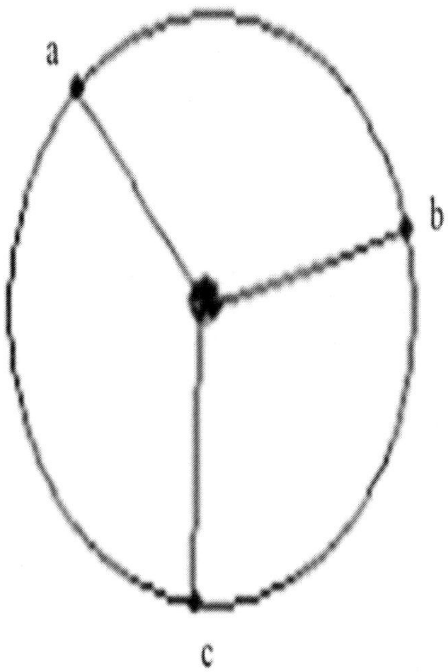

Figure 4. The orbits have been removed.

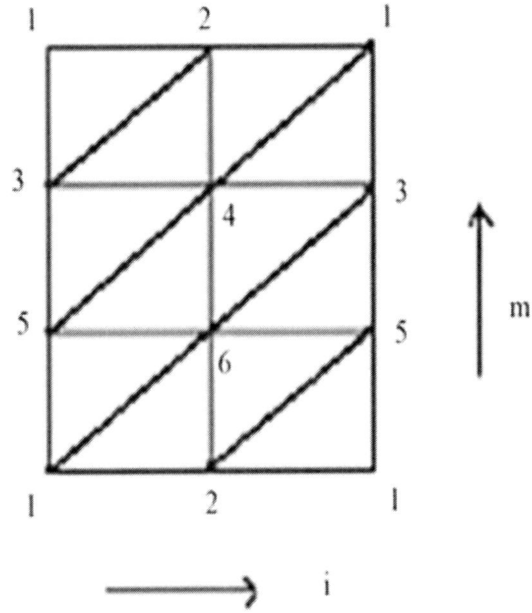

Figure 5. The diagonals marked in bold represent regular fibers.

To obtain tetrahedra from previously made triangulation we build on each one consisting of two adjacent triangles, three tetrahedra. Taking the first two rectangles on the rectangle 1,2,4,3 we do the following tetrahedra with vertices 3,2,4,7, 1,2,7,3 and 1,7,3,8.

The face (3,4,7,8) of the obtained triangular prism P_1 showed in Figure 6, must be identified (gluing) with the face (3,4,9,10) of a triangular prism P_2, which rises over 3,4,6,5-rectangle, thereby obtaining a central orbit.

Therefore the number of tetrahedra to triangular a toroidal neighbourhood of a singular orbit is:

$$N_{(\text{sing})} = 3a_1a_2a_3. \tag{13}$$

Now, consider a regular orbit S, the representation h of a regular orbit in ∂TS is,

$$h = l + am. \tag{14}$$

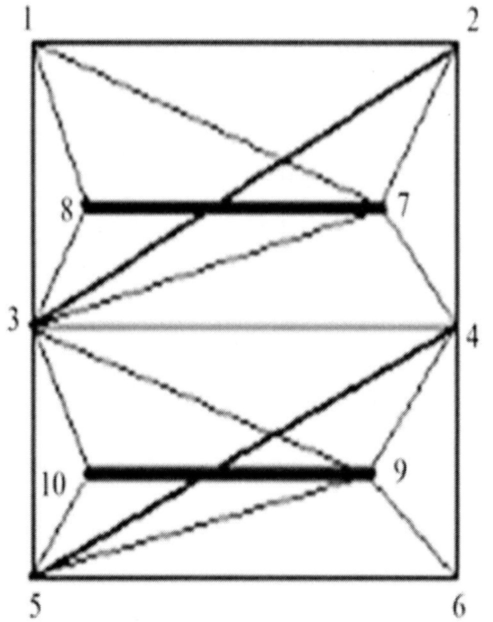

Figure 6. The lines marked in bold represent a central orbit.

As we can see in Figure 7.

In order to make the complete triangulation including both types of orbits we build on each rectangle consisting of two adjacent triangles, three tetrahedra. See Figure 8.

To have a central orbit we identify the line segment $\overline{bc}$ with $\overline{ad}$. By doing this, the face (e,c,f) of the tetrahedron d,e,c,f is identified with the opposite face. That is why we divide the tetrahedron in two parts. When we identify the faces of these tetrahedra, this induces the subdivision of a,b,c,e and a,c,d,e in each of the two tetrahedra.

4.3. Cusp Hyperbolic Space Triangulation

First we need to construct a cusp hyperbolic space with the following process.

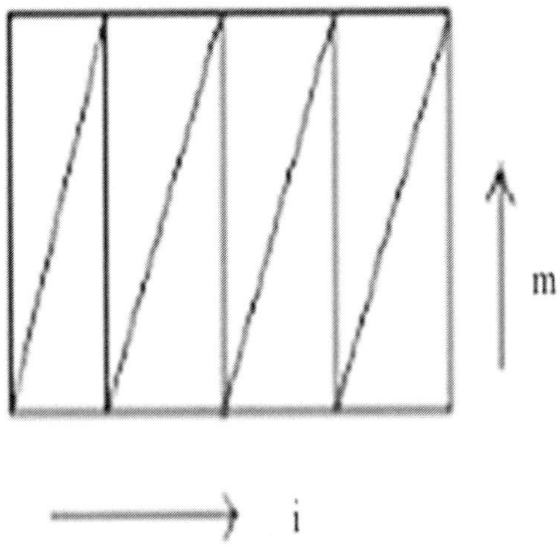

Figure 7. This is a T^2 triangulation, the diagonals of each rectangle represents a regular fiber h.

Take the 3-dimensional hyperbolic space $\mathbb{H}^3$ (each dimension of the 3-dimensional play the role of a spatial section) which is the upper half of $\mathbb{R}^3$ with

$$ds^2 = \frac{dx^2 + dy^2 + dz^2}{z^2} \tag{15}$$

and $\partial\mathbb{H}^3$ is identified with $\mathbb{C} \cup \{\infty\}$, the isometry group preserving orientation is $PSL(2,\mathbb{C})$. Hyperbolic planes are hemispheres and we have the concept of points at infinity. A fine feature of hyperbolic geometry is the fact that infinite objects can have finite volumes. This is the case of the ideal tetrahedron (one having one or more points at infinity). See [21] .

We now follow the Thurston geometrization conjecture [22] to construct our space.

We cut a manifold along essential spheres and tori. We take a connected component of the result of this process, completing spheres with balls and adding the "torus at infinity" in order to obtain the so-called "end" of the manifold. A graphical representation of this is the Figure 9.

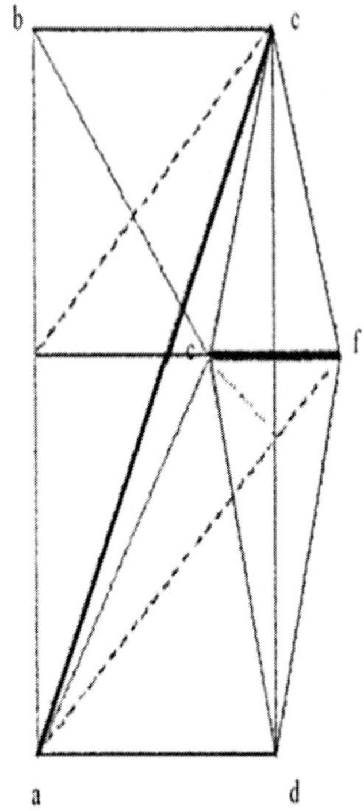

Figure 8. For example, the rectangle (a,b,c,d) to make the tetrahedra, b,c,e,a,c,d,e and d,e,c,f.

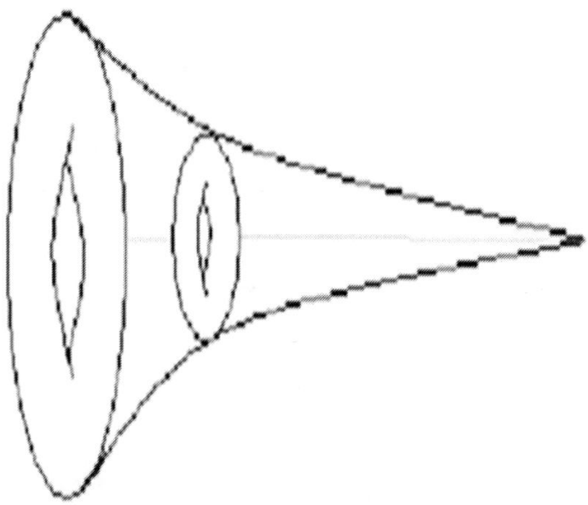

Figure 9. Cusp hyperbolic space.

Note: If we cut the cusp, we can visualize it as if we cut along a Euclidean plane in the upper half space model and see the previous movements associated to the toroidal boundary. This gives the eigenvalues $\lambda, \mu \in SL(2,\mathbb{C})$. From the last statement we have the abelian group $Z \oplus Z$. This is the acting group in the $\mathbb{H}^3$ manifold.

4.3.1. Cusp Hyperbolic Space

Here we construct our hyperbolic space with a single cusp and toroidal boundary.

Consider first the cube M which is composed as follows:

Gluing face A with A' and B to B' (following the direction of the arrows in the cube of the Figure 10), we obtain a cusp hyperbolic space and boundary homeomorphic to a 2-dimensional torus as we can see in Figure 11.

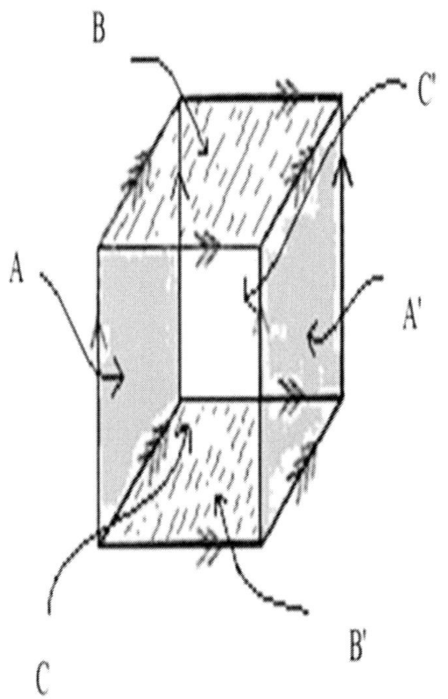

Figure 10. Where the C face has its vertices at infinity for $z \in \mathbb{C}$ and $z = 0$; the C' face has its vertices in the complex plane $z \in \mathbb{C}$ and $z = r$, for some $r \in \mathbb{R}$ and $r \neq 0$.

The above construction gives the definition of a cusp.

We need the two-dimensional torus, since in the next section we need to glue the boundary homeomorphically to a torus with either SF-sphere or SFH-sphere.

4.3.2. Cusp Hyperbolic Space Triangulation

We describe in this part a triangulation of a cusp hyperbolic space M, from a triangulation on the torus.

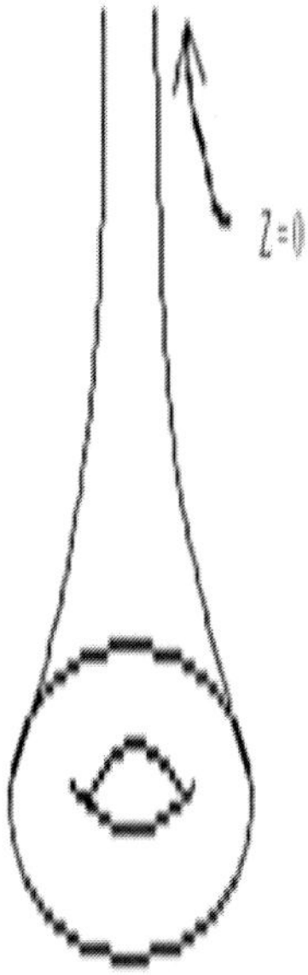

Figure 11. Cusp hyperbolic space with toroidal boundary.

Based on [8] and [9] (with the same ideology) do the following:

Take a partition from a toroidal boundary, with the number of rectangles $a_1a_2a_3$ (where $a_1a_2a_3$ are Seifert hyperbolic invariants).

The details are as follows:

We have first $a_1a_2 = 1$ and $a_3 = 1$, i.e. we have two triangles, then we will generalize in order to obtain a formula for the number of generated tetrahedra.

Take the rectangle with vertices $1,2,3,4$, in C'. We denote by η_i the points at infinity. Now we take the four points at infinity $\eta_1,\eta_2,\eta_3,\eta_4 \in C$ (we will call the resulting cube as Z_i) and construct the following tetrahedra $\eta_1,\eta_4,1,2$; $2,3,4,\eta_4$ $\eta_4,\eta_3,2,3$; $\eta_1,\eta_3,\eta_4,2$; $\eta_1,\eta_2,\eta_3,2$; and finally the tetrahedron formed by $1,2,4,\eta_4$. Similarly, as in SFH-triangulation we have the number of generated tetrahedra in this case 6 (see Figure 12).

As we need to generalize this process for $a_1a_2 = 2$ and $a_3 = 1$, we have, in this case, two glued cubes.

For the cube Z_i (showed in Figure 13) we have then 12 tetrahedra, namely $6a_1a_2a_3$.

Note that by gluing A with A' and B with B' the tetrahedra of the partition does not generate new vertices, since the figures have symmetrical form. Not having new vertices the gluing will not generate any more tetrahedra.

Generalizing this process for any a_1,a_2 and a_3, we have:

That is, in Figure 14 we see that for each cube Z_i we have 6 tetrahedra. If for example $a_1a_2 = 4$ and $a_3 = 3$ then we have 12 cubes, i.e. $12Z_i$ then we will have 72 tetrahedra. The following result is obtained by the above.

$$H = 6a_1a_2a_3 \qquad \text{tetrahedra.} \tag{16}$$

With this we triangulated our cusp hyperbolic space obtaining also the number of generated tetrahedra, given by H.

In Section 5 we will construct our model and we will use splice diagrams which are similar to graphs.

5. MODEL OF A UNIVERSE WITH A FUNDAMENTAL INTERACTION

In our topological model, each fundamental interaction is characterized by the following pair of parameters (n,ℓ), $(n,\ell \in \overline{0,4})$. These will be related to an assembly $E_{nh}^{(\ell)}$ of topological spaces composed of Seifert fibered spheres and a cusp hyperbolic space for each singular orbit. With the triangulations obtained in the preceding sections.

Each (n,ℓ) interaction is related to an assembly $E_{nh}^{(\ell)}$ of topological spaces $E_{nh}^{(\ell)}(\mathbb{R})$, which are interpreted as a set of allowable 3-spatial sections of the "universe" with one fundamental interaction. See [6] and [27] .

We begin by describing the assemblies $E_{nh}^{(\ell)}$. See [10] for the discrete volume.

We take a Seifert fibered sphere (SF-sphere), $\Sigma_0^{(\ell)}\left(1,a_{20}^{(\ell)},a_{30}^{(\ell)}\right)$ of the family

$$\left\{\Sigma_n^{(\ell)} := \Sigma^{(n-\ell)}\left(q_{2n-1}, p_{2n}, p_{2n+1}\right)\middle| n \in \overline{0,4}, \ell \in \overline{-4,4}\right\} \tag{17}$$

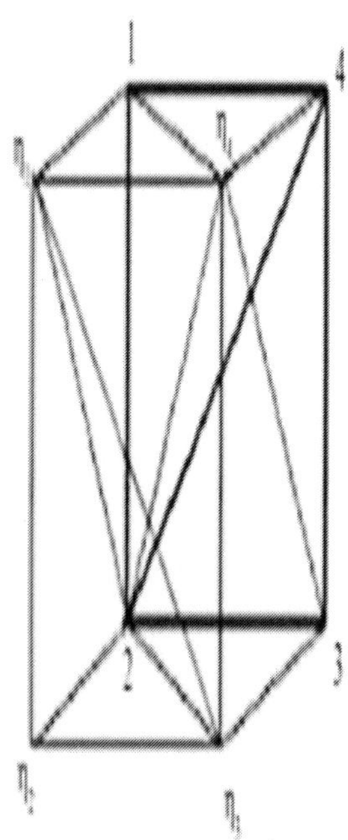

Figure 12. Note that each cube Z_i has inside five mixed tetrahedra, i.e. each tetrahedron may have only 2 or 3 or even 1 point at infinity.

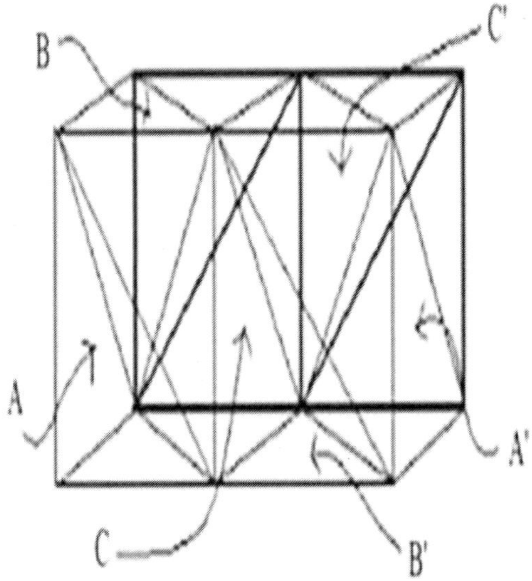

Figure 13. Where C has its vertices at infinity.

which contains two regular and two singular orbits as described in the Subsection 4.2 and we associate the following splice diagram.

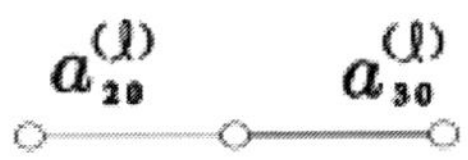

Notation: the second subscript denotes the level h, for $n = 0$, we have $a_{20}^{(\ell)}$ and $a_{30}^{(\ell)}$.

Removing the torus neighbourhoods of singular orbits and introducing the torus part of a cusp hyperbolic space by a splice operation (see [17]). Then we have constructed a new manifold denoted by $M_{0h}^{(\ell)}$

$$M_{0h}^{(\ell)} := h_0^{(\ell)} \underset{a_1^{(\ell)} \quad a_{20}^{(\ell)}}{\rule{2cm}{0.4pt}} \Sigma_0^{(\ell)} \underset{a_{30}^{(\ell)} \quad a_2^{(\ell)}}{\rule{2cm}{0.4pt}} h_0^{(\ell)} \tag{18}$$

and the corresponding splice diagram.

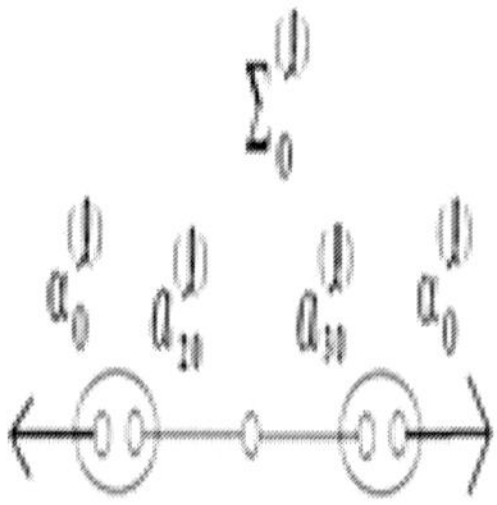

With this process we removed singular orbits and exchanged them by cusp hyperbolic spaces.

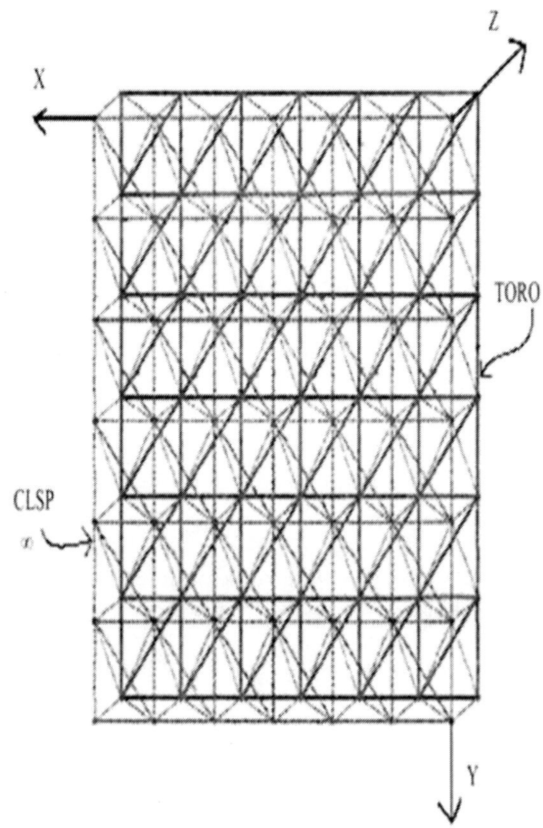

Figure 14. The cusp corresponds to $z=0$ and the back corresponds to $z=r$ for $z \in C$ and $r \in R, r \neq 0$.

Now, consider a SF-sphere $\Sigma_0^{(\ell)}\left(1,a_{20}^{(\ell)},a_{30}^{(\ell)}\right)$ containing two singular and two regular orbits (this is the zero level) and another SF-sphere $\Sigma_1^{(\ell)}\left(a_{10}^{(\ell)},a_{20}^{(\ell)},a_{30}^{(\ell)}\right)$, containing three singular and three regular orbits (this is level one).

In (18) we have glued two hyperbolic spaces, in order to glue in $\Sigma_0^{(\ell)}$. The end of $a_{20}^{(\ell)}$ with $\Sigma_1^{(\ell)}$; but in $a_{30}^{(\ell)}$ we must first remove the hyperbolic space $h_0^{(\ell)}$ and then we will glue $a_{30}^{(\ell)}$_ with $a_{21}^{(\ell)}$ applying a splice operation.

After that, we add two hyperbolic spaces (also level one) in the singular orbits $a_{31}^{(\ell)}$ and $a_{11}^{(\ell)}$ of level one SF-sphere. Here $\alpha_n := a_{1n}^{(\ell)},a_{2n}^{(\ell)},a_{3n}^{(\ell)}$.

For the construction of the assembly $\Sigma_{1h}^{(\ell)}$ we need to define the following manifold:

$$
\begin{array}{l}
M_{1h}^{(\ell)} := h_0^{(\ell)} \underset{a_1^{(\ell)}\,a_{20}^{(\ell)}}{\rule{2em}{0.4pt}} \Sigma_0^{(\ell)} \underset{a_{30}^{(\ell)}\,a_{21}^{(\ell)}}{\rule{2em}{0.4pt}} \Sigma_1^{(\ell)} \underset{a_{31}^{(\ell)}\,a_3^{(\ell)}}{\rule{2em}{0.4pt}} h_1^{(\ell)} \\
\qquad\qquad\qquad\qquad\qquad\quad \Big|\begin{smallmatrix} a_{11}^{(\ell)} \\ a_2^{(\ell)} \end{smallmatrix} \\
\qquad\qquad\qquad\qquad\qquad\quad h_1^{(\ell)}
\end{array}
$$

In other words, we are indicating that the splice operation was made in singular orbit S_3 of $\Sigma_0^{(\ell)}$(which corresponds to $a_{30}^{(\ell)}$ invariant), and $\Sigma_1^{(\ell)}$ (corresponding to $a_{21}^{(\ell)}$ invariant) as well as $a_{31}^{(\ell)}$ and $a_{11}^{(\ell)}$ invariants.

We remove one toroidal neighbourhood at each singular orbit and it is replaced by the corresponding $h_1^{(\ell)}$ space.

For this we associate the splice diagram below.

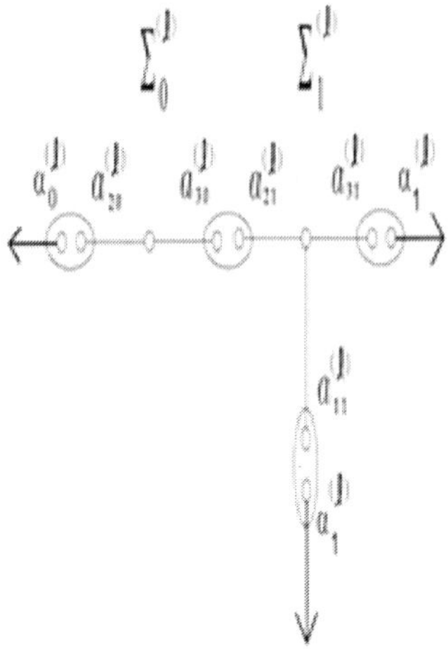

Then the assembly will be defined by:

$$E_{1h}^{(\ell)} := \left\{ M_{1h}^{(\ell)}(R) \middle| R \in \overline{0, V_{0h}^{(\ell)} - 2H_0^{(\ell)}} \right\}. \tag{19}$$

The $M_{1h}^{(\ell)}(R)$ components are obtained from $M_{0h}^{(\ell)}$ manifold using the connected sum R times between $M_{0h}^{(\ell)} \setminus TS_3$ sub manifold. Where TS_3 is the tubular neighbourhood of a singular orbit S_3 extracted from SFsphere $\Sigma_0^{(\ell)}$ using a splice operation. We remove singular orbits of $\Sigma_1^{(\ell)}$ and replaced them by a cusp hyperbolic spaces.

Generalizing this idea we get.

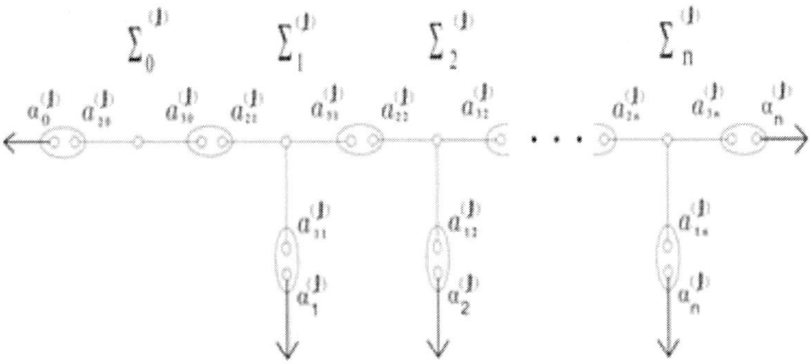

Finally, all singular orbits are removed and replaced by either another SFH-sphere or by a cusp hyperbolic space, fulfilling our goal of avoiding singular orbits (no singular points). See [10] for the discrete volume of this new manifold and a comparison with coupled constants.

6. CONCLUSIONS

In our construction we remove the tubular neighbourhoods of singular orbits and replace them by a hyperbolic space thus achieving two things. First, avoiding singularities from any space is closer to a physical representation. Usually it is hard to have a physical representation when in theory we have a singularity. Instead, we can imagine singular points as points at infinity. Second, we construct a different space than Efremov and Mitskievich did, introducing Thurston's theory and using the work done by Knesser, Milnor, Jaco, Shalen, Johanson. This space shows an improvement for the calculation of the hierarchy of the coupling constants, i.e. reproducing more closely the exponents of the coupling constants. This is important because the use of hyperbolic spaces instead the local use of hyperbolic geometry make a better approach.

Building the model of the universe with 3-dimensional spatial sections from Seifert fibered homology spheres and cusp hyperbolic spaces provides a description of the evolution at least for the hierarchical approach of the coupling constants by means of topological changes. We also use manifolds which contain a finite volume. Therefore, we can calculate volumes in these spaces. In other words, we have an infinite space with finite volume.

We have created here a new cusp hyperbolic space with fibers. This allows calculating the volume of this particular manifold. Once we obtain how to calculate the volume of each tetrahedra in the space, only what we need is to multiply for the number of tetrahedra contained in the cusp hyperbolic space. All of this is due to the space which was discretized. And this could be used in cosmology theories for calculating the volume density of the space.

As a final comment, we emphasize the importance of the approach presented here for the study of alternative cosmological models. Highlighting an important example, the standard Friedmann Singularity-Free (FSF) theories usually consider more complex structures than the simple spherically symmetric point like a Friedmann singularity (the big bang seed with null dimension and infinite density). Following this path, we point out here a possible application of the topological construction introduced in this paper to cosmological scenarios known as "Cusp Cosmology" which are based on the cusp-like geometries resulting from non-linear wave theory [29] .

ACKNOWLEDGEMENTS

Maria E. Mejía, Reinaldo R. Rosa Thanks to: Ignacio Barradas, Klaus-Peter Schröder, César Caretta, Hugo Garca, Stephanie Dunbar and Lawrence Nash. CONACyT, LAC-INPE, CAPES and PROAP.

REFERENCES

1. Hawking, S. (1988) Wormholes in Spacetime. Physical Review D, 37, 904.
2. Weinberg, S. (1989) The Cosmological Constant Problem. Reviews of Modern Physics, 61, 1.
3. Horowitz, G.T. (1991) Topology Change in Classical and Quantum Gravity. Classical and Quantum Gravity, 8, 587-601.
4. De Lorenci, V.A., Martin, J., Pinto-Neto, N. and Soares, I.D. (1997) Topology Change in Canonical Quantum Cosmology.
5. Kronheimer, P.B. and Mrowka, T.S. (1993) Gauge Theory for Embedded Surfaces. Topology, 32, 773-826.
6. Efremov, V.N. (1996) Flat Connection Contribution to Topology Changing Amplitudes in an Ensemble of Seifert fibered Homology Spheres. International Journal of Theoretical Physics, 35, 63-68.
7. Efremov, V.N. (1997) Siebenman-Type Cobordisms with Borders and Topology Changes by Quantum Tunnelling. International Journal of Theoretical Physics, 36, 1133-1151.
8. Efremov, V.N. and Mitskievitch, N.V. (2003) Discrete Model of Space-Time in Terms of Inverse Spectra of the T_0 Alexandroff Topological Spaces.
9. Efremov, V.N. and Mitskievitch, N.V. (2004) T Discrete Model of Space-Time in Terms of Inverse Spectra of the T_0 Alexandroff Topological Spaces Opology Changes in Terms of Proper Inverse Spectra of T_0-Discrete Spaces and Hierarchy of Fundamental Interactions in a Universe Glued Together of Seifert Homologic Spheres. In: Progress in General Relativity and Quantum Cosmology Research, Nova Sci. Publishers.
10. Mejia, M.E. (2009) Un Modelo del Universo y la Comparación de la Jerarquia de las Constantes de Acoplamiento Experimentales con las Constantes de Acoplamiento del Modelo. Memoria de los extensos del VI Encuentro Participación de la Mujer en la Ciencia, 1, 46-47.
11. Barrett, J.W. (1995) Quantum Gravity as Topological Quantum Field Theory. Journal of Mathematical Physics, 36, 6161.
12. Carlip, S. (2005) Quantum Gravity in 2+1 Dimensions: The Case of a Closed Universe. Living Reviews in Relativity, 8, 1.
13. Anderson, M., Carlip, S., Ratcliffe, J.G., Surya, S. and Tschantz, S.T. (2004) Peaks in the Hartle-Hawking Wave Function from Sums over Topologies Class. Quantum Gravity, 21, 729.
14. Bytsenko, A.A. and Guimaraes, M.E.X. (2008) Expository Remarks on Three-Dimensional Gravity and Hyperbolic Invariants.
15. Kneser, H. (1929) Jahresbericht der Deutshen. Mathematiker-Vereinigung, 38, 248-260.
16. Milnor, J. (1962) A Unique Decomposition Theorem for 3-Manifolds. American Journal of Mathematics, 84, 1-7.
17. Eisenbud, D. and Newmann, W.D. (1985) Three-Dimensional Link Theory and Invariants of Plane Curve Singularities. Ann. Math. Stud. Princeton University Press.
18. Johanson, K. (1979) Homotopy Equivalences of 3-Manifolds with Boundaries. Lecture Notes in Math. N 1-761. Springer-Verlag, Berlin.
19. Jaco, W. and Shalen, P.B. (1979) Seifert Fibered Spaces in 3-Manifolds. American Mathematical Society, 220.

20. Thurston W. P. (1982) Three-dimensional manifolds, Kleinian groups and hyperbolic geometry. American Mathematical Society. Bulletin. New Series, 6, 357-381.
21. Thurston, W.P. (1997) The Geometry and Topology of 3-Manifolds. Electronic Version 1.0.http://library.msri.org/books/gt3m/
22. Thurston, W. P. (1997) Three-Dimensional Geometry and Topology. Vol. 1, Princeton University Press, Princeton.
23. Jaco, W. (1980) Lectures on 3-Manifold Topology. Regional Conference Series in Mathematics. American Mathematical Soc.
24. Newmann, W. D. and Zagier, D. (1985) Volumes of Hyperbolic Three-Manifolds. Topology, 24, 307-332.
25. Aschenbrenner, M. Friedl, S. and Wilton, H. (2013) 3-Manifold Groups. e-print arXiv:1205.0202 [math.GT].
26. Rong,Y. (1992) Degree One Maps between Geometric 3-Manifolds. American Mathematical Society, 332, 411-436.
27. Hernándes, A.M. (2003) Cambios Topológicos y Jerarqua de las Interacciones Fundamentales en un Modelo Cosmológico T_0-Discreto del Universo Construido de Esferas Homológicas Fibradas de Seifert. Tesis para obtener el grado de maestro en ciencias en la Universidad de Guadalajara.
28. Milnor, J. (1975) On the 3-Dimensional Brieskorn Manifolds M (p,q,r).http://www.maths.ed.ac.uk/126aar/papersmilnbries.pdf
29. Rosa, R.R., Strieder, C. and Stalder. D.H. (2012) An Alternative Singularity-Free Cosmological Scenario from Cusp Geometries. AIP Conference Proceedings, 1483, 441-446.
30. Milnor, J. (1958) Review: Norman Steenrod, The Topology of Fibre Bundles. Bulletin of the American Mathematical Society, 64, 202-203.

INDEX